Lecture Notes in Mathematics

Edited by A. Dold and B. Eckmann

878

Numerical Solution of Nonlinear Equations

Proceedings, Bremen 1980

Edited by E. L. Allgower, K. Glashoff and H.-O. Peitgen

Springer-Verlag
Berlin Heidelberg New York 1981

Editors

Eugene L. Allgower
Mathematics Department, Colorado State University
Fort Collins, Colorado 80523, USA

Klaus Glashoff
Institut für Angewandte Mathematik
Universität Hamburg, Bundesstr. 55
2000 Hamburg 13, Federal Republic of Germany

Heinz-Otto Peitgen
Fachbereich Mathematik, FS „Dynamische Systeme"
Universität Bremen
2800 Bremen 33, Federal Republic of Germany

AMS Subject Classifications (1980): 35Q20, 35R35, 35R45, 47H10, 65-06, 65H10, 65H15, 65L10, 65L15, 65N20, 65N99, 65Q05

ISBN 3-540-10871-8 Springer-Verlag Berlin Heidelberg New York
ISBN 0-387-10871-8 Springer-Verlag New York Heidelberg Berlin

CIP-Kurztitelaufnahme der Deutschen Bibliothek
Numerical solution of nonlinear equations : proceedings, Bremen, 1980 / ed. by E. L. Allgower . . . – Berlin; Heidelberg; New York: Springer, 1981
(Lecture notes in mathematics; Vol. 878)
ISBN 3-540-10871-8 (Berlin, Heidelberg, New York)
ISBN 0-387-10871-8 (New York, Heidelberg, Berlin)
NE: Allgower, Eugene L. [Hrsg.]; GT

Printing and binding: Beltz Offsetdruck, Hemsbach/Bergstr.
2141/3140-543210

dedicated to the memory of

Emanuel Sperner

P R E F A C E

During July 21-25, 1980, a symposium on

"Numerical Solution of Nonlinear Equations"

was held at the University of Bremen under the sponsorship of the

Forschungsschwerpunkt "Dynamische Systeme",
Universität Bremen, Bremen,

and the

W. Blaschke Gesellschaft, Hamburg.

The organisation was due to *K. Glashoff* (Hamburg), *D. Hinrichsen* (Bremen), *K. Merten* (Bremen) and *H.O. Peitgen* (Bremen). This volume comprises the proceedings of the conference. The articles appearing herein have undergone an evaluation and reviewing process.

The aim of the conference was to furnish an opportunity for an exchange of problems and techniques for the numerical solution of nonlinear equations. Thus there were lectures on new sources

of nonlinear problems e.g. chaotic behaviour of discrete and continuous dynamical systems, image retrivial and nonlinear eigenvalue problems including bifurcation. Then there were lectures on the discretization of operator problems and the phenomenon of spurious solutions. Finally, there were lectures of the numerical aspect of solving a system of nonlinear equations. Included here were contributions to methods exploiting monotonicity, variants of Newton's method, continuation and simplicial methods. Invited lectures were given by: *E.L. Allgower* (Fort Collins), *E. Bohl* (Konstanz), *L. Collatz* (Hamburg), *K. Georg* (Bonn), *K.-P. Hadeler* (Tübingen), *K.-H. Hoffmann* (Berlin), *M. Kojima* (Tokyo), *H.-O. Kreiss* (Pasadena), *R. Saigal* (Evanston), *K. Schmitt* (Salt Lake City), *F. Stenger* (Salt Lake City), *M.J. Todd* (Ithaca) and *B. Werner* (Hamburg).
alltogether 37 lectures were held and approximately 65 participants from 7 countries attented the conference.

We take this opportunity to express our appreciation to all of the authors who have contributed to these proceedings and to the referees for their careful cooperation. Our special thanks go to the secretaries *I. Chromik*, *K. Limberg*, *E. Sieber* and *H. Siebert* whose conscientious efforts were responsible for the smooth running of the conference.

Many of the participants, and particularly the "simplicial school" members were saddened that Professor *E. Sperner* had passed away a few months prior to the conference. Professor *Sperner* had informed us that he was planning to attend and contribute to the conference. Indeed, in his Christmas greetings he communicated that he was looking forward to the meeting. He had also attented and made inspiring contributions at similar conferences at Bonn in 1978 and at Southampton in 1979. If *Herbert Scarf* is the father of the "simplicial methods", due to his significant paper in SIAM J. Appl. Math. (1967), then surely *Emanuel Sperner* is the grandfather, due to his highly useful lemma, published in Abh. Math. Sem. Univ. Hamburg (1928). Just as Sperner's lemma was used directly by *Knaster, Kuratowski* and *Mazurkiewicz* in Fund. Math. (1929) to give a new proof of Brouwer's fixed point theorem, it was also underlying the first algorithmical proofs by *Herbert Scarf* and *Harold Kuhn*. Professor *Sperner* was obviously pleased to see that approximately 50 years after its publication, his lemma had found yet another significant application. For his past and recent mathematical contributions, and especially for his pleasant personality, we would have greatly loved to have had Professor *Sperner* among us again in Bremen. Hence we dedicate this volume to Professor *Emanuel Sperner*.

April 1981

Eugene L. Allgower
Klaus Glashoff
Heinz-Otto Peitgen

PARTICIPANTS

ALLGOWER, E.L.	Fort Collins	MERTEN, K.	Bremen
ANSORGE, R.	Hamburg	MERZ, G.	Kassel
BÖRSCH-SUPAN, W.	Mainz	MITTELMANN, H.D.	Dortmund
BOHL, E.	Konstanz	OETTLI, W.	Mannheim
BRINK-SPALINK, J.	Münster	ORTLIEB, C.P.	Hamburg
BRÜBACH, W.	Göttingen	PEITGEN, H.O.	Bremen
CARMICHAEL, N.	Warwick	PETERS, H.	Bonn
COLLATZ, L.	Hamburg	PRITCHARD, A.J.	Warwick
COLONIUS, F.	Bremen	PRÜFER, M.	Bremen
CROMME, L.	Göttingen	SAIGAL, R.	Evanston
ECKHARDT, W.	Hamburg	SALAMON, D.	Bremen
EHRMANN, M.	Bremen	SAUPE, D.	Bremen
FORSTER, W.	Southampton	SCHEURLE, J.	Stuttgart
GEIGER, C.	Hamburg	SCHILLING, K.	Bayreuth
GEORG, K.	Bonn	SCHMIDT, G.	Bayreuth
GLASHOFF, K.	Hamburg	SCHMITT, K.	Salt Lake City
HACKBUSCH, W.	Köln	SEYDEL, R.	München
HADELER, K.-P.	Tübingen	SIEGBERG, H.W.	Bonn
HASS, R.	Hamburg	SIKORSKI, K.	Warszawa
HINRICHSEN, D.	Bremen	SKORDEV, G.	Sofia
HOFFMANN, K.-H.	Berlin	SPEDICATO, E.	Bergamo
HOLKENBRINK, B.	Mainz	STENGER, F.	Salt Lake City
HUI, G.	Hamburg	TALMAN, A.J.J.	Amsterdam
JEGGLE, H.	Berlin	TODD, M.J.	Ithaca
JONGEN, T.	Hamburg	UNGER, H.-J.	Münster
JÜRGENS, H.	Bremen	VÄTH, R.	Bayreuth
KEARFOTT, B.	Lafayette	VOSS, H.	Essen
KOJIMA, M.	Tokyo	WAGNER, H.D.	Hamburg
KREISS, H.-O.	Pasadena	WARBRUCK, H.W.	Bonn
LAAN, G.v.d.	Amsterdam	WEBER, H.	Dortmund
LEMPIO, F.	Bayreuth	WERNER, B.	Hamburg
LORENZ, R.	Stuttgart	WERNER, W.	Mainz
MC CORMICK, S.	Fort Collins	WILLE, F.	Kassel
		ZOWE, J.	Würzburg

LECTURES

ALLGOWER, E.L.	1. A survey of homotopy methods for smooth mappings 2. Discrete correction and mesh refinement for operator equations
BOHL, E.	On discrete models for reaction - diffusion - convection processes
COLLATZ, L.	Methoden der Approximationstheorie zur Berechnung von Fixpunkten
CROMME, L.	Remarks on the robustness and numerical stability of fixed-point algorithms
FORSTER, W.	Constructive versions of certain fixed point theorems and antipodal point theorems
GEORG, K.	Numerical integration of the Davidenko equation
HACKBUSCH, W.	Numerical solution of nonlinear equations by multi-grid methods of the second kind
HADELER, K.P.	Matrix inverse eigenvalue problems
HOFFMANN, K.H.	Fixpunktprinzipien und freie Randwertaufgaben
JÜRGENS, H.	SCOUT: A PL ´pathfinder` package
KEARFOTT, B.	A derivative-free continuation method
KOJIMA, M.	Continuous deformation of nonlinear programming problems

KREISS, H.-O.	Numerical methods for linear and nonlinear singular perturbation problems
LAAN, G.v.d.	Labelling rules and orientation: On Sperner's lemma and Brouwer degree
MERTEN, K.	Numerical solution of nonlinear boundary value problems by semi-iteration
MITTELMANN, H.D.	Fast solution of nonlinear free boundary problems
PEITGEN, H.O.	1. Global numerical PL-unfoldings of nonlinear eigenvalue problems (survey) 2. Nonlinear elliptic boundary value problems versus their finite difference approximations: Numerically irrelevant solutions (lecture + Film)
PETERS, H./SAUPE, D.	Numerical and finite dymensional approches to Nussbaum's conjecture
PRÜFER, M.	A PL-algorithm for a zero of an m-function
SAIGAL, R.	On computing one or several solutions to nonlinear problems
SAUPE, D.	Predictor - corrector methods and simplicial continuation algorithms
SCHMITT, K.	Global topological perturbations of nonlinear eigenvalue problems
SEYDEL, R.	On the discovery of branch points during continuation

SIEGBERG, H.W.	Chaotic approximations of stable fixed points
SIKORSKI, K.	Generalization of a bisection method for solving a system of n nonlinear equations
STENGER, F.	Ultrasonic tomography based on the Helmholtz equation
TALMAN, A.J.J.	On the structure of recent variable dimension algorithms to approximate a solution of nonlinear problems
TODD, M.J.	1. Simplicial methods for nonlinear constrained minimization problems 2. Exploiting structure in simplicial methods
WARBRUCK, H.W.	Zur Lösung konvexer Optimierungsaufgaben mit dem Kellog- Li -Yorke Algorithmus
WEBER, H.	On the numerical solution of secondary bifurcation problems
WERNER, B.	Simpliziale Methoden zur Berechnung von Lösungszweigen mit Umkehrpunkten
WERNER, W.	Some improvements of classical iteration methods for the solution of nonlinear equations

CONTENTS

E.L. ALLGOWER
A survey of homotopy methods for smooth mappings 1

E.L. ALLGOWER, K. BÖHMER and S. MC CORMICK
Discrete correction methods for operator equations 30

K.-H. BECKER, R. SEYDEL
A Duffing equation with more than 20 branch points 98

L. COLLATZ
Einschließungssätze für Fixpunkte 108

K. GEORG
A numerically stable update for simplicial algorithms 117

K. GEORG
Numerical integration of the Davidenko equation 128

K.-H. HOFFMANN
Fixpunktprinzipien und freie Randwertaufgaben 162

R.B. KEARFOTT
A derivative-free arc continuation method and a bifurcation technique 182

M. KOJIMA

An introduction to variable dimension algorithms for solving systems of equations 199

G.v.d.LAAN, A.J.J. TALMAN

Labelling rules and orientation: On Sperner's lemma and Brouwer degree 238

H.D. MITTELMANN

On the numerical solution of contact problems 258

H.O. PEITGEN, K. SCHMITT

Positive and spurious solutions of nonlinear eigenvalue problems 275

H. PETERS

Change of structure and chaos for solutions of $\dot{x}(t) = - f(x(t-1))$ 325

H.W. SIEGBERG

Chaotic mappings on S^1, periods one, two, three imply chaos on S^1 351

F. STENGER

An algorithm for ultrasonic tomography based on inversion of the Helmholtz equation 371

H. WEBER

On the numerical approximation of secondary bifurcation problems 407

W. WERNER

Some improvements of classical iterative methods for the solution of nonlinear equations 426

A SURVEY OF HOMOTOPY METHODS FOR SMOOTH MAPPINGS

BY

E.L. ALLGOWER

Mathematics Department
Colorado State University
Fort Collins, Colorado 80523
USA

1. Introduction

The numerical problem which we will survey here is the following. Suppose that a smooth curve C is implicitly defined as the set of points

$$C = \{(\lambda, x(\lambda)) | \lambda \in \Lambda_0 \subset \mathbb{R}^1\} \subset \mathbb{R}^{N+1}$$

such that

(1.1) $$H(\lambda, x(\lambda)) = 0 \text{ and } (\lambda_0, x_0) \in C.$$

(P) Numerically trace C assuming that $H: \Lambda \times \mathbb{R}^N \to \mathbb{R}^N$ is a smooth mapping and that (λ_0, x_0) is a regular point of H.

It follows from the implicit function theorem [41] that there exists an open interval Λ_0 containing λ_0 such that C is smooth on Λ_0. Applications for (P) have traditionally come from physical problems having extra parameters, nonlinear boundary value problems, the method of incremental loading in finite element methods, the solving of ill-posed problems via regularization, nonlinear programming, and in general zero-finding problems in which it is difficult to obtain an adequate starting point for classical iteration methods.

The survey we give here is organized as follows. In Section 2 we indicate how some classical applications may be formulated in the context of problem (P), and consider the question as to whether the curve C will be adequate to solve the application. In Section 3 we give a brief summary of classical embedding methods. This summary is not intended to be complete, but only to indicate the relationships to the present homotopy methods. In Section 4 we outline details of homotopy continuation algorithms for smooth maps. Elsewhere in these proceedings (see the papers of Georg and Kearfott) derivative-free continuation methods are discussed. In Section 5 we outline how continuation methods may be used to detect bifurcations along C and to switch to a new branch. In the final section we suggest several additional applications for homotopy continuation algorithms using the mesh size as a homotopy parameter.

2. Two classical applications

We shall discuss the details of two general applications of (P), which in essence subsume the particular applications listed in the introduction.

(A) Nonlinear eigenvalue problems.

In these applications λ is usually an intrinsic parameter, e.g. a Reynolds number, an oscillation constant or loading parameter, and a starting point (λ_0, x_0) is either trivially available or easily computed. As a simple example we may consider the eigenvalue problem

$$u'' + \lambda f(u) = 0 \qquad u(0) = 0 = u(1) \tag{2.1}$$

where $\lambda \in \mathbb{R}^1$ and f is smooth.

Numerically, we perform some discretization of (2.1) to a mesh of points $0 < x_1 < x_2 < \ldots < x_N < 1$ in [0, 1] and solve (P) for a discretized eigenvalue problem of the form

$$H(\lambda, \bar{u}) := A_N\bar{u} + \lambda B_N\bar{f}(\bar{u}) \tag{2.2}$$

where $\bar{u} = (u(x))^T$, $\bar{f}(\bar{u}) = (f(u(x)))^T$ for $x = x_i$, $i = 1, \ldots, N$ and A_N, B_N are N x N matrices associated with a consistent and stable discretization of (2.1).

Now an appropriate starting point for (P) might be $\lambda = 0$ and $\bar{u} = \bar{0} \in \mathbb{R}^N$. If, for example, a divided difference approximation is made on a uniform mesh, appropriate choices for A_N, B_N might be

$$(2.3) \qquad A_N = (N+1)^2 \begin{pmatrix} -2 & 1 & & 0 \\ 1 & \ddots & \ddots & \\ & \ddots & \ddots & 1 \\ 0 & & 1 & -2 \end{pmatrix}$$

(2.4) $B_N = I_N$ (the N x N identity matrix), or if f is sufficiently smooth,

$$(2.5) \qquad B_N = \frac{1}{12} \begin{pmatrix} 10 & 1 & & 0 \\ 1 & \ddots & \ddots & \\ & \ddots & \ddots & 1 \\ 0 & & 1 & 10 \end{pmatrix} .$$

(B) The zero-finding problem.

In these applications one generally wishes to solve a system of equations

$$(2.6) \qquad F(x) = 0$$

where $F: \mathbf{R}^N \to \mathbf{R}^N$ is a smooth mapping.

One resorts to the use of a continuation method when no adequate starting value for a fast iterative method such as a quasi-Newton method is available. One of the practical difficulties of the traditional iterative methods is that the zero-points of the map F may have domains of attraction (e.g. relative to Newton's method) which may be very difficult to hit with randomly chosen starting values.

The mapping H in these applications is usually artificially constructed, and may be chosen in order to exploit some property of the particular problem at hand. The following are examples of homotopies.

(α) Convex combination homotpy. (CH)

Here H is defined by

$$(2.7) \qquad H(\lambda, x) := (1 - \lambda)F_0(x) + \lambda F(x)$$

where F_0 is a smooth map chosen so that x_0 is a regular point of F_0 and $F_0(x_0) = 0$. For example, one may choose $F_0(x) = A(x - x_0)$ where

A is an N x N nonsingular matrix which remains free to be chosen. The starting point is $(0, x_0)$, and if the curve $C \subset H^{-1}(0)$ with $(0, x_0) \in C$ is defined and smooth for $\lambda \in [0, 1]$, then the points in the non-empty set $C \cap (\{1\} \times \mathbb{R}^N)$ are zero-points of F.

(β) Global homotopy. (GH)

Here H may be defined, for example, by

$$H(\lambda, x) := F(x) - e^{-\lambda}F(x_0) \tag{2.8}$$

where $(0, x_0)$ is the starting point.

If the curve $C \subset H^{-1}(0)$ is defined and smooth for $\lambda \in [0, \infty)$, then the limit points of C as $\lambda \to \infty$ are zero-points of F. If, furthermore, the zero-points of F are regular points, then C will have only one limit point as $\lambda \to \infty$.

Sometimes it is wished to find additional zero-points of F after a zero-point x_0 has already been determined or approximated. One means for doing this is to use a homotopy of the form

$$H_d(\lambda, x) := F(x) + \lambda d \tag{2.9}$$

where $d \in \mathbb{R}^N$ and $d \neq 0$, and $(0, x_0)$ is the starting point.

If the curve $C_d \subset H_d^{-1}(0)$ with $(0, x_0) \in C_d$ intersects $\{0\} \times \mathbb{R}^N$ at some new point $(0, x^*)$ with $x^* \neq x^0$, then x^* is also a zero-point of F.

The preceding general discussion leads naturally to the following questions.

a) What are sufficient conditions for C to be everywhere smooth in the zero-finding problem?

b) What are sufficient conditions for C to reach a solution to the zero-finding problem?

c) How can numerical implementations for numerically tracing C be made?

d) How can bifurcations in C be handled?

The first two questions can be answered by means of classical theoretical results.

a) The parametrized Sard's theorem.

This theorem, which we quote below ensures that C is smooth and consists of regular points of H (whether H is the convex or the global homotopy) for almost all $x_0 \in \mathbb{R}^N$ in the sense of N-dimensional Lebesgue measure. Similarly, $C_d \subset H_d^{-1}(0)$ is smooth and regular for almost all $d \in \mathbb{R}^N - \{0\}$.

(2.10) Parametrized Sard's theorem [1][13].

Let $U \subset \mathbb{R}^q$, $V \subset \mathbb{R}^m$ be nonempty open sets and $\psi: U \times V \to \mathbb{R}^p$ be a smooth map, and $p \leq m$. If 0 is a regular value of ψ, then for almost every $a \in U$, 0 is a regular value of the restricted map $\psi_a(\cdot) = \psi(a, \cdot)$.

In particular, by differentiating H with respect to both (λ, x) and x_0, one obtains that 0 is a regular value of $H|_{[0,1]\times\mathbb{R}^N}$ for almost all $x_0 \in \mathbb{R}^n$.

b) Basically, what we need here is that C (or C_d) should penetrate or approach the desired λ-level at least once. To ensure this the standard conditions from classical homotopy and degree theory are sufficient viz. a

(2.11) Bounding Condition:

There exists an open bounded neighborhood U of x_0 such that

i) x_0 is the only zero-point of H in $\{0\} \times U$.

ii) $H(\lambda, x) \neq 0$ for $(\lambda, x) \in [0, 1] \times \partial U$ for (CH)

for $(\lambda, x) \in [0, \infty) \times \partial U$ for (GH).

Thus, by (2.10), for almost all x_0, C is smooth and consists of regular points of H. By (2.11), C must reach the λ-level $\lambda = 1$ for (CH) or approach the λ-level ∞ for (GH).

An analogous bounding condition for H_d is

(2.12)

There exists an open bounded neighborhood U_d of x_0 such that

i_d) x_0 is the only zero-point of H_d in $\{0\} \times U_d$,

ii_d) $H_d(\lambda, x) \neq 0$ for $(\lambda, x) \in [0, \infty) \times \partial U_d$.

It follows again from (2.10) and (2.12) that for almost all x_0, C_d is smooth, consists of regular points and must be bounded above since

$$\lambda = -F_i(x)/d_i \quad \text{for } d_i \neq 0$$

and because F_i is bounded on the compact set $\bar{U}_d$. Hence C_d must "turn back" and penetrate $\lambda = 0$ at least once at some $x_* \neq x_0$ by the regularity of F at x_0.

Before proceding to numerical algorithms for tracing implicitly defined curves C, we present a few specific conditions which will yield bounding conditions.

(2.13) A coercivity condition.

Let U be a bounded open neighborhood of x_0 such that for every $x \in \partial U$ there exists a $v_x \in \mathbb{R}^N - \{0\}$ satisfying

$$v_x^T(x - x_0) > 0 \text{ and } v_x^T F(x) > 0.$$

Then for the convex homotopy we have

$$v_x^T H(\lambda, x) = (1 - \lambda)v_x^T(x - x_0) + \lambda v_x^T F(x) > 0$$

for all $\lambda \in [0, 1]$ and all $x \in \partial U$.

For the global homotopy an analogous bounding condition is that there exists a bounded neighborhood U of x_0 such that $DF(x)^{-1}F(x)$ always points into U or always points out of U for all $x \in \partial U$ ([32][58]).

The following are somewhat more special bounding conditions.

(2.14) The Leray-Schauder condition.

Let U be an open bounded neighborhood of 0 such that

$$G(x) := x - F(x) \neq tx$$

for all $t \geq 1$ and $x \in \partial U$. If we use the convex homotopy, then $x - F(x) \neq \frac{1}{\lambda}x$ for $\lambda \in [0, 1]$, $x \in \partial U$ implies $H(\lambda, x) = (1 - \lambda)x + \lambda F(x) \neq 0$ for $\lambda \in [0, 1]$, $x \in \partial U$.

(2.15) The Brouwer condition.

The hypothesis of the Brouwer fixed point theorem is $G: \mathbb{R}^N \to K \subset \mathbb{R}^N$ where G is continuous and K is convex and compact.

Suppose we let U be any bounded open neighborhood of K, let $x_0 \in K$ and $F(x) = G(x) - x$. Then the values of the convex homotopy

$$G(\lambda, x) = (1 - \lambda)x_0 + \lambda G(x)$$

lie in K since x_0 and $G(x)$ lie in K and K is convex. Thus the solution set to

$$\begin{aligned} H(\lambda, x) &= (1 - \lambda)x_0 + \lambda G(x) - x \\ &= (1 - \lambda)(x - x_0) + \lambda F(x) = 0 \end{aligned}$$

lies strictly inside of U for all $\lambda \in [0, 1]$ if $F(x) - x$ satisfies the Brouwer condition.

(2.16) Coercivity for H_d.

Suppose that there exists an open bounded neighborhood U of x_0 and a fector $v \in \mathbb{R}^N - \{0\}$ such that $v^T F(x) > 0$ for all $x \in \partial U$. If we choose $d \in \mathbb{R}^N$ such that $v^T d \geq 0$, then

$$v^T H_d(\lambda, x) = v^T F(x) + \lambda v^T d > 0$$

for all $x \in \partial U$ and $\lambda \in [0, \infty)$.

3. Classical embedding algorithms

We shall assume that H and (λ_0, x_0) are given as in Section 1, and for definiteness, we shall assume that it is wished to follow $C \subset H^{-1}(0)$ in the direction of increasing λ over $[\lambda_0, \lambda_*]$.

The following is a generical embedding algorithm in the sense that essentially each numerical step may be replaced by some similar step which may bring about an improvement or simplification, depending upon the objective in solving the problem (P).

(3.1) A generical embedding algorithm, $(H, \lambda_0, x_0$ given).

1. Start. Input positive real numbers $\Delta\lambda_0$, $\Delta\lambda_{min}$, $\Delta\lambda_{max}$, TOL and a positive integer j_*.
2. Predictor. Set $(\mu, z) := (\lambda_i + \Delta\lambda_i, x_i)$, $j := 0$. If $\mu > \lambda_*$, set $\mu = \lambda_*$.
3. Calculate $H(\mu,z)$ and $\partial_x H(\mu,z)$, or some appropriate nonsingular approximation $\hat{H}_x(\mu,z)$.
4. If $\|H(\mu,z)\| / \|\hat{H}_x(\mu,z)\| < TOL$, set $\lambda_{i+1} = \mu$, $x_{i+1} = z$ and STOP if $\lambda_{i+1} = \lambda_*$ (success). Otherwise set $\Delta\lambda_{i+1} = \min\{(\frac{j+2}{j+1})\Delta\lambda_i, \Delta\lambda_{max}\}$ and go to 2. Otherwise, continue.
5. Corrector. If $j < j_*$, solve for W in
$$\hat{H}_x(\mu,z)w = -H(\mu,z)$$
make the replacements = $z = z + w$, $j := j+1$ and go to 3. Otherwise, continue.
6. Decelerate. Replace $\Delta\lambda_i := \Delta\lambda_i/2$. If $\Delta\lambda_i < \Delta\lambda_{min}$, STOP (failure). Otherwise, go to 2.

Remarks. Numerical implementations of classical embedding methods appear to date back to Lahaye (1934) [34]. For surveys on embedding methods and their applications, see [44][62][63]. Figure (3.2) illustrates several steps of the generical algorithm.

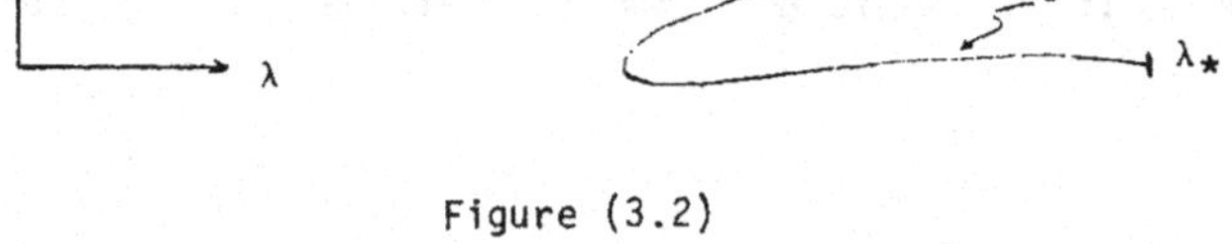

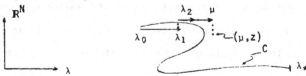

Figure (3.2)

It is evident from the Newton-Kantorovich theorem that if C is such a curve that

$$C \cap (\{\lambda\} \times \mathbb{R}^N)$$

consists of a single point $(\lambda, x(\lambda))$ for each $\lambda \in [\lambda_0, \lambda_*]$, then for sufficiently small $\Delta\lambda_{min}$, the algorithm (3.1) will attain the λ-level λ_* in finitely many steps. On the other hand, if C has a turning point at $\bar{\lambda} \in (\lambda_0, \lambda_*)$, then the algorithm (3.1) will generally fail to attain the level λ_*, because near $\bar{\lambda}$ the situation $\Delta\lambda_i < \Delta\lambda_{min}$ will eventually occur. The turning points of C can be described as the critical points of $\lambda(s)$, i.e., $\frac{d\lambda}{ds} = 0$ when C is regarded as being parametrized according to arc length.

Davidenko [16] observed that C is the solution curve to the initial value problem obtained by differentiating (1.1);

(3.3) $$\partial_x H \frac{dx}{d\lambda} + \partial_\lambda H = 0 \quad \text{and} \quad x(\lambda_0) = x_0$$

or

(IVP) $$\frac{dx}{d\lambda} = -\partial_x H^{-1} \partial_\lambda H \quad \text{and} \quad \lambda(\lambda_0) = x_0.$$

This fact has been implicitly made use of in the corrector step 5 of the embedding algorithm (3.1). With the formulation (3.3) it is now possible to solve the curve tracing problem (P) by means of solving (IVP) via the highly-developed methods for the numerical solution of initial value problems. If this is done, the predictor-corrector machinery of the IVP solver is utilized. In doing so, however, the useful fact that the points of C enjoy a contraction property is essentially discarded.

In recent years a considerable amount of research has been performed concerning the step size control (steps 4 and 5 in (3.1)) in embedding algorithms. It must be noted, however, that the step size strategy ought to be dictated by whether it is wished to follow C very reliably as, for example, in eigenvalue problems, or whether it is wished to merely attain a certain λ-level as rapidly as possible.

Some of the step size selection methods which have been developed are those of:

a) Deufelhard et al [17][18] in which it is attempted to make a maximum possible step in conformity with an empirically determined approximation to the local contraction factor of the Newton-Kantorvich theory.

b) Numerous authors, e.g. [31][33][38][50] in which a fixed contraction rate is uniformly maintained.

c) W. Schmidt [53] who maintains a successive error ratio formula connected with local quadratic convergence.

d) Wacker et al [25][60][61] attempt to minimize the total computational work in proceding from λ_0 to λ_*.

4. Homotopy continuation algorithms

As was seen in the preceding section, the corrector step in traditional embedding algorithms (step 5 in (3.1)) involves a condition constraining the iterates to a fixed λ-level, and this causes the algorithms to fail when C has a turning point with respect to the λ-parameter. Apparently Haselgrove [26] was the first to suggest a means for overcoming this difficulty and since then several authors have used similar ideas, e.g. [8, 11-14, 19, 20, 22-27, 31-33, 36-40, 49-54, 58-66].

In all of these methods no fixed variable is rigidly used for a special constraint in the corrector process. Generally, the predictor and corrector steps take the following form.

Tangent predictor.

Let $y_i = (\lambda_i, x_i) \in \mathbb{R}^{N+1}$ be a point which has been accepted as an approximating point for C. Choose as the predictor for the next approximating point

$$z_0 = y_i + \delta_i u_i \tag{4.1}$$

where:

a) $\delta_i > 0$ is the step length currently being used.

b) u_i is the solution to

$$\begin{pmatrix} u_i^T \\ H'(y_i) \end{pmatrix} u_i = \begin{pmatrix} 1 \\ \bar{0} \end{pmatrix} \tag{4.2}$$

satisfying

$$u_{i-1}^T u_i > 0. \tag{4.3}$$

Remarks. If y_i is a regular point of H, then the rank of $H'(y_i)$ is N and hence the solutions to (4.2) are the two unit vectors from the null space of $H'(y_i)$. The condition (4.3) assures that a constant direction of traversing C is maintained. The problem of solving the linear system (4.2) need not be discussed here except perhaps to recommend that stable numerical methods e.g. Gaussian methods with pivoting, Givens rotations or Householder transformations. Also if special structure e.g. sparseness or bandedness is present, these properties should be exploited.

If we regard C as being parametrized according to arc length, i.e.

$$C = \{y(s) = (\lambda(s), x(s)) \mid H(y(s)) = 0,\ y(0) = (\lambda_0, x_0),\ s \in [0, S]\},$$

then differentiation of $H(y(s)) = 0$ with respect to arc length yields

$$\begin{pmatrix} \dot{y}(s)^T \\ H'(y(s)) \end{pmatrix} \dot{y}(s) = \begin{pmatrix} 1 \\ \bar{0} \end{pmatrix} \quad \text{where } \dot{y} = \frac{dy}{ds}. \tag{4.4}$$

By comparing (4.2) and (4.4) we see that z^0 lies on a line parallel to a tangent to C at a point near y_i.

The matrix

$$A(y(s)) = \begin{pmatrix} \dot{y}(s)^T \\ H'(y(s)) \end{pmatrix}$$

arising in (4.4) will hereafter be referred to as the augmented Jacobian of H. As we have noted above, A(y(s)) will be nonsingular if C consists of regular points of H.

Adaptive corrector constraints.

Instead of the rigid corrector constraint used in step 5 of (3.1), a corrector process of the form:

Solve

(4.5) $\begin{pmatrix} r^T \\ H'(z_0) \end{pmatrix} W = -\begin{pmatrix} 0 \\ H(z) \end{pmatrix}$ and replace z by z + w.

is performed where initially $z = z_0$ and

a) $z_0 = y_i + \delta_i u_i$ is the tangent p

is the tangent predictor obtained via (4.1)-(4.3).

b) r^T is a fixed vector such that the hyperplane

$$r^T(y - z_0) = 0$$

is transversal to C near y_i. (See Figure 4.6.) Two reasonable choices for r are:

i) $r = u_i$. This choice often allows the reuse of the matrix in (4.2) in place of the matrix in (4.5). Thereby a recalculation of H' may be spared and if a factorization of the matrix in (4.2) has been performed, it may be reused. Choice i) has been used in [3, 5, 13, 36, 38, 65].

ii) $r = e^j$ where e^j is the standard $(N + 1)$-dimensional unit vector such that

$$u_i^T e^j = \max_{k=1,\ldots,n+1} |u_i^T e^k| .$$

This choice permits the immediate elimination of one variable in (4.5) and essentially maintains the structure of the matrix $H'(z^0)$ since the reduced problem to be solved now is

$$H_j'(z^0)\tilde{w} = -H(z)$$

where $H_j'(z_0)$ is the N x N matrix obtained by deleting the j-th row of $H'(z^0)$. Choice ii) has been used in [14][38][50].

Figure 4.6

In any case, because of the transversality condition (4.5)b) we can expect the corrector process to converge to C when y_i is near C and δ_i is sufficiently small. We may now outline the skeleton of a general-purpose homotopy continuation algorithm.

(4.7) A continuation algorithm. (H, y_0, direction of traversing are given.)

1. Start. Choose: $\delta_0 > 0$ (starting step size), $\varepsilon > 0$ (accuracy tolerance), $0 < \alpha_1 < \alpha_2 < 1$ (contraction factors to control step sizes), $0 \le \delta_{min}$, δ_{max} (minimum and maximum step sizes, respectively).
2. Predictor. Solve (4.2) and (4.3) and set $z_0 = y_i + \delta_i u_i$.
3. Corrector begin. Perform the corrector process (4.5) twice to obtain z_1 ($= z_0 + w_1$) and z_2 ($= z_1 + w_2$).
4. Step size selection. If:

   a) $\|w_2\| > \alpha_2 \|w_1\|$, replace δ_i by $\frac{\delta_i}{2}$. If $\frac{\delta_i}{2} < \delta_{min}$, exit (failure); otherwise, go to step 2 (step size decrease).

   b) $\|w_2\| < \alpha_1 \|w_1\|$, set $\delta_{i+1} = \min\{2\delta_i, \delta_{max}\}$ and continue (step size increase).

   c) $\alpha_1 \|w_1\| \le \|w_2\| \le \alpha_2 \|w_1\|$, set $\delta_{i+1} = \delta_i$ and continue.
5. Corrector finish. Perform (4.5) until

$$\|H(z)\| \le \varepsilon \|H'(z_0)\|$$

   then set $y_{i+1} = z$.
6. If the stopping criterion for tracing C is negative, go to 2; otherwise, stop.

<u>Remarks</u>. Very many variations upon the above are possible and many aspects remain to be explored and researched. Among these are:

i) The use of approximations to $H'(y_i)$ e.g. via differences or least change secant updates. Several authors have recently begun to to explore this aspect [23,24,29,52]. Also see the papers of Georg and Kearfott in these proceedings.

ii) Determination of best step size control for various purposes. Some of the results already cited in connection with classical embedding methods will apparently extend to the present methods at least for the corrector (4.5)b) (see e.g. [50]).

iii) The incorporation of higher order predictors. Haselgrove [26] suggests a higher order predictor which is based upon a formula of Shearing [56] relating chord length to arc length:

$$\|y_i - y_{i-1}\|^2 = (\Delta s_i)^2(\tfrac{5}{6} + \tfrac{1}{6}u_{i-1}^T u_i) + O((\Delta s_i)^6). \tag{4.8}$$

Since y_i, y_{i-1}, u_{i-1}, u_i are known, (4.8) can be used to calculate Δs_i, the arc length of C between y_{i-1} and y_i. Upon letting

$k = \dfrac{\Delta s_i}{\Delta s_{i-1}}$, the higher order predictor is given by

$$y_{i+1} = y_i - k^2(3k + 2)(y_i - y_{i-1}) + h(1 + k^2)[u_i + ku_{i-1}] + O((\Delta s_i)^6). \tag{4.9}$$

This formula does not, however, appear to have as yet been implemented and the aspect of incorporating higher order predictors remains unexplored.

Several authors [36,52,65] employ a highly sophisticated initial value problem integrator (e.g. [55]) to solve the Davidenko problem and thereby accept the sophisticated predictor-corrector machinery of the integrator. This is often a successful and relatively efficient expedient. However, the numerical results of Georg and Kearfott show the superiority of their experimental derivative-free algorithms.

To conclude this section we mention that the global Newton method introduced by Brannin [11] and since further researched and developed by several authors [22,27,32,58] is similar in nature to the homotopy methods discussed here. As noted before, an analogue of the Davidenko equation for the global homotopy is

$$F'(x)\dot{x} - \dot{\lambda}F(x^0) = 0$$
$$\dot{x}^T\dot{x} + (\dot{\lambda})^2 = 1.$$

5. Homotopy continuation methods and bifurcation

Let us now consider the final question (d) from Section 2 viz. How can we numerically handle bifurcations on C when we treat nonlinear eigenvalue problems in the context of continuation algorithms? Algorithm (4.7) will not in general detect a bifurcation point, but it can easily be modified to detect such bifurcation points on C at which the determinant of the augmented Jacobian changes sign.

Suppose that y^* is a point on C at which another arc $\tilde{C} \subset H^{-1}(\underline{0})$ intersects C. Then $\tilde{C}$ is tangent at y^* to a vector from the null space of $H'(y^*)$. Conversely, if the dimension of the null space of $H'(y^*)$ is $k \geq 1$, then at most $k - 1$ arcs can branch off from C at y^*. It is well known from bifurcation theory (see e.g. [15,48]) that if, for example, det $A(y(s))$ changes sign at y^*, then at least one curve $\tilde{C}$ will branch off from C at y^*. It turns out that it is a very minor additional calculation in step 2 of Algorithm (4.7) to determine det $A(y_i)$. Now if

$$\det A(y_i)\det A(y_{i-1}) < 0, \tag{5.1}$$

then a bifurcation point y^* lies on the arc of C between y_{i-1} and y_i. Successively improved approximations to y^* can be obtained by a bisection process applied to the step size. It may be possible to approximate y^* via a higher order method, but that is not of primary concern here.

Condition (5.1) is adequate for detecting the presence of simple bifurcations along C, but it merely signifies that between y_{i-1} and y_i an odd number of eigenvalues of $A(y(s))$ have changed sign. Thus it would seem desirable to try to monitor the actual eigenvalue structure at least near such y_i's where det $A(y_i)$ changes sign or becomes small in absolute value. The drawback to this is that $A(y(s))$ may often be a large matrix, and as formulated above is generally not symmetric. If $A(y)$ were symmetric, then Householder transformations could be used to obtain a symmetric tridiagonal matrix having the same eigenvalues as $A(y)$. Thus the familiar efficient methods for approximating the eigenvalues, e.g. Sturm sequences could be applied. It should be emphasized that it is only the sign structure of the eigenvalues which is really of interest. The choice of arc length parametrization which was previously made might very well be abandoned for another one which may offer more advantages.

In order to obtain a parametrization $y(t)$ of C which yields a symmetric $A(y(t))$, it is necessary that H_x be symmetric. This will be the case for appropriate discretizations of nonlinear elliptic eigenvalue problems (see e.g. (2.2)-(2.5)). If H_x is symmetric, we may take

$$A(y(t)) = \begin{pmatrix} k & H_\lambda^T \\ H_\lambda & H_x \end{pmatrix}$$

where k is a scalar function which remains to be determined. Thus under the parametrization $y(t) = (\lambda(t), x(t))$, we must require

$$\begin{pmatrix} k & H_\lambda^T \\ H_\lambda & H_x \end{pmatrix} \begin{pmatrix} \frac{d\lambda}{dt} \\ \frac{dx}{dt} \end{pmatrix} = \begin{pmatrix} \beta \\ \bar{0} \end{pmatrix}$$

where β must be a nonvanishing scalar function. Let us now consider the numerical aspects of solving (5.2).

At regular points of $C \subset H^{-1}(0)$, there must be a unique 1-dimensional set of vectors $\begin{pmatrix} \alpha \\ z \end{pmatrix} \in \mathbb{R}^{N+1}$ such that

$$(5.3) \qquad (H_\lambda \quad H_x)\begin{pmatrix} \alpha \\ z \end{pmatrix} = \bar{0} .$$

Equivalently,

$$(5.4) \qquad H_x z = -\alpha H_\lambda .$$

There are two cases to consider now.

i) If H_x is nonsingular, then (5.3) is solved by $\alpha\binom{1}{v}$ where $\alpha \neq 0$ and v is the solution to

(5.5) $$H_x v = -H_\lambda .$$

ii) If H_x is singular, then (5.3) is solved by $\alpha = 0$ and $z \in \mathcal{N}(H_x)$ (the null space of H_x). Say $z = \gamma w$ where w is a unit vector from $\mathcal{N}(H_x)$ and $\gamma \neq 0$.

The remaining equation which must be satisfied is

(5.6) $$(k,\ H_\lambda^T)\binom{\alpha}{z} = \beta .$$

In case i) we have

$$\alpha k + H_\lambda^T z = \beta$$

where $\alpha \neq 0$ and $z = -\alpha H_x^{-1} H_\lambda$. Thus

$$k = \frac{\beta}{\alpha} + H_\lambda^T H_x^{-1} H_\lambda$$

where we are free to choose $\alpha,\ \beta \neq 0$.

Numerous combinations of choices are possible and we outline only two of them below. For example, we could choose $k = \alpha$, which is somewhat in keeping with the arc-length parametrization. Then

(5.7) $$\alpha^2 - \alpha H_\lambda^T H_x^{-1} H_\lambda - \beta = 0$$

has a positive and a negative solution for any $\beta > 0$. Now two solutions for α and hence for z are possible. As in the arc length parametrization, we may choose the positive or negative value for α so that the direction of traversing is maintained by

(5.8) $$z_{i-1}^T z_i > 0 .$$

In case ii) we have that the solution to (5.3) is given by $\binom{0}{\gamma w}$ where $\gamma \neq 0$ and $w \in \mathcal{N}(H_x)$ is a unit vector. Again in keeping with $k = \alpha$, we have $k = 0$. Now from (5.6),

$$\gamma H_\lambda^T w = \beta .$$

If we take e.g. $\beta = |H_\lambda^T w|$, then $\gamma = \pm 1$ and the choice for γ is determined by (5.8). Of course, $H_\lambda^T w \neq 0$ since otherwise

$$\begin{pmatrix} H_\lambda^T \\ H_x \end{pmatrix} w = 0$$

would contradict the fact that (H_λ, H_x) must have rank n.

The choice of "symmetric parametrization" which we have made corresponds to using the augmented Jacobian

$$A(\lambda, x) = \begin{pmatrix} \frac{d\lambda}{dt} & H_\lambda^T \\ H_\lambda & H_x \end{pmatrix}$$

and the case when H_x is singular corresponds to the turning points in λ with respect to the parameter t.

Let us now briefly summarize the numerical steps involved in the above symmetric parametrization of C.

1. For any y_i, try to solve (5.5) via e.g. Householder transformations which reduce $H_x(y_i)$ to upper triangular form $QH_x(y_i)$.
2. If $H_x(y_i)$ is nonsingular, we obtain v_i. Using e.g. $\beta = |H_\lambda^T v_i|$ or $\beta = 1$, solve (5.7) to obtain an α.
3. Determine $z_i = \begin{pmatrix} \alpha_i \\ \alpha_i v_i \end{pmatrix}$ via (5.8) and set $k_i = \alpha_i$.
4. Take $u_i = \frac{z_i}{\|z_i\|}$ for the unit tangent to C at y_i relative to the symmetric parametrization.
5. If $H_x(y_i)$ is singular, solve for $w_i \in \mathcal{N}(H_x(y_i))$. Take $\beta = |H_\lambda^T w_i|$, $z_i = \pm w_i$ according to (5.8) and $k_i = 0$.
6. Evaluate $\det\left(\begin{array}{c|c} k_i & H_\lambda(y_i)^T \\ QH_\lambda(y_i) & QH_\lambda(y_i) \end{array}\right) = \det A(y_i)$ to monitor whether $\det A(y_i)$ is small or $\det A(y_{i-1}) \det A(y_i) < 0$. If so, a tridiagonalization of $A(y_i)$ via Householder transformations may be made in order to monitor the eigenvalue structure of $A(y_i)$.

The preceding is but one example of a symmetric parametrization. Another possibility is to set $\alpha \equiv 1$, $\beta \equiv 1$, and $k = 1 + H_\lambda^T H_x^{-1} H_\lambda$ for H_x nonsingular. In this case (5.7) and (5.8) become unnecessary. For the case (ii), k can be arbitrary and so we can simply omit monitoring det $A(y_i)$ at such points where $H_x(y_i)$ is singular.

Elsewhere in these proceedings Kearfott outlines a bisection algorithm to detect the solution branches (if there are any) at points where possibly several eigenvalues of A(y) change sign.

In the remainder of this section we discuss a method for numerically finding a new curve $\tilde{C}$ branching off from a point y* where a simple bifurcation takes place. This seems to be the most commonplace type of bifurcation. For this case the arc length parametrization is adequate and monitoring the signs of det A(y(s)) is sufficient to detect the bifurcation points.

In recent years numerous authors have given numerical methods for finding starting points on a new branch $\tilde{C}$ ([9,30,31,35,42,49,57,67]. It is beyond our present scope to survey these various methods (see e.g. [31][42]). It should be noted however, that many methods fail to work in practice for secondary bifurcations. Keller [31] and Rheinboldt [49] give methods which appear to work for secondary bifurcations.

Since it is in the spirit of the homotopy continuation methods under discussion here, and it has been shown to work in practice for secondary bifurcations, we will briefly outline the method of topological perturbations [23][28][45][46] in the context of continuation algorithms.

Suppose that in the process of performing a continuation algorithm successive points y_{i-1}, y_i have been encountered such that det $A(y_{i-1})$det $A(y_i) < 0$, and some point b has been obtained which approximates a bifurcation point y^* on C between y_{i-1} and y_i, possibly even rather crudely e.g. $b = (y_{i-1} + y_i)/2$. Now choose an $\varepsilon > 0$ e.g. $\varepsilon = \|y_{i-1} - y_i\|/2$ and define a perturbed H mapping e.g.

$$(5.9) \qquad \bar{H}(y) = \begin{cases} H(y) \text{ if } \|y - b\| \geq \varepsilon \\ H(y) + (\varepsilon - \|y - b\|)d \text{ otherwise} \end{cases}$$

where $d \in \mathbb{R}^N - \{0\}$.

Let us consider the curve $\bar{C} \subset \bar{H}^{-1}(0)$ with $y_{i-1} \in \bar{C}$. By the definition of $\bar{H}$ (5.9), $b \notin \bar{C}$. By the parametrized Sard theorem (2.10), for almost all $d \in \mathbb{R}^N - \{0\}$, $\bar{C}$ contains no singular point of $\bar{A}$, the corresponding augmented Jacobian for $\bar{H}$. Hence $\bar{C}$ must exit from the ball $\|y - b\| \leq \varepsilon$ at a point w having the same index as the entry point y_{i-1}. Hence, $w \notin C$ since the exit point has index different from that of y_{i-1}.

The topological perturbation device has also been used by Peitgen et al. [46] to artificially induce a bifurcation in order to reach other components of $H^{-1}(0)$ or to make short cuts in traversing C.

6. Mesh refinement and homotopies

In the numerical solution of nonlinear operator equations such as integral equations or boundary value problems, it is necessary to make some sort of projection to a finite dimensional problem such as the simple example discussed in Section 2. Recently mesh refining devices have been employed to improve the numerical efficiency of finite difference methods. (For references, see [7] elsewhere in these proceedings.) We now indicate briefly how mesh refining methods may be formulated in the context of vector homotopies. To illustrate this, let us recall the simple eigenvalue problem discussed in Section 2,

(6.1) $$u'' + \lambda f(u) = 0 \quad u(0) = 0 = u(1), \text{ f smooth.}$$

The finite difference analogue which was mentioned was

(6.2) $$H_N(\lambda, \bar{u}) = A_N\bar{u} + \lambda B_N f(\bar{u}) = 0$$

where $\bar{u}$ represents the restriction of u to the mesh $x_i = \frac{i}{N+1}$, $i = 1, \ldots, N$ and

(6.3) $$A_N = (N+1)^2 \begin{pmatrix} -2 & 1 & & \bigcirc \\ 1 & \ddots & \ddots & \\ & \ddots & \ddots & 1 \\ \bigcirc & & 1 & -2 \end{pmatrix}_{N\times N} \qquad B_N = \frac{1}{12}\begin{pmatrix} 10 & 1 & & \bigcirc \\ 1 & \ddots & \ddots & \\ & \ddots & \ddots & 1 \\ \bigcirc & & 1 & 10 \end{pmatrix}_{N\times N}.$$

The accuracy tolerance desired of the approximating solution and the order of accuracy of the truncation error determine how large N should be chosen. In mesh refining methods, roughly speaking, the largest corrections and most iterations are performed when these operations are least costly, i.e., when N is small. The approximations which are obtained for small N can be extrapolated to be used as starting values as N becomes larger.

We now give an example of a homotopy formulation for mesh refinement for the finite difference discretization (6.2), (6.3) of (6.1). The formulation is

(6.4) $$H_N(\lambda, \bar{t}, \bar{u}) = (2 + \sum_{i=1}^{N-1} t_i)^2 A_N(\bar{t})\bar{u} + \lambda B_N(\bar{t}) f(\bar{u}) = \bar{0}$$

where $\lambda \in \Lambda$, $\bar{t} = (t_1, \ldots, t_{N-1}) \in \prod_1^{N-1} [0, 1]$. For our present purposes we may regard λ as being held fixed. The matrices $A_N(\bar{t})$, $B_N(\bar{t})$ can be generated by e.g. a bordering process

$$A_N(\bar{t}) = \begin{pmatrix} -2 & t_1 & & \bigcirc \\ t_1 & -2t_1 & \ddots & \\ & \ddots & \ddots & t_{N-1} \\ \bigcirc & & t_{N-1} & -2t_{N-1} \end{pmatrix} \qquad B_N(\bar{t}) = \frac{1}{12}\begin{pmatrix} 10 & t_1 & & \bigcirc \\ t_1 & 10t_1 & \ddots & \\ & \ddots & \ddots & t_{N-1} \\ \bigcirc & & t_{N-1} & 10t_{N-1} \end{pmatrix}.$$

The t_i's can be allowed to vary from 0 to 1 in various combinations to generate various kinds of refinements. For example, the commonplace refinement by halving is obtained by successively letting

$$t_{2^J} = t_{2^J+1} = \ldots = t_{2^{J+1}-1} = t$$

and letting t vary from 0 to 1. The refinement obtained by letting t_1, t_2, t_3, etc., vary successively from 0 to 1 doesn't offer any numerical advantage, but it can be used to formulate a definition of the numerically relevant solutions of a discretization. Let us again take the example (6.4).

Two solutions $\bar{u}^J$ and $\bar{u}^{J+1}$ to

$$H_J(\lambda, \bar{u}) = \bar{0} \qquad \text{and} \qquad H_{J+1}(\lambda, \bar{u}) = \bar{0}$$

respectively, are said to <u>correspond</u> to one another if the points

$$\bar{z}^J = (\lambda, \prod_1^J\{1\} \times \prod_{J+1}^N\{0\}, \bar{u}^J \times \prod_{J+1}^N\{0\})$$

$$\bar{z}^{J+1} = (\lambda, \prod_1^{J+1}\{1\} \times \prod_{J+2}^N\{0\}, \bar{u}^{J+1} \times \prod_{J+2}^N\{0\})$$

belong to the same homotopy path $\mathcal{P}^{J+1}$ in

$$H_N^{-1}(\bar{0}) \cap (\{\lambda\}\times\prod_1^J\{1\} \times [0, 1] \times \prod_{J+2}^N\{0\} \times \mathbb{R}^{J+1} \times \prod_{J+2}^N\{0\}).$$

See Figure (6.5).

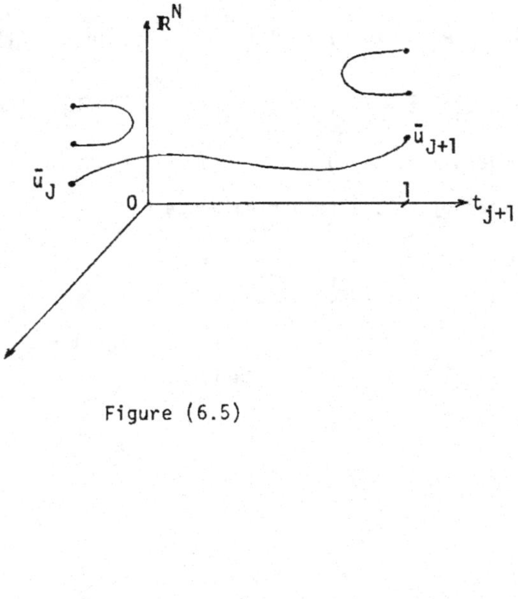

Figure (6.5)

A finite sequence $\{\bar{u}^j\}_{j=J}^N$ is called a <u>sequence of corresponding solutions</u> if $\bar{u}^j$ and $\bar{u}^{j+1}$ are corresponding solutions for $j = J, \ldots, N-1$.

If as $N \to \infty$ a convergent sequence of corresponding solutions is obtained, this sequence converges to a solution to the eigenvalue problem (6.1) for the fixed value λ.

A nonconverging sequence can be obtained in two ways (these are numerically irrelevant solutions):

i) There exists a subsequence $\{N_k\} \in \mathbf{Z}$ such that

$$\|\bar{u}^{N_k}\| \to \infty \quad \text{as } k \to \infty.$$

(Such solutions have been given by Gaines [21].)

ii) For some $J \geq 0$, the path $\mathcal{P}^{N+1}$ turns back to $t_{J+1} = 0$. (Such solutions have been given in [4][10] and have recently been studied by Peitgen, Saupe and Schmitt [47].)

As a final application of homotopy methods we mention the "eigenlength" problem which arises e.g. in neutron diffusion problems [43]. We shall again consider this problem in the context of our simple example of Section 2. The problem is: find b such that

$$\text{(6.6)} \qquad \begin{aligned} u'' + f(u) &= 0 \\ u(0) = 0 &= u(b) \end{aligned} \qquad (f \text{ smooth and } f(0) = 0)$$

has a nontrivial solution u.

By making the discretization $x_i = ih$, $h = \frac{b}{N+1}$ and using finite difference approximations as in Section 2, we have as in Section 2 the homotopy map

$$\text{(6.7)} \qquad H(h, \bar{u}) = h^{-2}A_N\bar{u} + B_N\bar{f}(\bar{u})$$

where A_N, B_N are as in (2.3)-(2.5). As a starting point we can choose $\bar{u} = \bar{0}$ and $h = 1$ and follow $H^{-1}(0)$ in the direction of decreasing h. From the bifurcation points $(h_k, \bar{0})$ on $[0, 1] \times \{\bar{0}\} \subset H^{-1}(\bar{0})$ we obtain the approximations

$$b_k = (N+1)h_k$$

to the eigenlengths.

REFERENCES

[1] R.Abraham and J.Robbin. Transversal Mappings and Flows, Benjamin, New York-Amsterdam, 1967.

[2] J.C.Alexander. The topological theory of an embedding method, Continuation methods, H.Wacker, ed., Academic Press, New York, 1978.

[3] J.C.Alexander and J.A.Yorke. The homotopy continuation method: Numerically implementable topological procedures, Trans.Amer.Math.Soc. 242 (1978),271-284

[4] E.L.Allgower. On a discretization of $y'' + \lambda y^k = 0$, Topics in Numerical Analysis II, J.J.H.Miller, ed., Academic Press, New York, pp. 1-15, 1975.

[5] E.L.Allgower and K.Georg. Simplicial and continuation methods for approximating fixed points and solutions to systems of equations, SIAM Review 22 (1980), 28-85.

[6] E.L.Allgower and K.Georg. Homotopy methods for approximating several solutions to nonlinear systems of equations, in Numerical Solution of Highly Nonlinear Problems, W. Förster, ed., North-Holland, 1980.

[7] E.L.Allgower, K.Böhmer and S.F.McCormick. Discrete correction methods for operator equations, these proceedings.

[8] P.Anselone and R.Moore. An extension of the Newton-Kantorovich method for solving nonlinear equations with an application to elasticity, J. Math.Anal.Appl., 13 (1966), 476-501.

[9] E.Bohl. Chord techniques and Newton's method for discrete bifurcation problems, Numer.Math., 34 (1980), 111-124.

[10] E.Bohl. On the bifurcation diagram of discrete analogues for ordinary bifurcation problems, Math.Meth.in the Appl.Sci., 1 (1979), 566-571.

[11] F.J.Branin,Jr. Widely convergent method for finding multiple solutions of simultaneous nonlinear equations, IBM J.Res.Develop. 16 (1972), 504-522.

[12] F.J.Branin,Jr. and K.S.Hoo. A method for finding multiple extrema of a function of n variables, in: Numerical Methods for Nonlinear Optimization, F.Lootsma, ed., Academic Press, pp. 231-327, 1972).

[13] S.N.Chow, J.Mallet-Paret and J.A.Yorke. Finding zeros of maps: Homotopy methods that are constructive with probability one, Math.Comput., 32 (1978), 887-899.

[14] L.O.Chua and A.Ushida. A switching-parameter algorithm for finding multiple solutions of nonlinear resistive circuits, IEEE Trans.Circuit Theory and Applications, 4 (1976), 215-239.

[15] M.G.Crandall and P.H.Rabinowitz. Bifurcation from simple eigenvalues, J.Func.Anal., 8 (1971), 321-340.

[16] D.Davidenko. On a new method of numerically integrating a system of nonlinear equations, Dokl.Akad.Nauk SSSR, 88 (1953), 601-604. (In Russian)

[17] P.Deufelhard. A modified continuation method for the numerical solution of nonlinear two-point boundary value problems by shooting techniques, Numer.Math. 26 (1976), 327-343.

[18] P.Deufelhard. A step size control for continuation methods and its special application to multiple shooting techniques, Numer.Math. 33 (1979), 115-146.

[19] F.-J.Drexler. Eine Methode zur Berechnung sämtlicher Lösungen von Polynomgleichungssystemen, Numer.Math., 29 (1977), 45-58.

[20] F.-J.Drexler. A homotopy method for the calculation of all zeros of zero-dimensional polynomial ideals, Continuation methods, H.Wacker, ed., Academic Press, New York, pp. 69-94, 1978.

[21] R.E.Gaines. Difference equations associated with boundary value problems for second-order nonlinear ordinary differential equations, SIAM J.Num. Anal., 11 (1974), 411-434.

[22] C.B.Garcia and W.I.Zangwill. Global continuation methods for finding all solutions to polynomial systems of equations in n variables, Int'l. Symp. on External Methods and Sys.Anal., Austin, TX, Univ. of Chicago, Dept. of Economics and Graduate School of Business, Report 7755, 1977.

[23] K.Georg. On tracing an implicitly defined curve by quasi-Newton steps and calculating bifurcations by local perturbations, to appear in SIAM J.Sci.Stat.Computing.

[24] K.Georg. Numerical integration of the Davidenko equation, these proceedings.

[25] H.Hackl, H.Wacker and W.Zulehner. Aufwandsoptimale Schrittweitensteuerung by Einbettungsmethoden, in Constructive Methods for Nonlinear Boundary Value Problems and Nonlinear Oscillations, Birkhauser, Basel, ISNM 48 (1979), eds. J.Albrecht, L.Collatz and K.Kirchgassner, pp. 48-67.

[26] C.Haselgrove. Solution of nonlinear equations and of differential equations with two-point boundary conditions, Comput.J. 4 (1961), 255-259.

[27] M.Hirsch and S.Smale. On algorithms for solving f(x) = 0, Comm.Pure Appl. Math., 32 (1979), 281-312.

[28] H.Jurgens, H.-O.Peitgen and D.Saupe. Topological perturbations in the numerical study of nonlinear eigenvalue and bifurcation problems, in Analysis and Computation of Fixed Points, ed. S.Robinson, Academic Press.

[29] R.B.Kearfott. A derivative-free arc continuation method and a bifurcation technique, preprint (also see these proceedings).

[30] J.P.Keener and H.B.Keller. Perturbed bifurcation theory, Arch.Rat.Mech.Anal. 50 (1973), 159-175.

[31] H.B.Keller. Numerical solution of bifurcation and nonlinear eigenvalue problems, in Applications of Bifurcation Theory, ed.: P.H.Rabinowitz, Academic Press, New York, pp. 359-384, 1977.

[32] H.B.Keller. Global homotopies and Newton methods, in Recent Advances in Numerical Analysis, eds: C.deBoor and G.H.Golub, Academic Press, New York, pp. 73-94, 1978.

[33] R.B.Kellogg, T.Y.Li and J.Yorke. A constructive proof of the Brouwer fixed point theorem and computational results, SIAM J.Numer.Anal., 4 (1976), 473-483.

[34] E.Lahaye. Une méthode de résolution d'une catégorie d'équations transcendantes, C.R.Acad.Sci., Paris, 198 (1934), 1840-1842.

[35] W.F.Langford. Numerical solution of bifurcation problems for ordinary differential equations, preprint, McGill University, Montreal (1976).

[36] T.Y.Li. Numerical aspects of the continuation method-flow charts of a simple algorithm, Proc. of Symp. on analysis and computation of fixed points, Madison, WI, ed. S.M.Robinson, Academic Press, New York, 1979.

[37] R.Menzel and H.Schwetlick. Über einen Ordnungsbegriff bei Einbettungsalgorithm zur Lösung nichtlinearer Gleichungen, Computing 16 (1976), 187-199.

[38] R.Menzel and H.Schwetlick. Zur Lösung parameterabhängiger nichtlinearer Gleichungen mit singulären Jacobi-Matrizen, Numer.Math., 30 (1978), 65-79.

[39] R.Menzel. Ein implementierbarer Algorithmus zur Lösung nichtlinearer Gleichungssysteme bei schwach singulärer Einbettung, Beiträge zur Numerischen Mathematik, 8 (1980), 99-111.

[40] G.Meyer. On solving nonlinear equations with a one-parameter operator imbedding, SIAM J.Numer.Anal., 5 (1968), 739-752.

[41] J.W.Milnor. Topology from the Differentiable Viewpoint, University Press of Virginia, Charlottesville, VA, 1969.

[42] H.D.Mittelman and H.Weber. Numerical treatment of bifurcation problems, University of Dortmund, preprint, 1979.

[43] Paul Nelson, Jr. Subcriticality for submultiplying steady-state neutron diffusion, in Nonlinear diffusion, ed. John Nohel, Research Notes in Math. 14, Pitman, London.

[44] J.M.Ortega and W.C.Rheinboldt. Iterative Solutions of Nonlinear Equations in Several Variables, Academic Press, New York-London, 1970.

[45] H.O. Peitgen and H.O. Walther, eds., Functional Differential Equations and Approximation of Fixed Points, Springer L.N.730

[46] H.O. Peitgen and M. Prüfer. The Leray Schauder continuation method is a constructive element in the numerical study of nonlinear eigenvalue and bifurcation problems, in [45], pp. 326-409.

[47] H.O. Peitgen, D. Saupe and K. Schmitt. Nonlinear elliptic boundary value problems versus their finite difference approximations ..., J. reine angew. Mathematik 322 (1981), 74-117.

[48] P.H.Rabinowitz. Some global results for nonlinear eigenvalue problems, J.Func.Anal., 7 (1971), 487-513.

[49] W.C.Rheinboldt. Numerical methods for a class of finite-dimensional bifurcation problems, SIAM J.Numer.Anal., 15 (1978), 1-11.

[50] W.C.Rheinboldt. Solution field of nonlinear equations and continuation methods, SIAM J.Numer.Anal., 17 (1980), 221-237.

[51] E.Riks. The application of Newton's Method to the problem of elastic stability, J.Appl.Mech.Techn.Phys., 39 (1972), 1060-1065.

[52] C.Schmidt. Approximating differential equations that describe homotopy paths, Univ. of Chicago School of Management Science Report 7931.

[53] W.F.Schmidt. Adaptive step size selection for use with the continuation method, Int'l. J.for Numer.Meths. in Engrg, 12 (1978), 677-694.

[54] H.Schwetlick. Ein neues Princip zur Konstruktion implementierbarer, global konvergenter Einbettungsalgorithmen, Beiträge Numer.Math., 4-5 (1975-6), 215-228; 201-206.

[55] L.F.Shampine and M.K.Gordon. Computer Solution of Ordinary Differential Equations: The Initial Value Problem, Freeman Press, San Francisco, 1975.

[56] G.Shearing. Ph.D. Thesis, Manchester (1960).

[57] R.Seydel. Numerische Berechnung von Verzweigungen bei gewöhnlichen Differentialgleichungen, TUM-Math-7736 Technische Universität München,1977.

[58] S.Smale. A convergent process of price adjustment and global Newton methods, J.Math.Econ., 3 (1976), 1-14.

[59] G.A.Thurston. Continuation of Newton's method through bifurcation points, J.Appl.Mech.Tech.Phys., 36 (1969), 425-430.

[60] H.Wacker. Minimierung des Rechenaufwandes für spezieller Randwertprobleme, Computing, 8 (1972), 275-291.

[61] H.Wacker, E.Zarzer and W.Zulehner. Optimal step size control for the globalized Newton methods, in Continuation Methods, ed. H.Wacker, Academic Press, New York, 1978, 249-277.

[62] H.Wacker, ed. Continuation Methods, Academic Press, New York, 1978.

[63] E.Wasserstrom. Numerical solutions by the continuation method, SIAM Review, 15 (1973), 89-119.

[64] L.T.Watson. An algorithm that is globally convergent with probability one for a class of nonlinear two-point boundary value problems, SIAM J.Num. Anal., 16 (1979), 394-401.

[65] L.T.Watson. A globally convergent algorithm for computing fixed points of C maps, Appl.Math. and Computation, 5 (1979), 297-311.

[66] L.T.Watson and D.Fenner. Chow-Yorke algorithm for fixed points or zeros of C^2 maps, ACM Trans. on Math. Software, 6 (1980), 252-259.

[67] H.Weber. Numerische Behandlung von Verzweigungsproblemen bei gewöhnlichen Differentialgleichungen, Numer.Math., 32 (1979), 17-29.

DISCRETE CORRECTION METHODS
FOR
OPERATOR EQUATIONS

BY

E.L. ALLGOWER,*
K. BÖHMER** AND
S. MC CORMICK*

*) Mathematics Department
Colorado State University
Fort Collins, Colorado 80523
USA

**) Fachbereich Mathematik
Universität Marburg
D-3550 Marburg

ABSTRACT

A numerical method is developed for approximating the exact solution of an operator equation on a certain finite grid to within a desired tolerance. The method incorporates discretizations which admit asymptotic expansions of the error, mesh refinement strategies and discrete Newton methods. An algorithm is given in which essentially the largest adequate mesh size is used. A homotopy method for obtaining good starting values for a Newton-type method applied to a coarse grid discretization and the connection that our approach has with multigrid methods are discussed. Numerical examples for two-point boundary value problems and elliptic boundary value problems are given.

1. Introduction

The aim of this paper is to develop a highly efficient numerical technique for solving nonlinear operator equations which are posed as follows

$$\mathcal{P} \quad \begin{cases} & Fy = 0 \\ \text{where} & F: \begin{array}{l} \mathcal{D} \subseteq E \to \hat{E} \\ y \mapsto Fy \end{array}, \\ & \text{and there exists a unique } z \in \mathcal{D} \text{ such that } Fz = 0. \end{cases}$$

We assume in $\mathcal{P}$ that E and $\hat{E}$ are Banach spaces and that F is a Frechét differentiable map.

The method developed in this paper incorporates aspects of two numerical techniques which have recently been successfully applied to the numerical solution of nonlinear two-point boundary value problems viz. the discrete Newton method (Böhmer [6]) and a mesh refinement method based upon a mesh independence principle for Newton's method (Allgower-McCormick [3] and Allgower, McCormick and Pryor [4]). Among the applications which the formulation $\mathcal{P}$ permits are boundary value problems for ordinary and partial differential equations and integral equations. Without making explicit assumptions concerning the smoothness of F, we shall implicitly assume that F is sufficiently smooth so as to admit certain asymptotic expansions that are given below. We also assume the validity of formulations of Newton's method

(1.1) $\qquad F'(y_{\ell-1})(y_\ell - y_{\ell-1}) = -Fy_{\ell-1}, \quad \ell = 1,2,\dots$

and of modifications of Newton's method such as

(1.2) $\qquad F^*(y_0)(y_\ell - y_{\ell-1}) = -Fy_{\ell-1}, \quad \ell = 1,2,\dots$.

Here $y_0 \in \mathcal{D}$ and $F^*(y_0)$ may be an approximation to $F'(z)$ of a type to be made more specific below. For example $F^*(y_0) = F'(y_0)$ may be used.

For the numerical solution of P, we introduce finite-dimensional discrete approximations P^h to P, for $h > 0$, of the form

$$P^h \begin{cases} F^h\eta^h = (\phi^h F)\eta^h = 0 \\ \text{where } F^h := \phi^h F \quad : \begin{cases} D^h \subseteq E^h \to \hat{E}^h \\ \eta^h \mapsto F^h\eta^h , \end{cases} \\ \text{and for sufficiently small h, there exists a unique } \zeta^h \in D^h \\ \text{such that } F^h\zeta^h = 0. \end{cases}$$

Here E^h, $\hat{E}^h$ are finite dimensional spaces of the same dimension corresponding to discrete approximations of operators (e.g., by means of finite differences, collocations or quadratures upon a uniform mesh of size $h > 0$). The local uniqueness of a solution ζ^h can always be established for P if $\phi^h F$ is a discretization which is stable and consistent with respect to F. Although there are many important instances in which it is desirable to employ nonuniform meshes, this aspect shall not fall within our present scope. In section 2 we shall give explicit examples of discretizations P^h for some boundary value problems P.

Especially for non linear boundary value problems in differential equations and Fredholm problems in integral equations, the solution ζ^h of $F^h\zeta^h = 0$ is usually computed via a Newton method

$$(1.3) \qquad (F^h)'(\xi^h_{\ell-1})(\xi^h_\ell - \xi^h_{\ell-1}) = -F^h\xi^h_{\ell-1} \qquad , \quad \ell = 1,2,\ldots .$$

The approach given below is especially worthwhile and efficient whenever (1.3) is used. Therefore and because of the equidistant step size, our method is not advisable for initial value problems nor for stiff problems when additional step size strategies are not incorporated.

A further assumption concerning P^h shall be that $\zeta^h - \Delta^h z$ admits a global asymptotic error expansion so that

$$(1.4) \qquad \zeta^h = \Delta^h\{z + \sum_{\nu=p}^{\bar{q}} h^\nu e_\nu(z)\} + O(h^{\bar{q}+\alpha})$$

for some $\alpha > 0$ and positive integers $\bar{q} \geq p$. Here

$$\Delta^h : E \to E^h$$

is an appropriate projection operator (e.g., the restriction of a function to a grid) and the coefficients $e_\nu(z) \in E$, $\nu = p,\ldots,\bar{q}$, are independent of h. Examples in which asymptotic expansions hold are studied in many papers (e.g., in Richardson [31], Stetter [44,45] Pereyra [28] and Böhmer [6,7,8]) and are briefly reviewed for two examples in section 3. The asymptotic expansions (1.4) shall be employed to obtain improved approximations to z of increasingly higher orders by discrete Newton methods. Note that we have assumed only the *existence* of an asymptotic expansion (1.4).

The discrete Newton methods are iterative procedures of the form

$$(1.5) \qquad (\Phi^h F)'(\zeta_0^h)(\zeta_\ell^h - \zeta_{\ell-1}^h) = -\Omega^h F z_{\ell-1}, \quad \ell = 1,2,\ldots$$

where ζ_0^h is, in practice, a sufficiently good approximation to ζ^h, Ω^h is a projection (e.g., Δ^h) and $z_{\ell-1}$ is a suitable extension of $\zeta_{\ell-1}^h$. We shall also allow the replacement of $(\Phi^h F)'$ by a suitable approximation $(\Phi^h F)^*$. All of these elements shall be more precisely defined below. The crucial property of the discrete Newton method (1.5) is that it generates iterates ζ_ℓ^h which admit an asymptotic expansion of the form

$$(1.6) \qquad \zeta_\ell^h = \Delta^h\{z + \sum_{\nu=(\ell+1)p}^{q_\ell} h^\nu e_{\nu,\ell}(z)\} + O(h^{q_\ell+\alpha_\ell}), \quad \ell = 1,2,\ldots$$

for some $\alpha_\ell > 0$ and positive integers $p \leq q_\ell \leq q$.

The aim in performing the discrete Newton iterations (1.5) is to obtain approximations to $\Delta^h z$ which have increasingly higher accuracy. By (1.6), this will only be accomplished when h is so small that the first terms of the summation in (1.6) are dominant. Since the coefficients $e_\nu(z)$ in (1.4) and $e_{\nu,\ell}(z)$ in (1.6) are independent of h, this will certainly hold for sufficiently small h. Thus, for the discrete Newton process we make use of the following

Fundamental Assumption:

$$
(1.7) \quad \begin{cases} \text{At any stage of the discrete Newton process (1.5), h is so small} \\ \text{that by rewriting (1.6) as} \\ \zeta_\ell^h - \Delta^h z = h^{(\ell+1)} \{\Delta^h e_{(\ell+1)p,\ell} + \Delta^h d_\ell(h)\}, \text{ we have} \\ \| \Delta^h e_{(\ell+1)p,\ell} \| > \| \Delta^h d_\ell(h) \| . \end{cases}
$$

This means that the leading terms in (1.6), those with smallest indices, determine the behaviour of the error. For computational reasons we will try to make h as big as possible without violating (1.7).

Unless it is otherwise specified, the uniform norm shall be used throughout this paper.

The discrete correction method which we develop here involves the blending of the discrete Newton method with monitoring of the mesh size based upon the mesh independence principle, so as to dynamically determine with high reliability a maximal mesh size h which is adequate to attain a desired accuracy in approximating z using the diescretization P^h. During this process, the mesh sizes are also reliably monitored to verify that the Fundamental Assumption (1.7) holds.

The organization of our paper is as follows:
In § 2 we discuss as examples two sample classes of problems P, corresponding discretizations P^h and the particular characteristics of their asymptotic expansions (1.4), (1.6). In § 3 the discrete Newton method is presented in detail and the role of the Fundamental Assumption

is discussed. In § 4 the discrete correction method is described. The primary novelty is that the corrections of the defect are not made with respect to approximations to z, but rather to the discrete solution on the next grid in the refinement sequence. In § 5 the process for the dynamical determination of the final mesh size h is presented. In § 6 methods for empirically monitoring the validity of the Fundamental Assumption and other assumptions are discussed. In § 7 the steps of a discrete correction algorithm are sketched. § 8 presents the necessary details of our algorithm for the examples treated here. Numerical examples are presented in § 9. Since the discrete correction method relies to some extent on the availability of a reasonably good approximation ζ_o^h on a relatively coarse mesh, we discuss in Appendix 1 how a ζ_o^h can be obtained by a continuation method. In Appendix 2, some aspects of the multigrid method are presented that relate to its use in connection with our present methods.

Before proceeding, it is perhaps worthwhile to discuss the relationship of the present discrete correction method with numerical methods which are currently extant. In the numerical methods which use a fixed order discretization, even if efficient mesh refinement methods are incoporated (c.f., Allgower and McCormick [3] and Allgower, McCormick and Pryor [4] or the multi-grid methods as in Brandt [10]), if it is desired to obtain a highly accurate approximation to z, it is usually necessary to solve the discretized problem on a correspondingly fine mesh. This results in the situation that the dimension of E^h is often very large and the process of solving the discretized problem requires substantial computational work. Furthermore, the approximation of z at a large numer of "co-ordinates" is usually not required.

The method of Richardson extrapolation [31] may yield approximations which have higher orders of accuracy on a relatively coarse mesh, but it also requires solving a lower order discretization on a sequence of refining meshes.

The methods of deferred corrections (Fox-Goodwin [14]) iterated deferred corrections (Pereyra [28]) error estimations (Zadunaiski [38,39,40]) and the latter's extension to iterated defect corrections (Stetter [34,35] Frank-Hertling-Ueberhuber [15,16,17,18,19]) are more closely related to the discrete Newton method in that, for the most part, the discretizations are used as much as possible on the same relatively coarse mesh. These methods essentially are obtained by adding appropriate perturbation terms to the discrete equation $F^h\zeta^h = 0$ to define $F^h_\ell\zeta^h_\ell = 0$. These solutions ζ^h_ℓ are usually different from the ζ^h_ℓ in (1.5), but they also satisfy the asymptotic relation (1.6) with different $e_{\nu,\ell}(z)$. So in every iteration step a system of nonlinear equations has to be solved for which an increasingly better initial guess, $\zeta^h_{\ell-1}$, is available. These nonlinear systems are usually solved with a Newton method. However, it is not known whether in computing ζ^h_ℓ from $\zeta^h_{\ell-1}$ one can choose the same matrix for all values of ℓ and for all iterations.

In discrete Newton methods only the linear systems (1.5) need to be solved where the same matrix is used for every ℓ and only one iteration is necessary in every transition from $\zeta^h_{\ell-1}$ to ζ^h_ℓ. Checking the validity of (1.7) is done by computing ζ^h_ℓ and $\zeta^{\bar{h}}_\ell$ for some $\bar{h} > h$. We use a mesh refinement strategy to compute, say, $\zeta^h = \zeta^h_o$ from a known $\zeta^{\bar{h}}_o$. In this paper we will see that even the iterative improvement of the $\zeta^{\bar{h}}$ to ζ^h can be achieved by the matrix $(\phi^{\bar{h}}F)'(\zeta^{\bar{h}}_o)$

in (1.5). This requires but one suitable matrix for every h and leads to considerable savings in computation.

2. Asymptotic error expansion and fundamental assumption.

Before presenting the main steps of our algorithm, we introduce some notation and recall basic properties concerning the asymptotic expansion of global discretization errors. We also discuss the meaning of an asymptotic expansion in the context of discrete Newton methods and discrete correction methods.

The elements of the infinite dimensional spaces $E;\hat{E}$ and their respective finite dimensional spaces $E^h;\hat{E}^h$ in (1.1) and (1.3) are denoted by e, $\bar{e}, f, .., u, y, z$; $\hat{e}, \hat{f}, .., \hat{y}, \hat{z}$ and respectively by greek letters $\varepsilon^h, \gamma^h, ..., \upsilon^h, \xi^h, \eta^h, \zeta^h$; $\varepsilon^h, \hat{\gamma}^h, .., \hat{\xi}^h, \hat{\eta}^h, \hat{\zeta}^h$, where we shall sometimes for convenience drop the superscripts h. Operators between E and $\hat{E}$ (we shall regard these spaces as corresponding to h=0) or between E^h and $\hat{E}^h$ (i.e., between spaces with the same h) are denoted by F,G,.. or F^h, G^h respectively. For operators on spaces for different values of h, we use capital greek letters, e.g.,

$$
(2.1)\qquad \left\{\begin{array}{l}
\Phi^h: (D \to \hat{E}) \to (D^h \to \hat{E}^h), \\
\Omega^h: \hat{d} \in \hat{E} \to \Omega^h \hat{d} \in \hat{E}^h, \\
\Delta^h: E \to E^h, \hat{\Delta}^h: \hat{E} \to \hat{E}^h, \\
\Delta_h^{\bar{h}}: E^h \to E^{\bar{h}}, \hat{\Delta}_h^{\bar{h}}\ \hat{E}^h \to \hat{E}^{\bar{h}}, \\
\Delta_h: E^h \to E, \text{ etc.}
\end{array}\right.
$$

The subscript denotes the domain, the superscript the range, and h = 0 is understood by its absence. The Δ^h are usually projections, e.g., restrictions to grids, while the Δ_h are extension operators, usually realized by interpolation or approximation.

Following the terminology of Stetter [34], a <u>discretization method</u> M, applicable to the original problem P, is a sequence $M = \{E^h, \hat{E}^h, \Delta^h, \hat{\Delta}^h, \Phi^h\}_{h \in \mathbb{H}}$, $\mathbb{H} \subset (0, h_0]$, $\inf \mathbb{H} = 0$, where $E \subseteq E_0, \hat{E}, E^h, \hat{E}^h$ are Banach spaces with $\dim E^h = \dim \hat{E}^h < \infty$, E continuously embedded into E_0; $\Delta^h, \hat{\Delta}^h$ are bounded linear operators, with $\lim_{h \to 0} \| \Delta^h y \|_{E^h} = \| y \|_E$ for $y \in E$, $\lim_{h \to 0} \| \hat{\Delta}^h \hat{y} \|_{\hat{E}^h} = \| \hat{y} \|_{\hat{E}}$ for $\hat{y} \in \hat{E}$, and $\Phi^h : C \to (E^h \to \hat{E}^h)$ where $C \subset (E \to \hat{E})$ and $F \in C$. Whenever the context indicates which norm has to be used, we write $\| \cdot \|$. A sequence of discretizations or discrete problems P^h is then formulated as

$$
(2.2) \begin{cases} P^h : = \{E^h, \hat{E}^h, F^h := \Phi^h F\}_{h \in \mathbb{H}}, \\ F^h : \mathcal{D}^h := \Delta^h \mathcal{D} \to \hat{E}^h, \ 0 \in F^h \mathcal{D}^h, \\ \text{find the unique solution } \zeta^h \in \mathcal{D}^h \text{ of } F^h \eta^h = 0. \end{cases}
$$

The usual assumptions, namely, consistency, stability of P^h and smoothness of P, guarantee then that $F^h \eta^h = 0$ is uniquely solvable in $\mathcal{D}^h$ and that $\lim_{h \to 0} \| \zeta^h - \Delta^h z \| = 0$ (convergence). As detailed examples of the general theory, we will discuss boundary value problems (B.V.P.s) for systems of ordinary differential equations (O.D.E.s) and for (elliptic) partial differential equations (P.D.E.s).

<u>Example 2.1:</u> <u>*Box-scheme for B.V.P.s in systems of O.D.E.s.*</u>

Let the subscript b indicate the box scheme and let

$$(2.3)\quad \begin{cases} E:=C^1([a,b],\mathbb{R}^n)\subset E_o:=C([a,b],\mathbb{R}^n):=\{y\mid y:[a,b]\to\mathbb{R}^n, y\in C[a,b]\}.\\ \hat{E}:=C([a,b],\mathbb{R}^n)\times\mathbb{R}^n,\ f\in C([a,b]\times F_1,\mathbb{R}^n), g\in C(F_2\times F_3),\\ F_i\subseteq\mathbb{R}^n, i=1,2,3, F_b y:=\{y'(\cdot)-f(\cdot,y(\cdot)), g(y(a),y(b))\}\\ \text{for } y\in \mathcal{D}\subseteq E \text{ and } (x,y(x))\in[a,b]\times F_1, y(a)\in F_2, y(b)\in F_3,\\ \mathcal{P}_b:=\{E,\hat{E},F_b\}, \text{ find the unique solution } z\in\mathcal{D}\\ \text{for } F_b y=0. \end{cases}$$

The unique solvability of $F_b y = 0$ in $\mathcal{D}$ is guaranteed by imposing certain conditions on $\mathcal{D}$, f, g which we do not discuss here (c.f., Keller [22,23]). As indicated in the introduction, we shall confine our discussion to uniform grids with constant mesh size h. The grid $\mathbb{G}^h$ is defined with $h:=(b-a)/m$ and the abbreviations $x_\nu:=a+\nu h$, $x_{\nu+1/2}:=a+(\nu+1/2)h$ as $\mathbb{G}^h: = \mathbb{G}:=\{x_o<x_1<\ldots<x_m\}, \mathbb{G}_o:=\mathbb{G}_o^h:=\mathbb{G}^h\setminus\{b\}$. With $\eta_\nu:=\eta^h(x_\nu)$ the box scheme is then given as:

$$(2.4)\quad \begin{cases} E^h:=\{\eta^h:\mathbb{G}^h\to\mathbb{R}^n\}, \hat{E}^h:=\{\hat{\eta}^h\mid\hat{\eta}^h=(\tilde{\eta}^h,\alpha):\tilde{\eta}^h:\mathbb{G}_o^h\to\mathbb{R}^n,\alpha\in\mathbb{R}^n\},\\ \Delta^h y:=y\,|_{\mathbb{G}^h}, \hat{\Delta}^h\hat{y}=\hat{\Delta}^h(\tilde{y},\alpha):=(y(\cdot+\frac{h}{2})|_{\mathbb{G}_o^h},\alpha) \text{ for } \tilde{y}\in C([a,b],\mathbb{R}^n),\\ F_b^h\eta^h:=(\Phi_b^h F_b)\eta^h:=\{\frac{\eta_{\nu+1}-\eta_\nu}{h}-f(x_{\nu+1/2},\frac{\eta_\nu+\eta_{\nu+1}}{2}),\\ \qquad\qquad \nu=0,1,\ldots m-1, g(\eta_o,\eta_m)\}\\ \mathcal{P}_b^h:=\{E_b^h,\hat{E}_b^h,F_b^h\}\ h\in\mathbb{H}, \mathbb{H}:=\{h:=\frac{b-a}{m}\mid m\in\mathbb{N}\},\\ \text{find the unique solution } \zeta^h\in\mathcal{D}^h=\Delta^h\mathcal{D} \text{ for } F_b^h\eta^h=0. \end{cases}$$

In (2.3) and (2.4) we take as norms for E_o, E, $\hat{E}^h$ the usual sup-, Sobolev-, and maximum norms. Under certain assumptions on the functions f, g in (2.3) the box scheme (2.4) provides a consistent and stable discretization for P_b and we have, for the solution ζ^h and z, that

$$\lim_{h\to 0} \|\zeta^h - \Delta^h z\| = 0. \tag{2.5}$$

For convenience we have omitted the subscript b for $E, \hat{E}, E^h$ etc. □

Example 2.2: *Modified five point star scheme for Dirichlet B.V.P.s in elliptic P.D.E.s on general regions.* For reasons of simplicity, we shall consider the following elliptic B.V.P., with the subscript s in F_s representing the modified five point star scheme:

$$F_s u := \begin{cases} -\Delta u(x,y) + f(x,y,u(x,y)) = -u_{xx}(x,y) - u_{yy}(x,y) \\ \quad +f(x,y,u(x,y)) \text{ for } (x,y) \in R, \\ u(x,y) - g(x,y) \quad \text{for } (x,y) \in \partial R. \end{cases} \tag{2.6}$$

A presentation for more general linear and nonlinear elliptic equations is given in Böhmer [7, 8]. Let $C(R)$, etc., stand for $C(R, \mathbb{R})$, etc., and let

$$\begin{cases} E := C^2(R) \subset E_o := C(R) \\ \hat{E} := C(R) \times C(\partial R), \ f \in C(R \times F), F \subseteq \mathbb{R}, \ g \in C(\partial R), \\ F_s u \text{ in (2.6) for } u \in D \subseteq E \text{ and } (x,y,u(x,y)) \in R \times F \\ P_s := \{E, \hat{E}, F_s\}, \text{ find the unique solution} \\ z \in D \text{ for } F_s u = 0 \end{cases} \tag{2.7}$$

In Bers [3], conditions for the unique solvability of $F_s u = 0$ are given. With a certain mesh size $h \in (0, h_o]$, we introduce the extended grid $\mathbb{E}^h$ and the grid lines $\mathbb{L}^h$ as

$$(2.8)\begin{cases} \mathbb{E}^h := \{(x,y)\in \mathbb{R}^2 \mid x=n_1h, y = n_2h,\ n_1,n_2 \in \mathbb{Z}\} \\ \mathbb{L}^h := \{(x,y)\in \mathbb{R}^2 \mid x=n_1h \text{ or } y = n_2h, n_1,n_2 \in \mathbb{Z}\}, \end{cases}$$

so one co-ordinate is held fixed in $\mathbb{L}^h$ (e.g., $x=n_1h$ and $y \in \mathbb{R}$ and vice versa). Since we are mainly interested in general (open) regions R having smooth boundary ∂R, we have to distinguish the sets

$$(2.9)\begin{cases} \mathbb{G}_r^h := \{(x,y)\in R \mid (x,y)\pm he_i \in R,\ i=1,2\} \cap \mathbb{E}^h \\ \quad = \{\text{regular grid points}\}, \\ \mathbb{G}_i^h := \{(x,y)\in R \mid \exists_i : (x,y)+he_i \notin R, \\ \quad\quad \text{or } (x,y)-he_i \notin R\} \cap \mathbb{E}^h \\ \quad = \{\text{irregular grid points}\}, \\ \mathbb{G}_b^h := \partial R \cap \mathbb{L}^h = \{\text{boundary grid points}\}, \\ \mathbb{G}^h := \mathbb{G}_r^h \cup \mathbb{G}_i^h \cup \mathbb{G}_b^h = \{\text{grid points}\}, \end{cases}$$

where e_1 and e_2 are the unit vectors in the x and y directions, respectively. For regular grid points we use the well known five point star discretization

$$(2.10)\begin{cases} h^2\ (\Phi_s^h\ F_s)\upsilon^h(x,y) := -\upsilon^h(x+h,y)\ -\upsilon^h(x-h,y) \\ -\upsilon^h(x,y+h)-\upsilon^h(x,y-h)+4\upsilon^h(x,y)+h^2 f(x,y,h(x,y)) \\ \text{for } (x,y) \in \mathbb{G}_r^h\ . \end{cases}$$

For an irregular grid point $(x,y) \in \mathbb{G}_i^h$, let, say, $(x+he_1,y)\notin R$. Then let h be so small and ∂R be so smooth that there is exactly one boundary point between $(x,y)\in R$ and $(x+he_1,y)\notin R$, namely, $(x+(1-s)h,y)\in \mathbb{G}_h^h$, where $0\leq s<1$. We want to avoid the value $\upsilon^h(x+h,y)$ for the noninterior point $(x+h,y)$. Therefore (see Pereyra-Proskurowski-Widlund [29] and Böhmer [8])

$$(2.11)\begin{cases} \text{for } (x,y)\in \mathbb{G}_i^h, \text{e.g. } (x+h,y)\notin R, \text{substitute} \\ \text{in (2.10) for } \upsilon^h(x+h,y) \text{ the value obtained} \\ \text{as the value of the polynomial } P_k \text{ of degree} \\ k \text{ which interpolates the } k+1 \text{ points} \\ (x+(1-s)h, \upsilon^h(x+(1-s)h,y)), (x-\nu h, \upsilon^h(x-\nu h,y))\ \nu=0,\dots k-1. \end{cases}$$

For $s=0$ we have $(x+h,y) \in \mathbb{G}_b^h$ and (2.11) reduces to $\upsilon^h(x+h,y)=g(x+h,y)$, see ((2.12)). A treatment with more complete formulas may be found in Böhmer [7,8]. For $s\approx 1$, one has to modify (2.11) as decribed in [7,8]. Finally, we add the boundary values

$$(2.12)\quad (\Phi_s^h F_s)\upsilon^h(x,y) := \upsilon^h(x,y) - g(x,y) \text{ for } (x,y)\in \mathbb{G}_b^h.$$

The operators Δ^h and $\hat{\Delta}^h$ are defined similarly to (2.4) as

$$(2.13)\begin{cases} \Delta^h u := u|_{\mathbb{G}^h} \quad \text{for } u \in C^2(R), \\ \hat{\Delta}^h \hat{u} := \hat{\Delta}^h(\tilde{u},\tilde{v}) := (\tilde{u}|_{\mathbb{G}_r^h \cup \mathbb{G}_i^h}, \tilde{v}|_{\mathbb{G}_b^h}), \\ \qquad \text{where } \tilde{u} \in C(R) \text{ and } \tilde{v} \in C(\partial R). \end{cases}$$

Under the conditions discussed in Böhmer [7,8], (2.9) - (2.13) defines a consistent and stable sequence of discretizations P_s^h for P_s and the solutions z and ζ^h of $F_s z = 0$ and $(\Phi_s^h F_s)\ \zeta^h = 0$, respectively, satisfy

$$(2.14)\quad \lim_{h\to 0} \|\ \zeta^h - \Delta^h z\| = 0. \quad \Box$$

3. Asymptotic error expansions and discrete Newton methods.

In many cases the discrete approximations P^h contain much more information than is given by (2.5) or (2.14). If the problem P is smooth enough and the discretization P^h is well enough suited to the problem (for precise conditions see Stetter [34]), then the global discretization error $\varepsilon^h := \zeta^h - \Delta^h z$ admits an asymptotic expansion such that

$$(3.1) \quad \zeta^h = \Delta^h\{z + \sum_{\nu=p}^{\bar{q}} h^\nu e_\nu\} + O(h^{\bar{q}+\alpha}), \quad p \le \bar{q} \le \bar{q}+\alpha,$$

or, equivalently,

$$(3.2) \quad \|\zeta^h - \Delta^h\{z + \sum_{\nu=p}^{\bar{q}} h^\nu e_\nu\}\|_{E^h} = O(h^{\bar{q}+\alpha}).$$

In (3.1) the $e_\nu = e_\nu(z)$ are certain fixed functions independent of h so that ζ^h approximates $\Delta^h z$ to an accuracy of order p. (3.1) is used in Richardson extrapolation, deferred and iterated defect corrections, and discrete Newton methods to obtain higher order methods.

Discrete Newton methods are based upon discretizations of equations of the form (1.2), that is, operators of the form

$$(3.3) \quad F_i + \hat{d} : y \mapsto F_i y + \hat{d} \text{ for } F_i \in L(E,\hat{E}), i=0,1, \hat{d} \in \hat{E}.$$

Here $L(E,\hat{E})$ denotes the space of linear transformations between E and $\hat{E}$. Since we are concerned only with bounded, additive discretization methods (see Böhmer [6]), then for $F_0+\hat{d}$, F_0, F_1. $F_1 \in \mathrm{dom}\ \Phi^h, F_i \in L(E,\hat{E}), i=0,1,$

we have

$$(3.4)\quad \begin{cases} \Phi^h(F_0+\hat{d})=\Phi^h F_0+\Omega^h\hat{d}, \text{ where} \\ \Phi^h F_0 \in L(E^h,\hat{E}^h) \text{ and } \Omega^h \in L\ \hat{E},\hat{E}^h) \text{ and} \\ \| \Phi^h F_0-\Phi^h F_1 \| \leq C \| F_0-F_1 \| \text{ with } C \in \mathbb{R}_+ \text{ independent of } h. \end{cases}$$

We shall require that the <u>extension operators</u> $\Delta_h: E^h \to E$, respectively, $\hat{\Delta}_h:\hat{E}^h \to \hat{E}$ shall be <u>asymptotic preserving</u>(see Böhmer [6]), that is,

$$(3.5)\quad \eta^h=\Delta^h\{y+\sum_{\nu=p}^{\bar{q}} h^\nu e_\nu^*\}+O(h^{\bar{q}+\alpha})$$

implies

$$(3.6)\quad y_h:=\Delta_h\eta^h:=\Delta_{q,h}\eta^h=y+\sum_{\nu=p}^{q_1} h^\nu e_\nu^* \ +O(h^{q_1+\alpha_1}),$$

where $q_1 \leq \min\{q,\bar{q}\}$ and $\alpha_1=\alpha_1(q,q_1,\alpha)$ depend on the operator $\Delta_h=\Delta_{q,h}$ and the respective norms in E and E_o.(See Example 3.1 below for a discussion of some operators $\Delta_{q,h}$ having the above properties.)

<u>Discrete Newton methods</u> are defined as iteration processes of the form:

$$(3.7)\quad \begin{cases} \text{(i) given the starting value } \zeta_o^h = \zeta^h \text{ such that } F^h\zeta_o^h=0, \\ \quad \text{where } F^h=\Phi^h F \text{ is introduced in (2.2), compute } \zeta_\ell^h \text{ from} \\ \text{(ii) } (\Phi^h F^*(\Delta_h\zeta_o^h))(\zeta_\ell^h-\zeta_{\ell-1}^h)= -\Omega^h F\Delta_h\zeta_{\ell-1}^h, \ell=1,2,\ldots\ . \end{cases}$$

Here F is the original operator, M is a bounded additive discretization method , Δ_h preserves asymptotics and

$F^*(z_o)$ is an <u>admissible approximation for $F'(z)$</u> (see Böhmer [6]).In the cases discussed in this paper, we use

$$(3.8)\quad \begin{cases} \phi^h F^*(z_o):=\phi^h F'(z_o) & \text{or} \\ \phi^h F^*(z_o):=(\phi^h F)'(\zeta_o^h). \end{cases}$$

Under suitable assumptions, which are satisfied for many discretization methods (arrising, e.g., from finite differences in differential equations or from quadratures in integral equations) the ζ_ℓ^h in (3.7) satisfy (Böhmer [6])

$$(3.9)\quad \begin{cases} \zeta_\ell^h=\Delta^h\{z+\sum_{\nu=(\ell+1)p}^{q_\ell} h^\nu e_{\nu,\ell}\}+O(h^{q_\ell+\alpha_\ell}) \\ q_\ell \leq q_{\ell-1}\leq\cdots\leq q_o \leq \overline{q},\ 0 < \alpha_\ell. \end{cases}$$

Hence ζ_ℓ^h is an approximation to $\Delta^h z$ of order $(\ell+1)p$.

<u>Example 3.1:</u> <u>*Asymptotic preserving extension operators*</u> $\Delta_h=\Delta_{q,h}$: Let in (3.5) $y \in C_L^{\overline{q}+r}([a,b],\mathbb{R}^n)$, $e_\nu^* \in C_L^{\overline{q}+r-\nu}([a,b],\mathbb{R}^n)$ (the index L means that the highest derivative is still Lipschitz-continuous). This is usually true in a system of O.D.E.s of order r with defining functions $\overline{q}$-times Lipschitz-continuously differentiable. Let $\|\cdot\|_E$ be the usual Sobolev-norm in $W_\infty^r([a,b],\mathbb{R}^n)$. Then define $\Delta_{q,h}$ in one of the following two ways (see Böhmer [6]):

$$(3.10)\quad \begin{cases} \text{Let } m=\frac{b-a}{h}=kq,\ m,k,q \in \mathbb{N}, \text{ and} \\ \Delta_{q,h}\eta^h|(x_{\mu q},x_{(\mu+1)q}) \in \Pi_q \ \text{(polynomials of degree q)} \\ (\Delta_{q,h}\eta^h)(x_\nu)=\eta^h(x_\nu),\ \nu=0,\ldots,m. \end{cases}$$

$$(3.11)\quad \begin{cases} \text{Let } m=k'(q-2s), s \in \mathbb{N} \text{ for } \mu=0,\dots,k'-1 \\ P_\mu \in \Pi_q, P_\mu(x_\nu)=\eta^h(x_\nu), \nu=\mu(q-2s)-s,\dots,(\mu+1)(q-2s)+s, \\ (\Delta_{q,h}\eta^h - P_\mu)\big|_{(x_{\mu(q-2s)}, x_{(\mu+1)(q-2s)})} \equiv 0. \end{cases}$$

Then both $\Delta_{q,h}$ satisfy (3.6) with

$$(3.12)\qquad q_1=\min\{q,\bar{q}\}-r, \alpha_1 \leq \min\{q-q+\alpha, 1\}.$$

In (3.10), close to the gridpoints $x_{\mu q}$, the polynomials usually start to oscillate. These oscillation are avoided in (3.11), especially if it is possible to extend the functions y, e_ν^* and η^h, and to define the overlapping polynomials in (3.11) on a larger interval but restrict their use to [a,b].□

In many cases it is possible, and sometimes even necessary, to totally avoid the application of the interpolation in (3.10), (3.11) by using admissible local approximations for $\Omega^h F z_{\ell-1}$. We present them in the special case of our Examples 3.2 and 3.3 and refer to Böhmer [6] for the general case.

Example 3.2: *Error asymptotic and discrete Newton methods for the box scheme:* Let in (2.3), $f \in C_L^{2\bar{q}}([a,b]\times F_1, \mathbb{R}^n)$, $g \in C_L^{2\bar{q}+1}(F_2 \times F_3)$, be such that all the partials $f^{(\mu,\nu)}, g^{(\mu,\nu)}$ for $\mu+\nu \leq 2\bar{q}$ and $\mu+\nu \leq 2\bar{q}+1$, respectively, are Lipschitz continuous. Then ζ^h, defined in (2.4), admits the asymptotic expansion

$$(3.13)\quad \begin{cases} \zeta^h = \Delta^h\{z + \sum_{\nu=1}^{\bar{q}} h^{2\nu} e_{2\nu}\} + O(^{2\bar{q}+1}) \\ \text{with } z \in C_L^{2\bar{q}+1}([a,b],\mathbb{R}^n), e_{2\nu}=e_{2\nu}(z) \in C_L^{2(\bar{q}-\nu)+1}([a,b],\mathbb{R}^n). \end{cases}$$

The operator Ω^h in (3.7) reduces in this case to $\Omega_b^h = \hat{\Delta}^h$ in (2.4). With

$$(3.14)\qquad f^*(\cdot,z_o(\cdot)) := \begin{cases} f^{(0,1)}(\cdot,z_o(\cdot)) & \text{or} \\ f^{(0,1)}(\cdot,\dfrac{z_o(\cdot+h/2)+z_o(\cdot-h/2)}{2}) & \end{cases}$$

and for one of the operators $\Delta_{q,h}$ in Example 3.1, we have then the discrete Newton method for the box-scheme as

$$(3.15)\qquad \begin{cases} F_b^h \zeta_o^h = 0 \quad \text{from (2.4) and with } z_{\ell-1} := \Delta_{q,h}\zeta_{\ell-1}^h, \delta_\nu := (\zeta_\ell^h - \zeta_{\ell-1}^h)(x_\nu) \\ F_b^{h'}(\zeta_o^h)(\zeta_\ell^h - \zeta_{\ell-1}^h) = \{\dfrac{\delta_{\nu+1}-\delta_\nu}{h} - f^*(x_{\nu+1/2}, z_o(x_{\nu+1/2}))\dfrac{\delta_{\nu+1}+\delta_\nu}{2}, \\ \qquad \nu=0,\dots,m-1, g^{(1,0)}(\zeta_o^h(x_o),\zeta_o^h(x_m))\delta_o \\ \qquad + g^{(0,1)}(\zeta_o^h(x_o),\zeta_o^h(x_m))\delta_m\} \\ = -\{z'_{\ell-1}(x_{\nu+1/2}) - f(x_{\nu+1/2}, z_{\ell-1}(x_{\nu+1/2})), \nu = 0,\dots,m-1, \\ \qquad g(\zeta_{\ell-1}^h(x_o), \zeta_{\ell-1}^h(x_m))\}. \end{cases}$$

Then the ζ_ℓ^h from (3.15) satisfy (for $q \le \bar{q}$)

$$(3.16)\qquad \zeta_\ell^h = \Delta^h\{z + \sum_{\nu=\ell}^{q-1-[\frac{\ell-2}{2}]} h^{2\nu} e_{2\nu,\ell}\} + O\,(h^{2q-\ell+1}).$$

In (3.15) we may replace $z'_{\ell-1}(x_{\nu+1/2})$ and $z_{\ell-1}(x_{\nu+1/2})$ in the next to last line by symmetric divided differences and by linear combinations of the $\zeta_{\ell-1}^h(x_\mu)$ of order 2ℓ respectively. Close to the endpoints a and b, either unsymmetric formulae should be used, or the problem must be extended to a larger interval. Then the ζ_ℓ^h again satisfy (3.16) with q replaced by $\bar{q}$.□

<u>Example 3.3:</u> *<u>Error asymptotics and discrete Newton methods for the modified five point star:</u>* Let, in (2.6), (2.7), $f \in C^{2q+\alpha}(R \times F)$ $\partial R \in C^{2q+2+\alpha}$, $g \in C^{2q+2+\alpha}(\partial R)$, $0<\alpha<1$, and the solution z be in $C^1(R)$. Then $z \in C^{2q+2+\alpha}(R)$ and for nonlinear problems (2.6), (2.7), the ζ^h from (2.9)-(2.12) admits the expansion (see Böhmer [7])

$$(3.17)\begin{cases}\zeta^h=\Delta^h\{z+h^2e_2+h^4e_4\}+O(h^{k-1+\frac{1}{\sigma}}),\, k \le 6, \\ \text{where } \sigma=2, \text{ or } \infty \text{ corresponding to } \|\cdot\|_2 \text{ or } \|\cdot\|_\infty \text{ in } E^h.\end{cases}$$

For linear problems (2.6) we even have (see Böhmer [8])

$$(3.18)\begin{cases}\zeta^h=\Delta^h\{z+\sum_{\nu=1}^{q} h^{2\nu}e_{2\nu}\}+O(h^{\min\{k+1,2q+\alpha\}}) \\ \text{and } \|\cdot\|_\infty \text{ in } E^h \text{ for } k \in \mathbb{N}.\end{cases}$$

The modification of the usual five point star close to the boundary is unavoidable, since Wasow [36] has proved that for $k = 0$ and $k = 1$ there is no error asymptotic. Since the replacement described in (2.11) only occurs in $\upsilon^h(x\pm h,y)$ or $\upsilon^h(x,y\pm h)$ and never in $f(x,y,\upsilon^h(x,y))$, we have

$$(3.19)\begin{cases}\Phi_s^h F_s'(z_o) = (\Phi_s^h F_s)'(\zeta_o^h), \text{ where we} \\ \text{have defined } z_o \text{ by the } \Delta_{q,h} \text{ in Example 3.1 along} \\ \text{the grid lines.}\end{cases}$$

If we denote by $\Delta_{q,h}^x$ and $\Delta_{q,h}^y$ the application of $\Delta_{q,h}$ in the x- and y-directions respectively, the discrete Newton equation corresponding to (2.10) for $\ell = 0$ has the form ($\delta^h := \zeta_1^h - \zeta_0^h$)

$$(3.20)\begin{cases} -\delta^h(x+h,y)-\delta^h(x-h,y)-\delta^h(x,y+h)-\delta^h(x,y-h)+4\delta^h(x,y) \\ +h^2 f^{(0,0,1)}(x,y,\zeta_0^h(x,y))\delta^h(x,y) = \\ -\{-(\Delta_{q,h}^x \zeta_0^h)_{xx}(\cdot,\cdot)-(\Delta_{q,h}^y \zeta_0^h)_{yy}(\cdot,\cdot)+f(\cdot,\cdot\ \zeta_0^h(\cdot,\cdot)\}(x,y) \\ \text{for } (x,y) \in \mathbb{G}_r^h. \end{cases}$$

For irregular points, we again must replace the corresponding $\delta^h(x\pm h,y)$, $\delta^h(x,y\pm h)$ as indicated in (2.11). Then Pereyra-Proskurowski-Widlund [29] proved for $-\Delta z-f=0$ in R, $z-g=0$ on ∂R and Böhmer [7, 8] for more general cases that

$$(3.21)\qquad \zeta_1^h = \Delta^h z + O(h^{3.5}) \text{ for } k = 6.$$

Again, it is possible to replace in (3.20) the $\Delta_{q,h}^x, \Delta_{q,h}^y$ by local approximations analogous to Example 3.2.□

4. Discrete correction method

The crucial aim of this paper is the incorporation of a mesh refinement strategy into a discrete Newton method. The strategy reduces the computations involved in achieving an acceptable approximation to the discrete solution $\zeta^h = \zeta_0^h$ to P^h in E^h. Newton's method uses the error asymptotic (3.1) for ζ^h by applying (3.7) in E^h, to obtain better and better approximations for $\Delta^h z$. Therefore we need to know ζ^h much more accurately than mere truncation error $O(h^p)$. Specifically, ζ^h must be known sufficiently well so that the last correction, say $\zeta^h_{\bar{\ell}}$, based on the knowledge of ζ^h, is still worthwhile. Thus we must first improve the approximation to ζ^h and not to $\Delta^h z$. The reward for this patience is that an excellent approximation to ζ^h allows us, via (3.7), to compute an excellent approximation to $\Delta^h z$. Again, it is possible to obtain the approximation for ζ^h, which must be much better here than that required in Allgower-McCormick [3], by using the original mesh refinement strategy. That this is possible is easily verified by checking through the development in Allgower-McCormick [3]. The essence of the strategy is to start with a rather coarse step size $h_c = h_0$ and, proceeding via $h_0 > h_1 > \ldots > h_i > \ldots > h_\mu = h$, to interpolate an approximation $\tilde{\zeta}^{h_{i-1}} \approx \zeta^{h_{i-1}}$ to obtain a good starting value $\tilde{\zeta}_0^{h_i}$ for Newton's method (1.3). Linear interpolation is usually adequate for such purposes. This usually results in the need to perform fewer and fewer iterations in (1.3) as h_i decreases . For the $h_i \approx h_\mu = h$, often one iteration in (1.3) is sufficient (see §5).

Further, for small h_i, since the matrices in (1.3) and (3.7) are large, it is usually more efficient to solve many systems defined by the same matrix than to solve a few defined by different matrices. We therefore consider the transition from $P^{h_{i-1}}$ to P^{h_i} in a similar way as from P^h to P and introduce a discrete Newton method as in (3.7), which now has to be applied between the discrete spaces $E^{h_{i-1}}$ and E^{h_i}. To avoid lengthy subscripts, we use $\bar{h} := h_{i-1}$ and $h := h_i$ and introduce the following <u>discrete correction method:</u>

$$
(4.1)\quad \begin{cases} \quad \text{Given the starting value } \xi_0^{\bar{h}} = \zeta^{\bar{h}} \text{ such that} \\ \text{(i)} \quad F^{\bar{h}} \zeta^{\bar{h}} = 0 , \qquad \text{compute } \xi_\ell^{\bar{h}} \text{ from} \\ \text{(ii)} \quad (F^{\bar{h}})'(\xi_0^{\bar{h}})(\xi_\ell^{\bar{h}} - \xi_{\ell-1}^{\bar{h}}) = -\hat{\Delta}_h^{\bar{h}} F^h \Delta_{\bar{h}}^h \xi_{\ell-1}^{\bar{h}}, \ell = 1,2\ldots . \end{cases}
$$

For many important discretization methods, we have $(F^h)'(\zeta^h) = \Phi^h \overset{*}{F}(z_0)$, where $\overset{*}{F}(z_0)$ is an admissible approximation for $F'(z)$. Hence the $\xi_\ell^{\bar{h}}$ in (4.1) and the $\zeta_\ell^{\bar{h}}$ in (3.7) should be rather close if $\hat{\Delta}_h^{\bar{h}} F^h \Delta_{\bar{h}}^h \zeta_{\ell-1}^{\bar{h}}$ and $\Omega^{\bar{h}} F \Delta_{q,\bar{h}} \zeta_{\ell-1}^{\bar{h}}$ are close. This is certainly true for $\ell = 1$ and, to a lesser extent, for $\ell > 1$.

In (3.7), the choice of a modified Newton method with $\Phi^h \overset{*}{F}(z_0)$ replacing $\Phi^h F'(z_{\ell-1})$ is important, since for a $z_0 = z + O(h^p)$, as used in (3.7), and for the fixed discretization method of order p, the change to $\Phi^h F'(z_{\ell-1})$ would not only cost much more computational time, but would asymptotically not have improved the convergence, since at most h^p can be gained in every iteration. For the $P^{\bar{h}} - P^h$ transition, this restriction is no longer valid, but other arguments become important. Since the matrix $(F^{h_i})'$ is larger than $(F^{h_{i-1}})'$ for $h_{i-1} > h_i$, it is usually cheaper to compute several iterations in (4.1ii) for $\bar{h} = h_{i-1}$ instead of one h_i- iteration in (1.3). So in our process from $h_o = h_c$ down to $h_\mu = h$ it is worthwhile, especially for small h_{i-1}, to use the last matrix for h_{i-1} in (1.3) to compute a better starting value for $\xi_o^{h_i}$ via (4.1ii). Further, for the final $h_\mu = h$ we need to check via an additional $\bar{h}>h$, if the Fundamental Assumption (1.7) is satisfied. For this purpose we must use the matrix $\varphi^{\bar{h}} F^*(\Delta_{\bar{h}} \zeta^{\bar{h}}) = (F^{\bar{h}})'(\zeta^{\bar{h}}) \approx (F^{\bar{h}})'(\xi_o^{\bar{h}})$ in (3.7) and may as well use it in (4.1ii) to obtain a starting value $\tilde{\zeta}_o^h = \Delta_{\bar{h}}^h \, \xi_\ell^{\bar{h}} \approx \zeta^h$. This is often good enough to avoid another iteration in (1.3) and may be used as starting value in (3.7).

In §5 we will see that always $h_{i-1} = c\, h_i$ with a fixed constant $c > 1$. So we assume $\bar{h} = c\, h$, $c > 1$ fixed in the following asymptotic results, which are concerned with the limiting process $h \to o$. Besides $\bar{h} = h_{i-1}$ and $h = h_i$ we will use, especially for the last $h = h_i = h_\mu$, an $\bar{h}$ as a "controlling" step size to check if the Fundamental Assumption (1.7) is satisfied or not. Again, we want to point out that the results in Theorems 4.1 and 4.2 only indicate for small enough h that the leading terms are dominating. Hence, neglecting the higher terms for $h \approx h_c$ may lead to bad estimations, whereas for $h \approx h_\mu$ these estimations will usually be reliable. For the following asymptotics, let $\bar{h}, h \in \mathbb{H}$ tend to zero as elements of null sequences such that $\bar{h} = c\,h$, $c > 1$ fixed. The results show that the asymptotic properties of the ζ_ℓ^h and the ξ_ℓ^h are rather similar, especially for $\ell = 1$. Since we use the result for $\ell = 1$ much more than for $\ell > 1$ and since the technical conditions for $\ell=1$ are much simpler, we give the two results in separate theorems (for the local error function Λ_h see Stetter [34], p 22 ff.):

<u>Theorem 4.1:</u> *Suppose that, for a fixed* $c>1$, $h \in \mathbb{H}$ *and* $\bar{h} = c\,h \in \mathbb{H}$ *tend to zero. Let* F, $F^h = \Phi^h F$ *be Frechét differentiable,* $(F^{\bar{h}})'(\zeta_o^{\bar{h}}) = \Phi^{\bar{h}} F^*(z_o)$, *where* $F^*(z_o) = F'(z) + O(h^p)$ *and* $\Phi^{\bar{h}} F^*(z_o)$ *is stable. Further, let* (3.1) *hold with* $\bar{q}+\alpha \geq 2p$ *and let the local error function* Λ_h *be* $(p+1,p)$*-smooth. Finally let the operators* $\Delta^{\bar{h}}$, $\Delta^h_{\bar{h}}$, Δ^h, $\Delta^{\bar{h}}_h$, $\hat{\Delta}^h$, $\hat{\Delta}^{\bar{h}}_h$ *and* Ω^h *satisfy the relations*

$$(4.2)\quad \begin{cases} \Delta^{\bar{h}}_h\,\Delta^h = \Delta^{\bar{h}} + O(h^{2p}), \quad \Delta^h_{\bar{h}}\,\Delta^{\bar{h}} = \Delta^h + O(h^{2p}), \\ \hat{\Delta}^{\bar{h}}_h\,\hat{\Delta}^h = \hat{\Delta}^{\bar{h}} + O(h^{2p}) \ \text{and}\ \Omega^h = \hat{\Delta}^h + O(h^p). \end{cases}$$

Then the following relations hold:

$$(4.3)\quad \zeta^{\bar{h}} = \Delta^{\bar{h}}_h\{\zeta^h + \Delta^h \sum_{\nu=p}^{2p-1} \bar{h}^{\,\nu}\,(1-(\tfrac{h}{\bar{h}})^\nu)\,e_\nu\} + O(h^{2p}),$$

$$(4.4)\quad \xi_1^{\bar{h}} - \xi_o^{\bar{h}} = (1-(\tfrac{h}{\bar{h}})^p)(1+O(h))\,(\zeta_1^{\bar{h}} - \zeta_o^{\bar{h}}).$$

If ζ^h *admits an* h^2 *expansion in* (3.1) *and if* Λ_h *is* $(p+2,p)$ *smooth,* $O(h)$ *in* (4.4) *may be replaced by* $O(h^2)$.

Proof: By (3.1) we have

$$\zeta^{\bar h} = \Delta^{\bar h}\{z + \sum_{\nu=p}^{\bar q} \bar h^{\nu} e_{\nu}\} + O(h^{\bar q+\alpha})$$

and with $\Delta_h^{\bar h}\Delta^h = \Delta^{\bar h} + O(h^{2p})$ then

$$(4.5) \qquad \Delta_h^{\bar h}\zeta^h = \Delta^{\bar h}\{z + \sum_{\nu=p}^{2p-1} h^{\nu} e_{\nu}\} + O(h^{2p}) .$$

Hence (4.3) is satisfied. To prove (4.3) we need the $O(h^{2p})$ in (4.2) and $O(h^p)$ in $F^*(z_o) = F'(z) + O(h^p)$, whereas for the proof of (4.4), the $O(h^{2p})$ and $O(h^p)$ could be relaxed to $O(h^p)$ and $O(h)$, respectively.
To show (4.4) we have to compare

$$\Omega^{\bar h} F \Delta_{\bar h} \zeta^{\bar h} \text{ and } \Delta_h^{\bar h} F^h \Delta_{\bar h}^h \zeta^{\bar h} .$$

Now, by (4.2),

$$(4.6) \qquad \begin{cases} \Omega^{\bar h} F \Delta_{\bar h} \zeta^h = \Omega^h F(z + \sum_{\nu=p}^{2p-1} \bar h^{\nu} e_{\nu} + O(h^{2p})) \\ = \hat\Delta^h F'(z) \sum_{\nu=p}^{2p-1} \bar h^{\nu} e_{\nu} + O(h^{2p}) \\ = \bar h^p \hat\Delta^{\bar h} \{F'(z)\, e_p + O(h)\} . \end{cases}$$

In the rest of the proof we need the local error function and its asymptotic expansion, which we write in the form

$$\Lambda_h u = \sum_{\nu=1}^{\bar q} h^{\nu} \lambda_{\nu} u + O(h^{\bar q+\alpha})$$

where Λ_h and λ_{ν}: $E \to \hat E$.

We use this notation, which is different from (2.1), to stay somewhat consistent with Stetter [34] and Böhmer [6]. From Stetter [34], chapter 1, and from (4.5) with h and $\bar h$ interchanged, we find that

$$(4.7)\quad \begin{cases} F^h\Delta^h_{\bar h}\zeta^{\bar h} = F^h\Delta^h\{z+\bar h^p e_p + O(h^{p+1})\} \\ =\hat\Delta^h\{(F+\Lambda_h)(z+\bar h^p e_p + O(h^{p+1})\} \\ =\hat\Delta^h\{h^p(F'(z)e_p+\lambda_p z)+(\bar h^p-h^p)F'(z)e_p+O(h^{p+1}))\} \\ =\hat\Delta^h(\bar h^p-h^p)F'(z)e_p+O(h^{p+1}), \end{cases}$$

since the $F'(z)e_p+\lambda_p z=0$ by definition of e_p.
Now (4.2) and (4.7) imply that

$$(4.8)\qquad \hat\Delta^{\bar h}_h F^h\Delta^h_{\bar h}\zeta^{\bar h} = h^p\hat\Delta^{\bar h}\{\{(\tfrac{\bar h}{h})^p-1\}F'(z)e_p+O(h)\}\ .$$

By comparing (4.6), (3.7) with (4.8), (4.1), we obtain (4.4).□

<u>Remark:</u> In (4.3), $\bar h^\nu$ may be changed into h^ν provided $(1-(\frac{h}{\bar h})^\nu)e_\nu$ is replaced by $((\frac{\bar h}{h})^\nu-1)e_\nu$. □

The next result is concerned with the behavior of the $\xi^{\bar h}_\ell-\xi^{\bar h}_{\ell-1}$ for $\ell > 1$. To formulate and prove it completely would require many technical details developed in Böhmer [6].However, the proof and the conditions for Theorem 4.2 are totally analogous to the corresponding proof and conditions for the validity of (3.9) as given in [6]. That is due to the fact that we have assumed in [6] that the terms defining $e_{\nu,\ell}$ in (3.9) via

$$F'(z)e_{\nu,\ell}= \sum_\mu g_{\mu,\nu,\ell}(z)$$

are elements of certain linear spaces and that the $e_{\nu,\ell}$ again are elements of suitable linear spaces. So, if the $g_{\mu,\nu,\ell}(z)$ are replaced by multiples, $e_{\nu,\ell}$ is only replaced by another element of the same space. To avoid the introduction of a lengthy formulation, we therefore assume that (3.9) has been proved as in Böhmer [6] using the assumptions on linear spaces indicated above. In Theorem 4.2 this is formulated as "let (3.9) be proved as in Böhmer [6]".

<u>Theorem 4.2:</u> *Let the conditions in* Theorem 4.1 *be satisfied with* 2p *replaced by* $q+\alpha$, *let* $F^*(z_o)$ *be an admissible approximation for* $F'(z)$, *let* Λ_h *be* (q,p)*- smooth and let* (3.9) *be proved as in* Böhmer [6]. *Then we have, with* $e^*_{\nu,\ell}$ *independent of* h *and for* $(\ell+1)p<q_\ell, \alpha_\ell>0$, *that*

$$\text{(4.9)} \qquad \xi_\ell^{\bar h} = \Delta_h^{\bar h}\{\zeta^h + \sum_{\nu=(\ell+1)p}^{q_\ell} \bar h^\nu \Delta^h e^*_{\nu,\ell}\} + O(h^{q_\ell+\alpha_\ell}).$$

<u>Proof:</u> (4.9) is, for $\ell=0$, an immediate consequence of (4.3) since $\xi_o^{\bar h}=\zeta^{\bar h}$. As in the foregoing proof we have to compute

$$F^h\Delta^h_{\bar h}\xi^{\bar h}_{\ell-1} =$$

$$F^h\Delta^h_{\bar h}\,\Delta_h^{\bar h}\{\zeta^h + \sum_{\nu=\ell p}^{q_{\ell-1}} \bar h^\nu \Delta^h e^*_{\nu,\ell-1} + O(h^{q_{\ell-1}+\alpha_{\ell-1}})\}$$

$$= \quad F^h\,\Delta^h[z + \sum_{\nu=p}^{q_\ell} h^\nu \tilde e_\nu + O(h^{q_\ell+\alpha_\ell})]\}\ \text{(by (4.2), (4.9))}$$

where $\tilde e_\nu = e_\nu, \nu=p,\ldots,\ell p-1, \tilde e_\nu = e_\nu + (\frac{\bar h}{h})^\nu e^*_{\nu,\ell},\ \nu \geq \ell p$,

$$= \hat\Delta^h\{(F+\Lambda_h)(z + \sum_{\nu=p}^{q_\ell} h^\nu \tilde e_\nu + O(h^{q_\ell+\alpha_\ell})\}$$

$$= \hat\Delta^h[F'(z)\sum_{\nu=p}^{q_\ell} h^\nu \tilde e_\nu + \sum_{\nu=p}^{q_\ell} h^\nu \lambda_\nu z + \sum_{\nu=1}^{q_\ell} h^\nu \lambda'_\nu(z)\sum_{k=p}^{q_\ell} h^\nu \tilde e_\nu$$

$$+ \sum_{m=2}^{[q_\ell/p]} \frac{1}{m!}\{F^{(m)}(z) + \sum_{\nu=1}^{q_\ell-mp} h^\nu \lambda_\nu^{(m)}(z)\}(\sum_{k=p}^{q_\ell} h^\nu \tilde e_\nu)^m] + O(h^{q_\ell+\alpha_\ell}).$$

Since $\tilde{e}_\nu = e_\nu, \nu=p,\dots,\ell p-1$, all coefficients of $h^\nu, \nu=p,\dots,\ell p-1$, vanish since the e_ν have been defined just by this property (see Stetter [34]; for Lipschitz-continuous F^h this is an immediate consequence of $\xi^{\bar{h}}_{\ell-1}=\Delta^{\bar{h}}_h \zeta^{\bar{h}} + O(h^{\ell p}))$.

Thus we finally have

$$F^h \Delta^h_{\bar{h}} \xi^{\bar{h}}_{\ell-1} = \hat{\Delta}^h \sum_{\nu=\ell p}^{q_\ell} h^\nu \hat{e}_{\nu,\ell-1} + O(h^{q_\ell+\alpha_\ell}),$$

where the $\hat{e}_{\nu,\ell-1}$ are obtained exactly as the $\bar{\bar{f}}_\mu$ in the proof of Lemma 3.3 in Böhmer [6]. Hence we can argue as in Theorem 3.7 in [6] using Assumption 2.4 with the linear space structure of the admissible sets and ranges. Thus, (4.9) is proved. □

Our main aim in (4.1) is the approximation of ζ^h by $\Delta^h_{\bar{h}} \xi^{\bar{h}}_\ell$. The degree to which $\Delta^h_{\bar{h}} \xi^{\bar{h}}_\ell$ approximates ζ^h depends of course on how close $\bar{h}$ is to h as well on the size of h itself. In particular, we have for $\bar{h}/h \approx 1$ that $e^*_{\nu,\ell} \approx 0$ in (4.9). Certainly if $\bar{h}$ is much larger that h, $\Delta^h_{\bar{h}}\xi^{\bar{h}}_\ell$ may be quite far from ζ^h no matter how big ℓ is. Thus one should take into account some knowledge of the convergence characteristics of (4.1), especially by investigating $\hat{\Delta}^{\bar{h}}_h F^h \Delta^h_{\bar{h}} \xi^{\bar{h}}_{\ell-1}$ or even $F^h \Delta^h_{\bar{h}} \xi^{\bar{h}}_{\ell-1}$ to estimate the quality of $\Delta^h_{\bar{h}}\xi^{\bar{h}}_\ell \approx \zeta^h$. If $\bar{h}$ satisfies the Fundamental Assumption (1.7) well enough, we have from the previous theorem that

$$(4.10) \quad \begin{cases} \| \Delta^h_{\bar{h}} \xi^{\bar{h}}_\ell - \zeta^h \| \approx \bar{h}^{p(\ell+1)} \varepsilon_\ell \\ \text{where } \varepsilon_\ell := \| \Delta^h e^*_{p(\ell+1),p} \| . \end{cases}$$

<u>Remarks:</u> 1) The result in (4.9) reflects the fact that, for a modified Newton method as (4.1), we can only expect linear convergence instead of a quadratic convergence by using $(F^{\bar{h}})'(\xi^{\bar{h}}_{\ell-1})$. However, the reduced computational costs for additional iterations more than compensates for the slower convergence rate. Moreover, if we expect that only a few iterations are to be used, as may be the case for differential equations, then the concept of quadratic convergence is not very relevant. We shall further discuss the advantage of (4.1) in sections 5 and 6, especially in connection with P.D.E.s.

2) The conditions required in Theorem 4.1 and 4.2 are satisfied for the two examples treated in this paper (see Böhmer [6,7,8]).

5. Dynamical determination of h.

Suppose that the task is to determine in E^h an approximation to $\Delta^h z$ such that the estimated error, which is not necessarily an upper bound for the error, is less than a given tolerance "tol". If we can determine $\zeta^h_{\bar{\ell}}$ precisely, where $\zeta^h_{\bar{\ell}}$ is the last iterate from (3.7) in E^h, we must then require that

(5.1) $$\| \zeta^h_{\bar{\ell}} - \Delta^h z \| \leq \text{tol}.$$

As discussed in § 1, these iterates $\zeta^h_{\bar{\ell}}$ are only useful if the Fundamental Assumption (1.7) is satisfied. Hence we need to describe numerical means for checking the validity of (1.7) (see §6).Since at the beginning of our computations we may not know how small h has to be to meet (5.1), a guess may be necessary. Theoretical aspects and numerical experience often suggests upper bounds $\bar{\ell}_o$ for the $\bar{\ell}$ in (5.1). For example, with B.V.P.s in elliptic P.D.E.s, $\bar{\ell} \leq \bar{\ell}_o := 1$ is appropriate (see Example 3.3).For B.V.P.s in O.D.E.s, as given in Example 3.2, $\bar{\ell}$ depends on the choice of $\Delta_{q,h}$. Thus, $\Delta_{q,h}$ in (3.10) admits only $\bar{\ell} \leq \bar{\ell}_o := 3$, $8 \leq q \leq 12$, $\Delta_{q,h}$ in (3.11) admits $\bar{\ell} \leq \bar{\ell}_o := 4$, and perhaps $\bar{\ell} \leq \bar{\ell}_o := 5$ for $10 \leq q \leq 20$. If one uses the local approximations as indicated at the end of Example 3.2, even $\bar{\ell} \leq \bar{\ell}_o := 7$ is possible for small enough h. In any event, we will assume for the moment that $\bar{\ell}_o$ is known approximately. For more problem specific ways, see the end of this section.

For many problems, the coefficients $e_{(\ell+1)p,\ell}$ in (3.9) do not increase too much with ℓ so that we often have

$$\| e_{(\ell+1)p,\ell} \| \leq \| e_p \|^{\ell+1} \text{ for } \ell \leq \bar{\ell} .$$

If both sides are approximately equal, (3.1) and (3.9) imply that

$$(5.2) \qquad \| \zeta_\ell^h - \Delta^h z \| \approx \| \zeta_o^h - \Delta^h z \|^{\ell+1} \text{ for } \ell \leq \bar{\ell}$$

as soon as h is small enough to satisfy the Fundamental Assumption (1.7). (5.2) suggests that we attempt to determine h so as to guarantee that ζ_o^h does not differ from $\Delta^h z$ in norm by more than tol$^{\frac{1}{\bar{\ell}+1}}$. Now, to account for the variation in $\| e_{(\ell+1)p,\ell} \|$ over ℓ, and for the possibility that (1.7) might not be satisfied too well, and that the ζ_ℓ^h will be determined only approximately, we incorporate a security factor $c \approx 1$ leading to the requirement that

$$(5.3) \qquad \| \zeta_o^h - \Delta^h z \| \approx c \; (\text{tol})^{\frac{1}{\bar{\ell}+1}} .$$

The following consideration allows an interpretation for c: We obtain, instead of (5.2), an equality by

$$\| \zeta_\ell^h - \Delta^h z \| = \frac{\| \zeta_\ell^h - \Delta^h z \|}{\| \zeta_o^h - \Delta^h z \|^{\ell+1}} \cdot \| \zeta_o^h - \Delta^h z \|^{\ell+1} \text{ for } \ell \leq \bar{\ell} .$$

In particular, for $\ell = \bar{\ell}$ with (5.1)-(5.3), c satisfies

$$(5.4) \quad c \approx \tilde{c} := \| \zeta_o^h - \Delta^h z \| / \| \zeta_\ell^h - \Delta^h z \|^{\frac{1}{\ell+1}} \text{ for } \ell \leq \bar{\ell} .$$

The question remaining now is how to get an estimation for $\| \zeta_o^h - \Delta^h z \|$ in terms of h so that h may be determined to satisfy (5.3). To this end, note that from (3.1), (3.9) and Assumption (1.7), we have

$$\zeta_o^h - \zeta_1^h = \Delta^h h^p e_p + O(h^{p+1})$$
$$= \zeta_o^h - \Delta^h z + O(h^{p+1}).$$

Hence,

(5.5) $$\| \zeta_o^h - \zeta_1^h \| \approx \| \zeta_o^h - \Delta^h z \| \approx h^p \| \Delta^h e_p \|.$$

By (5.3), we must then find an h such that

(5.6) $$\| \zeta_o^h - \zeta_1^h \| \approx c\ (\mathrm{tol})^{\frac{1}{\bar{\ell}+1}}.$$

For this purpose we start with a rather coarse h_c and compute

$$\| \zeta_o^{h_c} - \zeta_1^{h_c} \| \approx h_c^p \| \Delta^{h_c} e_p \|.$$

Comparing this result with (5.5), (5.6) we must then choose h such that

(5.7) $$\begin{cases} c(\mathrm{tol})^{\frac{1}{\bar{\ell}+1}} \approx h^p \| \Delta^h e_p \| \approx (\frac{h}{h_c})^p h_c^p \| \Delta^{h_c} e_p \| \\ \approx (\frac{h}{h_c})^p \| \zeta_o^{h_c} - \zeta_1^{h_c} \|. \end{cases}$$

We therefore estimate h according to

(5.8) $$h = h_c \sqrt[p]{\frac{c(\mathrm{tol})^{1/(\bar{\ell}+1)}}{\| \zeta_o^{h_c} - \zeta_1^{h_c} \|}}.$$

In the derivation of (5.2) and of (5.7), we have assumed

that

$$(5.9) \qquad \| e_{(\ell+1)p,\ell} \| \approx \|e_p\|^{\ell+1}$$

and

$$(5.10) \qquad \| \Delta^h e_p \| \approx \| \Delta^{h_c} e_p \|,$$

respectively. Either or both of these conditions might in fact be violated. When this happens, (5.8) will represent a bad estimation for h. In particular, h will be too small or too large accordingly as $\| e_{(\ell+1)p,\ell} \| << \| e_p \|^{\ell+1}$ or $>> \| e_p \|^{\ell+1}$. If we have $E^h = \{\eta^h : \mathbb{G}^h \to \mathbb{R}^n\}$ where $\mathbb{G}^h$ is a grid as in Example 2.1 and 2.2 and if $\mathbb{G}^{h_c} \subset \mathbb{G}^h$ and (1.7) is well enough satisfied for h_c and for $\ell=1$, we then have that $\| \Delta^h e_p \| \geq \| \Delta^{h_c} e_p \|$. Hence, the choice of c via (5.7) might result in an h smaller than the right hand side in (5.8). However, since we must in any case compute the differences $\zeta_0^h - \zeta_1^h$, then (5.10) can be easily checked via

$$(5.11) \qquad \| \Delta_h^{h_c} (\zeta_0^h - \zeta_1^h) \| \approx \| \zeta_0^h - \zeta_1^h \| \text{ for } \Delta_h^{h_c} \Delta^h \approx \Delta^{h_c}.$$

Moreover, (5.9) may be verified using the higher corrections for h as

$$(5.12) \qquad \| \zeta_\ell^h - \zeta_{\ell+1}^h \| \approx \| \zeta_0^h - \zeta_1^h \|^{\ell+1}.$$

We will return to these relations in §6.

A second reason for a potentially inadequate h arising from (5.8) is that often, for the coarse h_c, Assumption (1.7) will only be very roughly true even for $\ell=0$. Therefore, the estimation for h should be updated as described below.

From (5.8), we have at least a rough estimate of the size of h that is needed in order to satisfy (5.1). The next step is therefore to compute ζ^h more exactly. That is done by a method based upon the Allgower-McCormick [3] mesh independence principle, which is stated somewhat loosely for B.V.P.s in ordinary second order scalar differential equations as follows:

Let $\varepsilon > 0$ be a given tolerance and $y_0 \in C[a,b]$ be a given continuous function . Then, for rather general discretization methods, there exists an $h_c > 0$ such that, if it takes μ iterations in Newton's method (1.3), starting with $\tilde{\zeta}_0^{h_c} := \Delta^{h_c} y_0$, to compute, for the coarse step size h_c, an approximation $\tilde{\zeta}_\mu^{h_c}$ for ζ^{h_c} so that $\| \zeta^{h_c} - \tilde{\zeta}_\mu^{h_c} \| \leq \varepsilon$, then the same number of iterations is sufficient to compute $\tilde{\zeta}_\mu^h$ within $\| \zeta^h - \tilde{\zeta}_\mu^h \| \leq \varepsilon$ for every $h \leq h_c$, again starting with $\tilde{\zeta}_0^h := \Delta^h y_0$. We have intentionally used μ as number of iterations in (1.3) for h_c since we will see below that it coincides with μ in the sequence $h_0 = h_c > h_1 > \ldots > h_\mu = h$.

According to this principle, we should iterate in (1.3) for h_c and h_μ respectively, μ times using different matrices in (1.3) for each iteration. For h_c this causes no difficulty, since the dimensions of corresponding matrices $(F^{h_c})'(\tilde{\zeta}_{\ell-1}^{h_c})$ are rather small. For small h_μ ,however, such computations are no longer inexpensive so it is important to try to reduce the number of iterations for $h = h_\mu$. When such iterations are unavoidable, we modify the methods by using the same matrix for every iteration. As noted earlier, this is the motivation for introducing (4.1). This is quite compatible with the Allgower-McCormick [3] modified principle:

Define a mesh refinement factor r as

$$(5.13)\quad \begin{cases} r := [(h_c/h)^{1/\mu}] \text{ and let} \\ h_o := h_c, h_i := h_{i-1}/r = h_o/r^i, \ldots i=1,\ldots,\mu-1, h_\mu = h. \end{cases}$$

Here $[\alpha]$ denotes the integer part of $\alpha \in \mathbb{R}$. Then $\mu-i$ Newton steps in (1.3) are required for every h_i, where the starting value $\xi_o^{h_i}$ is obtained by linear or higher interpolation of $\xi_{\mu-i+1}^{h_{i-1}}$.

This modified principle appears also to be true for more general operator equations and their discretization such as the examples treated above.

In practice we start with a rather coarse h_c, which may not be small enough to satisfy the mesh indipendence principle. Suppose μ_c iterations in (1.3) are necessary to obtain $\| \xi_{\mu_c}^{h_c} - \zeta^{h_c} \| \leq$ tol. We then use the last matrix for h_c needed in (1.3), namely, $(F^{h_c})'(\xi_{\mu_c-1}^{h_c})$, to compute $\zeta_1^{h_c} - \zeta_o^{h_c}$ via (3.7). h is then estimated from (5.8) and r is computed via (5.13). For large h_i the same steps are performed except that $\xi_o^{h_i} := \Delta_{h_{i-1}}^{h_i} \xi_{\mu_{i-1}}^{h_{i-1}}$ is used to start the iteration (1.3), where $\mu_{i-1} \approx \mu-i+1$ is such that $\| \xi_{\mu_{i-1}}^{h_{i-1}} - \zeta^{h_{i-1}} \| <$ tol. For small values of h_i, we use an appropriate combination of (1.3), (3.7) and (4.1). With the actual μ_{i-1} let $\xi_o^{h_{i-1}} := \xi_{\mu_{i-1}}^{h_{i-1}}$. Then (4.1) is used to obtain $\xi_1^{h_{i-1}} - \xi_o^{h_{i-1}}$. For every second or third step size h_i (see n = 2, 3 in <1> section 7), $\zeta_1^{h_{i-1}} - \zeta_0^{h_{i-1}}$ is estimated via (4.4) and h is updated in (5.8) and r in (5.13). The higher iterates in (4.1) are then used to define a good starting value $\tilde{\zeta}_0^{h_i} := \Delta_{h_{i-1}}^{h_i} \xi_\ell^{h_{i-1}}$ for (1.3). We then perform μ_i iterations to achieve $\|\tilde{\zeta}_{\mu_i}^{h_i} - \zeta^{h_i}\|_{h_i} < \text{tol}$. We iterate (4.1) relatively often to obtain a $\tilde{\zeta}_0$ which is used to approximate $\zeta_0^{h_i}$. The suitability of this approximation may be checked by computing the defect $F^{h_i}\tilde{\zeta}_0^{h_i}$. If by one step in (1.3) $\|F^{h_i}\zeta_0^{h_i}\| < c_i$ tol, $c_i \leq 1$, we use $\tilde{\zeta}_0^{h_i}$ as $\zeta_0^{h_i} = \zeta^{h_i}$. Otherwise, we perform a certain number of iterations, usually just one, in (1.3) to suitably approximate $\zeta_0^{h_i}$. The reliability of the results for the final step size is checked by comparing these results with $\bar{h}$-results for some $\bar{h} > h$.

There is yet another concern with respect to the choice of $\bar{\ell}$ relative to the accuracy required. High $\bar{\ell}$, that is, high-order methods, are only worthwhile for high accuracy. Thus, if "tol" in (5.1) is large or, more precisely, if the relative error is large, it is sensible to choose only small values of $\bar{\ell}$. Since a proper relation between tol and $\bar{\ell}$ depends very much on the specific problem, we do not place much restriction upon $\bar{\ell}$ with respect to tol. In any event, the checks discussed in section 6 will restrict $\bar{\ell}$ in a much more problem-related way. Here we simply require that

$$(5.14) \qquad \bar{\ell} \leq \min\{(\log_{10}(\text{tol}/\|\zeta^{h_c}\|) - 1, \bar{\ell}_0\},$$

where $\bar{\ell}_0$ is introduced following (5.1).

6. Verification of the basic assumptions

Throughout this paper we have made three essential assumptions which must be checked during the computation. We discuss them in the order which is naturally indicated by the availability of the results. We need the first correction to check (5.11) and higher corrections to check (1.7) and (5.2) or (5.1). We assume throughout this section that

$$(6.1) \qquad \Delta_h^{\bar{h}} \Delta^h \approx \Delta^{\bar{h}} \text{ for any pair } \bar{h}, h \in \mathbb{H},\ \bar{h} > h.$$

In (5.11) we postulated that

$$(6.2) \qquad c_1(h) := \| \zeta_o^h - \zeta_1^h \| \,/\, \| \Delta_h^{h_c} (\zeta_o^h - \zeta_1^h) \| \approx 1$$

for a reliable estimation of h. In (6.2), h_c means the smallest step size which we have used in our updating process, described at the end of § 5, to obtain h. For $\Delta_h^{h_c} \Delta^h = \Delta^{h_c}$ we always have $c_1(h) \geq 1$. But $c_1(h) \gg 1$ suggests that the behavior of the error $\zeta^{h_c} - \Delta^{h_c} z$ is not well estimated in E^{h_c} and that the computed h must therefore be replaced by a different value. We recommend

$$(6.3) \qquad \begin{cases} \text{for } c_1(h)^{1/p} \geq 2 \text{ or } c_1(h)^{1/p} \leq 0.5 \\ \text{replace h in (5.8) by } h/c_1(h)^{1/p}. \end{cases}$$

Because of (6.1), $c_1(h)$ will in practice almost never be much smaller than 1. Moreover, h will not be changed if $1 < c_1(h) < 2^p$, but if $2 < c_1(h) < 2^p$, we should be more sensitive to change h if any other source (e.g.,(6.4), (6.6) and (6.7) below) suggest we do so.

The estimation of h in (5.8) is based upon the fact that Assumption (1.7) is satisfied with $\ell = 0$ for h and the last h_c which we have used in our updating. This may be verified. by examining the condition

$$(6.4) \qquad c_2(h) := \frac{h_c^p \; \| \Delta_h^{h_c} (\zeta_o^h - \zeta_1^h) \|}{h^p \; \| \zeta_o^{h_c} - \zeta_1^{h_c} \|} \approx 1.$$

This is very easily done since, with the exception of h_c^p and h^p, all other evaluations in (6.4) have already been made via either in (6.2) or (5.8). When $c_2(h) \not\approx 1$, one should then replace $c_1(h)$ in (6.3) by $c_1(h) \cdot c_2(h)$.

By comparing the corrections δ_ℓ^h obtained from (3.7) via

$$(6.5) \qquad \delta_\ell^h := -(\zeta_{\ell-1}^h - \zeta_\ell^h) \approx -(\zeta_{\ell-1}^h - \Delta^h z)$$

for different values of ℓ and of h, we are then able to verify (1.7) and (5.1). Since the estimation of h in (5.8) is occasionally too small and since much less work is involved to compute the $\bar{h}$ - than the h-approximations (since $\bar{h} > h$), we always start with the $\bar{h} := 2h$ computations. We compute $\delta_1^{\bar{h}}$, define $c_1(\bar{h})$ in (6.2) and $c_2(\bar{h})$ in (6.4), and monitor if we have to change h corresponding to (6.3). If so, we redefine h and $\bar{h}$ and return to (6.2) - (6.4).

Because of (6.5), in place of (5.2) we more directly verify (5.1) in the form

$$(6.6) \qquad \| \zeta_{\bar{\ell}}^h - \zeta_{\bar{\ell}+1}^h \| \approx \| \zeta_{\bar{\ell}}^h - \Delta^h z \| \leq \text{tol}.$$

We will describe below a way to test how reliable $\| \zeta^h_{\bar{\ell}} - \zeta^h_{\bar{\ell}+1} \|$ estimates $\| \zeta^h_{\bar{\ell}} - \Delta^h z \|$. Since (5.2) for $\ell \leq \bar{\ell}$ was mainly an estimation to obtain (5.3) and, for $\ell = \bar{\ell}$, the size of h via (5.8), we would not have to check (5.2) for $\ell < \bar{\ell}$ to verify (6.6). Nevertheless, we test (5.2) in the form

$$(6.7) \quad \begin{cases} \| \zeta^h_\ell - \Delta^h z \| \approx \| \zeta^h_\ell - \zeta^h_{\ell+1} \| \approx \| \zeta^h_o - \Delta^h z \|^{\ell+1} \approx \| \zeta^h_o - \zeta^h_{\ell_o} \|^{\ell+1} \\ \text{for } \ell \leq \bar{\ell} \text{ and } \ell_o \leq \ell + 1. \end{cases}$$

Usually we take $\ell_o = 1$. Only if $\bar{\ell}$ is rather large, $\| \delta^h_2 \| \geq 0.1 \| \delta^h_1 \|$ and we need an excellent estimation, is $\ell_o = 2$ appropriate. If (6.7) is strongly violated for a small $\ell < \bar{\ell}$, this indicates that h is chosen too small or large to satisfy (6.6). To avoid the use of too small an h we again start as indicated above, with the $\bar{h}$ - computations, and check (6.6) and (6.7) at first for $\bar{h}$ instead of h. If (6.6) is satisfied for $\bar{h}$ already, we use $h := \bar{h}$ and $\bar{h} := 2h$ to test (1.7). Otherwise we guess our final h once more by the following consideration. If for $\bar{h}$ the Fundamental Assumption (1.7) is satisfied for $\ell = \bar{\ell}$, we have

$$(6.8) \qquad \varepsilon := \| \zeta^{\bar{h}}_{\bar{\ell}} - \Delta^{\bar{h}} z \| \approx \| \zeta^{\bar{h}}_{\bar{\ell}} - \zeta^{\bar{h}}_{\bar{\ell}+1} \| \approx \bar{h}^{(\bar{\ell}+1)p} \| \Delta^{\bar{h}} e_{(\bar{\ell}+1)p,\bar{\ell}} \|.$$

By replacing in (6.8) $\bar{h}$ by h, we want to be able to replace ε by tol. For that purpose we define

$$(6.9) \qquad c_3 := (\varepsilon/\mathrm{tol})^{\frac{1}{p(\bar{\ell}+1)}}$$

and determine h by the requirement that

$$(6.10)\qquad h = \bar{h}([c_3-\delta]+1),\ 0 \le \delta < 1.$$

We usually choose $\delta=0$. The presence of $\delta \ge 0$ allows for the possibility that tol in (5.1) was determined loosely. A $\delta > 0$ may effect a reduction in computation at the sacrifice of some accuracy. Again (6.10) is reliable if (1.7) is satisfied suitably well for $\bar{\ell}$, h and $\bar{h}$.

If the tolerance tol in (5.1) is met for $\bar{h}$ and for an $\ell<\bar{\ell}$ already, that $\|\delta_\ell^{\bar{h}}\| <$ tol, or if $\|\delta_\ell^{\bar{h}}\|$ for $\ell \le \bar{\ell}$ is within the range of round-off-errors, we use again $h:=\bar{h}$ and $\bar{h}:=2h$ and must again test the reliability later on.

As discussed before, the corrections ζ_ℓ^h and the estimations for the final error $\zeta_{\bar{\ell}}^h-\Delta^h z \approx \zeta_{\bar{\ell}}^h-\zeta_{\bar{\ell}+1}^h$ for the final h via (6.8) - (6.10) are all only worthwhile if Assumption (1.7) is satisfied well enough. As a necessary condition for (1.7), we find that

$$(6.11)\qquad \|\Delta_h^{\bar{h}}\delta_\ell^h\| \approx \left(\frac{h}{\bar{h}}\right)^{\ell p} \|\delta_\ell^{\bar{h}}\| \text{ for } \ell=0,1,\dots,\bar{\ell}+1.$$

This condition usually will be satisfied for small ℓ and will not be satisfied very well for $\ell\approx\bar{\ell}$. So we admit, analogously to Daniel-Martin [13] in a similar situation, that

$$(6.12)\qquad \left|\ln \frac{\|\Delta_h^{\bar{h}}\ \delta_\ell^h\|}{(h/\bar{h})^{\ell p}\ \|\delta_\ell^{\bar{h}}\|}\right| \le 0.15\,\ell p.$$

Admitting increasing numbers in (6.12) is suggested by the

fact that for increasing ℓ the leading terms in (3.9) no longer dominate the higher terms as strongly as for small ℓ and that, for higher ℓ, (1.7) might no longer be well satisfied for $\bar{h}$ even if it is for h. Violations of (6.12) will be caused by one of two reasons: either h and $\bar{h}$ are still too large and (1.7) is not yet satisfied, or h and $\bar{h}$ are already too small, with respect to the computer in use, and δ_ℓ^h and $\delta_\ell^{\bar{h}}$ are essentially influenced by round-off effects. The latter case is recognized immediately by comparing $\| \delta_\ell^h \|$ and $\| \delta_\ell^{\bar{h}} \|$ to machine accuracy. In the former case we replace h by h/2 and restart the process.

All the conditions discussed before are necessary for (1.7) and (5.1) or (5.2) and (5.10). One would obtain conditions closer to being sufficient if the norms in (6.2)-(6.12) were replaced by the corresponding function values in E^h and $E^{\bar{h}}$. Sometimes one is interested in the properties of the solution z only on a few given points. Then one should replace the norms in (6.2) - (6.12) by the function values in or close to the points of interest.

To verify (6.11) or (6.12) we do not necessarily need $\bar{h}/h \in \mathbb{N}$. However, this is usually advisable since we avoid interpolation errors if we use restrictions instead of interpolations. Interpolation is unavoidable for $\bar{h}/h \notin \mathbb{N}$. Further, if we have $\bar{h}/h \in \mathbb{N}$, we may then use Richardson extrapolation to improve the approximation in $E^{\bar{h}}$. With

$\mu:=(\ell+1)p, e_\mu:=e_{(\ell+1)p,\ell}, e_{\mu+1}:=e_{(\ell+1)p+1,\ell}$ we have

$$(6.13) \qquad \zeta_\ell^h = \Delta^h\{z+h^\mu e_\mu + h^{\mu+1} e_{\mu+1} + \ldots\}.$$

With

$$(6.14) \qquad \kappa_\ell^{\bar h} := \frac{\Delta_h^{\bar h}\zeta_\ell^h - \zeta_\ell^{\bar h}}{(h/\bar h)^\mu - 1} = \Delta^{\bar h}\{\bar h^\mu e_\mu + \bar h^{\mu+1} \frac{(h/\bar h)^{\mu+1}-1}{(h/\bar h)^\mu - 1} e_{\mu+1} + \ldots\}.$$

we obtain the Richardson extrapolation value as

$$(6.15) \qquad \rho_\ell^{\bar h} := \zeta_\ell^{\bar h} - \kappa_\ell^{\bar h} = \Delta^h\{z + \bar h^{\mu+1} \frac{(h/\bar h)^\mu - (h/\bar h)^{\mu+1}}{(h/\bar h)^\mu \quad -1} e_{\mu+1} + \ldots\}.$$

An estimation for the error in $\rho_\ell^{\bar h}$ is found by combining the error estimations $\delta_{\ell+1}^h = \zeta_{\ell+1}^h - \zeta_\ell^h$ and $\delta_{\ell+1}^{\bar h} = \zeta_{\ell+1}^{\bar h} - \zeta_\ell^{\bar h}$

$$(6.16) \qquad \sigma(\rho)_{\ell+1}^{\bar h} := \delta_{\ell+1}^{\bar h} - \Delta_h^{\bar h} (h/\bar h)^\mu \delta_{\ell+1}^h$$

$$= \Delta^{\bar h}\{\bar h^{\mu+1}(h/\bar h - 1)e_{\mu+1} + \ldots\}$$

or, comparing (6.13) - (6.16),

$$(6.17) \qquad \rho_\ell^{\bar h} - \Delta^{\bar h} z = \sigma(\rho)_{\ell+1}^{\bar h} \cdot \frac{(h/\bar h)^\mu - (h/\bar h)^{\mu+1}}{((h/\bar h)^\mu - 1)(h/\bar h - 1)} =: \delta(\rho)_{\ell+1}^{\bar h}.$$

If ζ^h in (1.4) admits an h^2 expansion, the $\mu+1$ in (6.13)-(6.17) must be replaced by $\mu+2$.

Because of the possibility of improving the $\zeta_{\bar\ell}^{\bar h}$, we do not need to satisfy (6.6) but only

$$(6.18) \qquad \| \rho_{\bar\ell}^{\bar h} - \Delta^h z \| \leq \text{tol}.$$

Therefore, we may replace tol by $d \cdot$tol in (6.6) and (6.9), where $2 \leq d \leq 1000$ depends on the basic discretization method and on $\bar{\ell}$.

During the mesh refinement process we want to ensure especially for small step sizes that we need only one or at most two matrices in the linear equations that are solved for every step size. This effect is often attainable by using the discrete correction method (4.1) to provide a good approximation $\tilde{\zeta}^h := \Delta^h_{\bar{h}} \zeta^{\bar{h}}_{\ell}$ for ζ^h. $\tilde{\zeta}^h$ is suitable essentially if $\| F^h \tilde{\zeta}^h \| < \text{tol}$, see § 7, and we use $\tilde{\zeta}^h$ to approximate ζ^h in (3.7). Otherwise, we must perform a few corrections in (1.3) but with $(F^h)'(\tilde{\zeta}^h)$ replacing $(F^h)'(\tilde{\zeta}^h_{\ell-1})$.

If we determine that the final $h = h_\mu$ must be decreased, we can, instead of the discrete corrections 4.1 indicated above, as well use the techniques described in Daniel-Martin [13]. In [13], the last reliable $\zeta^{\bar{h}}_{\ell}$ should be extended and then interpreted as a $\zeta^h_{\ell_o}$, where $\ell_o \geq 0$ is determined by special considerations. Since we must in any case solve the system in (3.7) with a matrix $(F^h)'(\zeta^h_{\ell_o})$, we prefer to start again with $(F^h)'(\zeta^h_o)$ since this usually amounts to solving only one or two additional linear equations with the same matrix. This is considerably less work than is required to factor $(F^h)'(\zeta^h_o)$ or to start an SOR iteration method with a proper relaxation parameter.

7. Description of the algorithm

The aim of this section is to give an idea for the organization of the algorithm. To avoid excessive indices, we write $\bar{h}$ for h_{i-1}, h for h_i, and h_e for the estimated values of the "end" step size h.

The different steps described here are motivated by the discussions in sections 5 and 6. We attempt to be more specific here by inserting comments where appropriate.

<1> Start with h_c; let n $\mathbb{N}$ be fixed (say $n = 2$ or $n = 3$); define $j := 0$; $\bar{h} := h_c$;

Comment: In <2> - <5> we estimate the "end" stepsize h_e to meet (5.1), μ and r to meet (5.13) and $\|\tilde{\zeta}_\mu^{h_c} - \zeta^{h_c}\| < \text{tol}$ and $\bar{\ell}$ to meet (5.14) based on h_c results. For small values of h_i, we want to update h_3, μ, r, $\bar{\ell}$ in every second or third step ($n = 2$ or 3). j counts the number of these steps (see p. 36).

<2> use (1.3) to compute $\tilde{\zeta}_\ell^{h_c}$, $\ell = 0, \ldots, \mu$ such that $\|\tilde{\zeta}_\mu^{h_c} - \zeta^{h_c}\| < \text{tol}$; use $\tilde{\zeta}^{h_c}$ to estimate the relative error $\text{tol}/\|\tilde{\zeta}^{h_c}\|$ and determine an upper bound for $\bar{\ell}$ via (5.14);

<3> use the last matrix for h_c in (1.3) $(F^{h_c})'(\tilde{\zeta}^{h_c}_{-1})$, to compute $\tilde{\zeta}_1^{h_c} - \zeta_0^{h_c}$ from (3.7);

<4> with $\zeta_1^{h_c} - \zeta_0^{h_c}$, estimate h_e from (5.8), where we use h_e for h;

<5> compute r from (5.13), where we use h_e for h;

<6> $h: = \bar{h}/r$;

<7> if $h \approx h_c$, that is, if h is relatively large, use $\zeta_\mu^{\bar{h}}$ to obtain a starting value for (1.3) as $\tilde{\zeta}_0^h: = \Delta_{\bar{h}}^h \tilde{\zeta}^{\bar{h}}$ and goto <13>;

Comment: For relatively large h, so $h \approx h_c$, the iterations in (1.3) are very cheap, so we start with a rather crude $\tilde{\zeta}_0^h: = \Delta_{\bar{h}}^h \tilde{\zeta}_\mu^{\bar{h}}$. For small values of h we try to avoid many steps in (1.3) and apply (4.1) to get a much better starting value, using always the same matrix in (4.1). This is achieved in <8>, <11>, <12>. In <9>, <10> we must estimate whether (5.1) may already be obtained with $h = \bar{h}/2$;

<8> use (4.1) with $\xi_0^{\bar{h}}: = \tilde{\zeta}_\mu^{\bar{h}}$ to make discrete corrections;

<9> use $\ell = 1$ and $\xi_1^{\bar{h}} - \xi_0^{\bar{h}}$ to estimate $\zeta_1^{\bar{h}} - \zeta_0^{\bar{h}}$ with (4.4);

<10> if $\|\zeta_1^{\bar{h}} - \zeta_0^{\bar{h}}\| < 2^p \, tol^*$, with $tol^*: = c \sqrt[\ell+1]{tol}$, goto <18>;

11 make further corrections in (4.1) until

$$\|F^h \Delta_{\bar{h}}^h \xi_{\ell^*-1}^{\bar{h}}\| < \sqrt{tol};$$

Comment: Since (4.1) converges only linearly, but (1.3) quadratically, it is enough to use $\sqrt{tol}$ in <11> to continue with <13>;

<12> obtain a starting value for (1.3) as

$\tilde{\zeta}_0^h := \Delta_{\bar{h}}^{h} \xi_{\ell*}^{\bar{h}}$;

<13> use (1.3) to compute $\tilde{\zeta}_\ell^h, \ell=1,\ldots,\ell^*$ until

$\| \tilde{\zeta}_{\ell*}^h - \zeta^h \| <$ tol and define $t^*(h)$ such that

$\| F^h \tilde{\zeta}_{\ell*}^h \| < t^*(h)$ tol (that is, use (1.3) for the

smaller step size h); define $\mu := \ell^*$;

<14> j: = j+1;

<15> if j < n and $h \approx h_c$ (see <7>) goto <17>;

Comment: Only for j=n and small values of h, so $h \not\approx h_c$, are the h_e, μ, r and $\bar{\ell}$ are updated. This is done in <16>. In <17> the reduction step (5.13) with $h_i = h_{i-1}/r$ is performed;

<16> otherwise use $\| \zeta_1^{\bar{h}} - \zeta_0^{\bar{h}} \|$ from <9>, μ from <13> and h from <6> to update h_e via (5.8), where we use $\bar{h}$ for h_c and h_e for h, and update r via (5.13) and $\bar{\ell}$ via (5.14), where we use $\bar{h}$ for h_c and h_e for h; define j:=0;

<17> define $\bar{h}$: = h; goto <6> ;

<18> With the actual $\bar{h}$ in <6> and μ in <13> ,perform more correction steps in (1.3), so starting with $(F^{\bar{h}})'(\tilde{\zeta}_{\mu}^{\bar{h}})(\tilde{\zeta}_{\mu+1}^{\bar{h}}-\tilde{\zeta}_{\mu}^{\bar{h}}) = -F^{\bar{h}}\tilde{\zeta}_{\mu}^{\bar{h}}$,to guarantee an excellent approximation $\|\tilde{\zeta}_{\sigma}^{\bar{h}}-\zeta^{\bar{h}}\| < tol$, for $\sigma \geq \mu$;

Comment: If the higher approximations $\zeta_{\ell}^{\bar{h}}$ and ζ_{ℓ}^{h} behave as expected, h will be the final h_e. So we need $\zeta^{\bar{h}}$ and ζ^{h} very accurately to get a good enough basis for our checks below;

Comment: In the following steps we must ensure that different assumptions are satisfied (see § 6);

thus in what follows the step sizes $\bar{h}$ and h are used to check these assumptions and are adjusted if some of them are not satisfied.

<19> compute $c_1:=c_1(\bar{h})$ and $c_2:=c_2(\bar{h})$ in (6.2) and (6.4), respectively;

<20> define, for the different values of $(c_1c_2)^{1/p}$, h and $\bar{h}$ in the following way:

$h:=\bar{h}$, $\bar{h}:= 2h$ for $(c_1c_2)^{1/p} \leq 0.6$; if the actual r in <6> is 2 or 3 we use $\bar{h}:=rh$;

$h:=\bar{h}/2$; $\bar{h}:=\bar{h}$ for $0.6 < (c_1c_2)^{1/p} < 1.5$;

$h:=\bar{h}/3$; $\bar{h}:=\bar{h}$ for $1.5 \leq (c_1c_2)^{1/p} < 1.9$;

$h:=\bar{h}/4$; $\bar{h}:=\bar{h}/2$ for $1.9 < (c_1c_2)^{1/p}$;

Comment: Since we have computed already the results for $\bar{h}$, we use $\bar{h}$ when possible either as the value for h or $\bar{h}$, comparing the results with those for $h = \bar{h}/2$ or $\bar{h}/3$. Only for much too large an $\bar{h}$ do we cut back $h: = h/4$ and $\bar{h}: = \bar{h}/2$;

<21> if $\bar{h}$ has been increased in <20>, use <10>, <13>, <19> to update $c_1(\bar{h})$, $c_2(\bar{h})$; if $\bar{h}$ has been decreased, use <8>, <10>, <13>, <19> to update $c_1(\bar{h})$, $c_2(\bar{h})$;

<22> if $(c_1c_2)^{1/p} < 0.6$ or $(c_1c_2)^{1/p} \geq 1.5$, goto <20>;

<23> use <18> to get an excellent approximation for $\zeta^{\bar{h}}$;

<24> compute $\zeta_{\ell}^{\bar{h}}$, $\ell = 1, \ldots, \bar{\ell} + 1$ from (3.7);

<25> if $\|\zeta_{\ell}^{\bar{h}} - \zeta_{\ell+1}^{\bar{h}}\| \leq$ tol or $\leq \varepsilon\text{ps}\cdot\tilde{d}$ for an $\ell \leq \bar{\ell}$, define $h: = \bar{h}$ and $\bar{h}: = 2h$, resp., rh (see <20>) where εps is the accuracy of the computer and $\tilde{d} > 1$ is a security factor;

Comment: <25> shows that either tol or essentially machine accuracy, $\tilde{d}\varepsilon$ps, is obtainable already using $\bar{h}$. So we choose $h: = \bar{h}$ and $\bar{h}: = 2h$ and must compute $\tilde{\zeta}_{\nu}^{\bar{h}}$ and $\zeta_{\ell}^{\bar{h}}$ in <26> for the new value of $\bar{h}$. The factor $\tilde{d}$ in $\tilde{d}\varepsilon$ps depends on the problem to be solved. For a very well-behaved problem we choose, say, $d \approx 10$ and for an unpleasant one we choose, say, $d \approx 10^5$ or even larger;

<26> with $\bar{h}$, h from <25> and $\tilde{\zeta}_0^{\bar{h}} := \Delta_h^{\bar{h}} \zeta_0^h$, use (1.3) to compute an approximation $\| \tilde{\zeta}_\nu^{\bar{h}} - \zeta^{\bar{h}} \| < tol$; compute $\zeta_\ell^{\bar{h}}$, $\ell=1,\ldots,\bar{\ell}+1$, using $\delta_{\ell-1}^h$ from (3.7); goto <33>;

<27> with $\varepsilon := \| \zeta_{\bar{\ell}}^{\bar{h}} - \zeta_{\bar{\ell}+1}^{\bar{h}} \|$ let $c_3 := (\varepsilon/tol)^{\frac{1}{p(\bar{\ell}+1)}}$ and define h by $\bar{h} = h\,([c_3 - \delta]+1)$, where $\delta > 0$ is a fixed small number and $[c_3-\delta]$ the integer part of $c_3-\delta$;

Comment: Using $1 \geq \delta > 0$ allows for an h which might not quite provide us with accuracy tol. So $\delta > 0$ should only be used if a slightly larger error is admitted in favor of less computational time;

<28> if $[c_3-\delta]+1 > 3$ define $\bar{h} := 2h$; interpolate the approximation for $\zeta^{\bar{h}}$ in <23> or <26>, respectively, to obtain a $\tilde{\zeta}_0^{\bar{h}}$ for the actual $\bar{h}$ and use (1.3) as in <26> ; goto <24> ;

Comment: $\bar{h}$ and h are now determined to satisfy (6.6); we are left with the problem of computing ζ_ℓ^h, $\ell=0,\ldots,\bar{\ell}+1$, and of determining whether or not (1.7) is satisfied;

<29> use (4.1) to obtain a good estimation $\zeta_0^h \approx \tilde{\zeta}_0^h := \Delta_{\bar{h}}^h \xi_\ell^{\bar{h}}$, where $\| F^h \Delta_{\bar{h}}^h \dot{\xi}_{\ell-1}^{\bar{h}} \| < tol$;

<30> Let $t^*(h)$ be an extrapolated value from the $t^*(h)$ in <13> for the actual h; with $\tilde{\zeta}_0^h$ in <29> iterate in (1.3) until $\| F^h\tilde{\zeta}_\nu^h\| \leq t^*(h)tol$; estimate $\| \zeta^h-\tilde{\zeta}_\nu^h\| \approx \| \tilde{\tilde{\zeta}}_1^h-\tilde{\zeta}_\nu^h\|$, where

(*) $(F^h)'(\tilde{\zeta}_{\nu-1}^h)(\tilde{\tilde{\zeta}}_1^h-\tilde{\zeta}_\nu^h) = -F^h\tilde{\zeta}_\nu^h$.

($\tilde{\zeta}_{\nu+1}^h$ is obtained from a modified Newton step in (1.3)) for $\| \tilde{\tilde{\zeta}}_1^h-\tilde{\zeta}_\nu^h \| > tol$ iterate (*) to obtain a $\| \zeta^h-\tilde{\tilde{\zeta}}_\sigma^h\| < tol$ and use $\zeta^h := \tilde{\tilde{\zeta}}_\sigma^h$;

<31> define $\ell := 0$, $\upsilon := 0$;

Comment: υ will be used to count how severely (6.12) will be violated, (see <34> , <35>);

<32> compute δ_ℓ^h from (3.7);

<33> if $\| \delta_\ell^h \| <$ tol or $< \tilde{d}\cdot\varepsilon ps$, goto <40> ;

Comment: For $\tilde{d}$ see the sentences following (6.18);

<34> if (6.12) is satisfied for δ_ℓ^h and $\delta_\ell^{\bar{h}}$, define $\upsilon := 0$;

<35> if (6.12) is satisfied with $.15(\ell+1)p$ replaced by $.25(\ell+1)p$, define $\upsilon := \upsilon+1$; otherwise $\upsilon := \upsilon+2$;

Comment: If (6.12) is well satisfied, δ_e^h and $\delta_e^{\bar{h}}$ are reliable, so we stay with $\upsilon:=0$. If (6.12) is mildly violated, we replace υ by $\upsilon+1$. One strong violation ($\upsilon=2$) or two mild violations ($\upsilon= 0+1+1 = 2$) will cause a smaller step size $\bar{h}: = h$, $h: = h/2$ in <36> ;

<36> if $\upsilon = 2$, define $\bar{h}:=h$, $h:=h/2$ and goto<29>;otherwise, if $\ell < \bar{\ell} + 1$, define $\ell:=\ell+1$ and goto <32>;otherwise, if (6.6) is satisfied, goto <40>; otherwise, if (6.6) is satisfied with tol replaced by d^*tol, goto <37> ,otherwise, define $\bar{h}:=h;h:=h/2$ and goto <29> ;

Comment: Replacing tol by d*tol facilitates the achievement of the final accuracy tol by using Richardson extrapolation. This is performed in <37> - <39> . d* depends on ℓ and may be chosen larger if we have an h^2 instead of an h expansion, since for small ℓ and for h^2 expansions, the Richardson extrapolation is much more effective than if otherwise is. Thus, for example, we may define

$$d^*: = \frac{0.8}{\ell+2} \text{ for } h^2 \text{ and } d^*: = \frac{0.3}{\ell+2} \text{ for h expansions;}$$

<37> compute $\rho_{\bar{\ell}}^{\bar{h}}$ and $\delta(\rho)_{\bar{\ell}+1}^{\bar{h}}$ from (6.15) and (6.17), respectively;

<38> if $\| \delta(\rho)^{\bar{h}}_{\bar{\ell}+1} \| <$ tol, goto <39> ; otherwise, define $\bar{h} := h$ and $h := h/2$ and goto <29> ;

<39> print $\rho^{\bar{h}}_{\bar{\ell}}$ as an approximation to $\Delta^{\bar{h}} z$ and $\delta(\rho)^{\bar{h}}_{\bar{\ell}+1}$ as an estimation for the error $\| \rho^{\bar{h}}_{\bar{\ell}} - \Delta^{\bar{h}} z \|$; goto <41> ;

<40> print ζ^{h}_{ℓ} as an approximation for $\Delta^{h} z$ and $\delta^{\bar{h}}_{\ell+1}$ as an estimation for the error $\| \zeta^{h}_{\ell} - \Delta^{h} z \|$;

<41> end.

8. Algorithm modifications for the examples

The main lines of the algorithm presented in § 7 are independent of the special problem. Nevertheless, there are some difficulties specific to the examples treated in this paper.

Example 8.1: *Discrete corrections for B.V.P.s in O.D.E.s.*

The difficulties arising here depend essentially on the choice of $\Delta_{q,h}$ introduced in Example 3.1. If we use $\Delta_{q,h}$ in (3.10) and (3.11), the algorithm in § 7 is unproblematic as long as we ensure that $(b-a)/h_i = k_i q$ and $(b-a)/h_i = k_i'(q-2s)$, respectively, with $k_i, k_i', q, s \in \mathbb{N}$. If k_i or $k_i' \notin \mathbb{N}$, one must define the corresponding polynomial pieces P_μ for $\mu = 0,\dots,[k_i]$ or $\mu = 0,\dots,[k_i']$, respectively, and must, close to the endpoint b, use some of the points $x_{[k_i]q-\nu}$, $\nu > 0$, twice to define the missing polynomial piece for the interval $(x_{[k_i]q}, b)$.

There are four reasons for starting the algorithm with a coarse h_c as indicated before. First, it is rather cheap to compute ζ^{h_c} via (1.3) for a relatively large h_c. This iteration (1.3) allows us to obtain μ. Further, we use $\zeta_0^{h_c}-\zeta_1^{h_c}$ to estimate h_e via (5.8), and r from μ, h_c and h_e via (5.13). Since we are interested only in the first correction $\zeta_0^{h_c}-\zeta_1^{h_c}$, we do not need the defect $\Omega_b^{h_c} F_b \Delta_{h_c} \zeta_0^{h_c}$ more accurately than order $2p = 4$ for the box scheme if we do not prefer to compute $\zeta_0^{h_c}-\zeta_1^{h_c}$ using the relation (4.4). So it is enough to use polynomial pieces of degree 4 with h_c, whereas for the smaller h_i, where we have to compute higher iterates $\zeta_{\ell}^{h_i}$, $\ell > 1$, we must use polynomials of correspondingly higher degree. If we use the local approximations, as indicated at the end of Example 3.2, we must use one additional point to the left of a and right of b, respectively, to compute $\Omega_b^{h_c} F_b \Delta_{h_c} \zeta_o^{h_c}$ of order 2p = 4. For the smaller step sizes and higher corrections up to the order $\bar{\ell}$, we need $\bar{\ell}(\bar{\ell}+1)/2$ outer points to the left of a and right of b. □

Example 8.2: *Discrete corrections for elliptic B.V.P.s in P.D.E.s.*

If we are interested in good accuracy over all of R^o, we must use k > 2 in Example 3.3 typically with k = 4 or 6. In this case, the matrices or nonlinear equations are rather widely banded, which is a somewhat severe but unavoidable disadvantage. In this case the first correction provides an excellent improvement and the second correction may be used for error estimation. For the computation of the defects of order 4 or 6, we use either symmetric, or, close to the boundary, unsymmetric, formulas. Using extrapolated values for additional outer points in symmetric formulas is essentially equivalent to the use of unsymmetric formulas. □

If we take into account these remarks for our two examples, we can immediately use the algorithm described in § 7. Analogous difficulties will arrise in other examples, but may be solved by similar considerations.

APPENDIX 1

In the foregoing we have assumed the availability of an approximate solution $\zeta_o^{h_c}$ which will lie within the domain of convergence $\mathcal{B}^{h_c}(\zeta^{h_c}) \subset \mathcal{D}^{h_c}$ of whichever discrete Newton method is being used for the discretization P^{h_c}. Here h_c is usually a relatively coarse starting mesh size having say, n mesh points.

Practical numerical experience shows that if a particular solution $\zeta^{h_c} \approx \Delta^{h_c} z$ is sought, and if $F^{h_c} y = 0$ has several solutions, then the determination of an adequate $\zeta_o^{h_c}$ may be a difficult matter if one merely attempts to make starting guesses in a region where a solution is expected to lie. Thus it is worthwhile to briefly indicate here how an appropriate starting guess may be obtained by a homotopy or deformation method. A discussion of the details and aspects of several homotopy and deformation algorithms is available in Allgower-Georg [1] and in corresponding references cited therein. Hence, we shall only indicate here how homotopy or deformation methods may be applied to the present problem.

The problem is: For a given $\delta>0$, find an approximation $\zeta_o^{h_c}$ to ζ^{h_c} where $F^{h_c}\ \zeta^{h_c} = 0$ such that $\|\zeta^{h_c}-\zeta_o^{h_c}\| < \delta$.

There are fundamentally two types of algorithms for homotopy or deformation methods :

<u>I Continuation methods.</u>

In these methods, (see e.g. Allgower-Georg [1],

Kellog-Li-Yorke [26], Chow-Mallet-Paret-Yorke [12]) it is generally assumed that F^{h_c} is a smooth map and a homotopy such as

$$\text{(A1.1)} \qquad H(t,y)=(1-t)F_o(y)+t\,F^{h_c}(y), \quad (t,y) \in \mathbb{R} \times \mathbb{R}^n$$

is constructed where F_o is a chosen mapping having a given zero-point y^o. Under conditions of regularity, there is a smooth curve C emanating from (o,y^o). The curve C may be followed within a tubular neighborhood of diameter δ by solving the corresponding Davidenko Initial Value Problem

$$\text{(A1.2)} \quad \begin{cases} H'(t,y)\dot{y} = 0, \\ ((\dot{t},\dot{y}),(\dot{t},\dot{y}))_2 = 1, \text{ where } (\cdot,\cdot)_2 \text{ indicates the Euclidean norm in } \mathbb{R}^{n+1}, \\ (t(0),y(0)) = (0,y^o) \end{cases}$$

by some predictor-corrector method. If in the process of following C, a value of $t \geq 1$ is attained, an approximate solution $\zeta_o^{h_c}$ has been encountered.

<u>II Simplicial methods.</u>

In these methods (c.f., Allgower-Georg [1]) one may again use a continuous homotopy such as (A1.1) or a discontinuous deformation such as

$$\text{(A1.3)} \; H(t,y) = \begin{cases} F_o(y) \text{ for } t = 0 \\ t\,F^{h_c}(y) \text{ for } t \in (0,1]. \end{cases}$$

The distinction from I is that a piecewise linear approximation

C_T to C defined by $H^{-1}(0)$ (either for (A1.1) or (A1.3) relative to some triangulation T of $[0,1)\times\mathbb{R}^n$) is followed instead of C. The process of following C_T consists of a pivoting between simplices as is done in linear programming algorithms.

To illustrate how these methods may be applied to the present problem, let us consider the following specific example. We want to find a solution $\lambda > 0$, $\bar{y} > \bar{0}$, $\bar{y} \in \mathbb{R}^n$ satisfying

$$(A1.4) \qquad F^{h_c}(\lambda,\bar{y}) := M_n\bar{y} - \lambda\bar{f}(\bar{y}) = \bar{0}.$$

Here we shall make the following assumptions

(i) M_n is a nonsingular $n\times n$ matrix and $M_n^{-1} > 0$.

(ii) $\bar{y} = (y_1,\ldots,y_n)^T$, $\bar{f}(\bar{y}) = (f(y_1),\ldots,f(y_n))^T$

where $f\colon \mathbb{R}^1 \to \mathbb{R}^1$ is a positive smooth function.

The problem (A 4) frequently arises from a finite difference discretization (e.g. via central differences) of a nonlinear elliptic boundary value problem. In general $\bar{f}(\bar{y})$ represents the discretization of a Nemytskii operator. The positivity assumption on f may be relaxed (c.f., Georg [21]), and we make it here primarily to simplify the following discussion.

If we wish to use a homotopy method to obtain a solution, we might define

$$(A1.5) \qquad \begin{cases} H(t,\bar{y}) := (1-t)M_n\bar{y} + t(M_n\bar{y} - \bar{f}(\bar{y})) \\ \qquad = M_n\bar{y} - t\bar{f}(\bar{y}) \text{ for } (t,\bar{y}) \in \mathbb{R}\times\mathbb{R}^n. \end{cases}$$

Since $M_n^{-1} > 0$, we can conclude by the implicit function theorem that, for some $\alpha > 0$, there is a smooth curve

$C_\alpha=\{(t,\bar{y}(t)): t \in(-\alpha,\alpha)\}$ passing through $(0,0)$ with

(A 1.6) $$H(t,\bar{y}(t)) = \bar{0}$$

and that $\bar{y}(t)>\bar{0}$ for $t>0$ and $\bar{y}(t)<\bar{0}$ for $t<0$.

The curve C_α may be "followed" from $(0,\bar{0})$ in the positive t-direction by either a continuation algorithm (Method I) or a simplicial algorithm (Method II). Recent works in which these approaches have been used are Keller [24], Rheinboldt [30], and Watson [37] (Method I), and Allgower-Jeppson [2], Jürgens-Peitgen-Saupe [22] and Peitgens-Prüfer [27] (Method II).

Now let us suppose that a solution $(\lambda_0,\bar{y}^0)>\bar{0}$ to (A1.4) has been located or approximated for some fixed $\lambda_0>0$. Suppose that it is wished to determine other solutions (λ_0,z) to (A1.4) for the same value of the parameter λ, or to determine whether there are any other such solutions. One approach to attempt to solve this problem would be to merely follow C_α in (A1.6) to determine whether C_α contains other solutions $(\lambda_0,\bar{z})$. However, it may very well occur that the solution set to (A1.4) may contain separated components and in this event the above approach may fail to yield solutions $(\lambda_0,\bar{z})$ which are in fact present. The following homotopy method may be used in this event. Suppose that there exist

(A1.7)
(i) $\bar{d} \neq \bar{0} \in \mathbb{R}^n$
(ii) a bounded open neighborhood $\mathcal{U}$ of $\bar{y}^0$ such that $H_d(t,\bar{y}):=t\bar{d}+F^{h_c}(\lambda_0,\bar{y}) \neq \bar{0}$ for all $\bar{y} \in \partial\,\mathcal{U}$ and $t \geq 0$.

Then it is clear that the smooth curve C_d with $(0,\bar{y}^0) \in C_\alpha \subset H_\alpha^{-1}(0)$ is also bounded in the positive t-direction and hence penetrates the t = 0 hyperplane a positive even number of times. An example where this device is used for (A1.4) occurs in Allgower-Georg [1] for the case that f(y) is asymptotically superlinear as $y \to \infty$. Recently, a simplicial homotopy algorithm incorporating topological perturbations has been applied in Jürgens-Peitgen-Saupe [22] to approximate additional solutions (λ_0, z) to problems such as (A.1.4).

APPENDIX 2

The discrete correction, DC, method (4.1) is closely related to the multigrid method, MG (cf., [10] and [11]), at least formally. In particular, the equation defining the MG coarse grid correction $\delta_\ell^{\bar h}$ to the fine grid approximation $\xi_{\ell-1}^h$ can be written in the present notation as

(A2.1) $$F^{\bar h}[\Delta_h^{\bar h}\xi_{\ell-1}^h + \delta_\ell^{\bar h}] - F^{\bar h}\Delta_h^{\bar h}\xi_{\ell-1}^h = -\hat{\Delta}_h^{\bar h}F^h\xi_{\ell-1}^h .$$

The new fine grid approximation is then given by

(A2.2) $$\xi_\ell^h = \xi_{\ell-1} + \Delta_{\bar h}^h\delta_\ell^h .$$

Since the two methods differ fundamentally by the grid on which they store their approximations (coarse for DC and fine for MG), the relationship between (4.1) and (A2.1) is not very precise. However, if $\xi_{\ell-1}^h$ happens to satisfy

(A2.3) $$\Delta_{\bar h}^h\Delta_h^{\bar h}\xi_{\ell-1}^h = \xi_{\ell-1}^h$$

(this would be unusual since the operator in (A2.3) is far from being the identity), then setting $\xi_{\ell-1}^{\bar h} = \Delta_h^{\bar h}\xi_{\ell-1}$ and rewriting (A2.1) we have

(A2.4) $$F^{\bar h}[\xi_{\ell-1}^{\bar h} + \delta_\ell^{\bar h}] - F^{\bar h}\xi_{\ell-1}^{\bar h} = -\hat{\Delta}_h^{\bar h}F^h\Delta_{\bar h}^h\xi_{\ell-1}^{\bar h} .$$

The relationship between (4.1) and (A2.4) now is fairly clear. For example, if (1) is a linear problem, that is, if $F^{\bar h}$ is a linear variety, then (4.1) and (A2.4) are equivalent. Thus, (4.1) can in general be thought of as a single Newton iteration step applied to (A2.4).

The differences between DC and MG are fundamental. First, as noted above, the grids on which they maintain approximate solutions are different. While DC attempts to approximate the grid h solution as closely as possible on grid $\bar{h}$, MG more typically attemps to approximate this solution on grid h. Thus,DC maintains a grid h approximation only by computing (but not storing) $\Delta_{\bar{h}}^{h}\xi_{\ell-1}^{\bar{h}}$ at each stage. Second, they are based upon different philosophies. MG views the residual error $F^h(\xi_{\ell-1}^h)$ in terms of its Fourier components, resolves the low frequency components on the coarse grid, and iterates by some relaxation process to resolve the high frequency ones on the fine grid. DC, on the other hand, views the actual error $\xi_{\ell-1}^h-\Delta^h z$ in terms of an asymptotic expansion and resolves this expansion by computing the residual on the fine grid and iterating on the coarse grid by a Newton-like method. A third distinction is that, while MG is a complete process,DC can be viewed as an outer loop iteration defined by (3.7) and (4.1). Each of these iterations requires a linear equation solver that might be based upon either a direct method such as Choleski or LU factorization or, in the case of partial differential equations, an inner loop iteration such as MG. We now turn our discussion to a brief account of the use of MG in this way.

The multigrid method that should be used as an inner loop iterative method for solving equations of the form

$$A^h\delta^h = f^h, \qquad \text{(A2.5)}$$

where $A^h = (F^H)'(\xi)$, is the correction scheme, CSMG(cf. [10]), used to solve linear systems. (Note that each equation defining our outer loop algorithm is of this form). CSMG has some small advantages over the more general MG in that slightly less computation work is required and the process is easier to conceptualize.

As a final note, we remark that the evaluation $-\hat{\Delta}_h^{\bar{h}} F^h \Delta_{\bar{h}}^h \xi_{\ell-1}^{\bar{h}}$ is not really necessary (at least for linear problems) provided F^h satisfies the homogenization condition

$$F^{\bar{h}}\xi^{\bar{h}} = \hat{\Delta}_h^{\bar{h}} F^h \Delta_{\bar{h}}^h \xi^{\bar{h}}, \qquad \xi^{\bar{h}} \in E^{\bar{h}} . \tag{A2.6}$$

That is, suppose $F^h(\xi^h) = A^h\xi^h - f^h$ where $A^h\colon E^h \to E^h$ is linear

$$A^{\bar{h}} = \hat{\Delta}_h^{\bar{h}} A^h \Delta_{\bar{h}}^h . \tag{A2.7}$$

(See [26] for more detail on this condition and situations where such a condition is met in practice.) Then the evaluation $\hat{\Delta}_h^{\bar{h}}(A^h\Delta_{\bar{h}}^h\xi^{\bar{h}} - f^h)$ is equivalent to $A^{\bar{h}}\xi^{\bar{h}} - f^{\bar{h}}$ provided $f^{\bar{h}} = \hat{\Delta}^{\bar{h}} f^h$; that is, it is unnecessary to appeal to grid h to compute the residual for grid $\bar{h}$ since it agrees with the residual evaluation on this grid. Thus, the discrete correction method is most useful in cases where (A2.7) is not satisfied. Such cases dominate applications,however, include all five-point discretizations of Laplace's equation in two dimensions, for example.

Literature

[1] Allgower, E.L. and Georg, K.: Simplicial and continuation methods for approximating fixed points and solutions to systems of equations, SIAM Review, 22 (1980)

[2] Allgower,E.L. and Jeppson, M.M.: The approximation of solutions of nonlinear elliptic boundary value problems with several solutions, Springer Lecture Notes in Mathematics, 333, 1-20 (1973).

[3] Allgower, E.L. and Mc Cormick, S.F.: Newton's method with mesh refinements for numerical solution of nonlinear boundary value problems, Numerische Math., 29, 237-260 (1978).

[4] Allgower, E.L.,McCormick, S.F., and Pryor, D.V.: A general mesh independence principle for Newtons method applied to second order bondary value problems, to appear in Computing.

[5] Bers, L.: On mildly-nonlinear partial differential equations of elliptic type, J.Res.Nat.Bur. Standards, 51, 229-236 (1953).

[6] Böhmer,K.:Discrete Newton methods and Iterated defect corrections, I.General theory, II. Proofs and applications to initial and boundary value problems, submitted to Numer.Math.

[7] Böhmer, K.: High order difference methods for quasilinear elliptic boundary value problems on general regions, University of Wisconssin Madison MRC Report.

[8] Böhmer, K.: Asymptotic expansions for the discretization error in linear elliptic boundary value problems on general regions, to appear in Math.Zeitschrift.

[9] Böhmer, K., Fleischmann, H.-J.: Self-adaptive discrete Newton methods for Runge-Kutta-methods,to appear in ISNM, Basel 1979/80.

[10] Brandt,A.: Multi-level adaptive solutions to boundary value problems, Math.Comp. 31, 333-390 (1977).

[11] Brandt, A.: Multi-level adaptive techniques (MLAT) for partial differential equations, Mathematical Software III, Academic Press, New York, 277-318, (1977).

[12] Chow, S.N., Mallet-Paret, J., and Yorke, J.: Finding zeros of maps: Homotopy methods that are constructive with probability one, Math.Comp. 32, 887-899 (1978).

[13] Daniel, J.W. and Martin, A.J.: Numerov's method with deferred corrections for two-point boundary value problems, SIAM J.Num.Anal., 14, 1033-1050 (1977).

[14] Fox, L., and Goodwin, F.T.: Some new methods for the numerical integration of ordinary differential equations. Proc.Camb. Phil. Soc. 45, 373-388 (1949).

[15] Frank, R.: Schätzungen des globalen Diskretisierungsfehlers bei Runge-Kutta-Methoden, ISNM 27, 45-70 (1975).

[16] Frank, R.: The method of iterated defect-correction and its application to two-point boundary value problems, Part I., Numer. Math. 25, 409-419 (1976), Part II, Numer. Math. 27, 407-420 (1977).

[17] Frank, R., Hertling, J., and Ueberhuber, C.W: An extension for the applicability of iterated deferred corrections Math. Comp. 31, 907-915 (1977).

[18] Frank, R., Hertling, J., and Ueberhuber, C.W.: Iterated defect correction based on estimates of the local discretization error, Report Nr. 18/76 des Instituts für Numerische Mathematik, Technische Universität Wien (1976).

[19] Frank, R., and Ueberhuber, C.W.: Iterated defect correction of the efficient solution of systems of ordinary differential equations, BIT 17, 146-159 (1977).

[20] Georg, K.: A simplicial algorithm with applications to optimization, variational inequalities and boundary value problems, to appear in Proceedings of Symposium on Fixed Point Algorithms and Complementarity, Southampton, England, 1979.

[21] Jürgens, H., Peitgen, H.-O., and Saupe, D.: Topological perturbations in the numerical study of nonlinear eigenvalue and bifurcation problems, in Proceedings of Symposium on Analysis and Computation of Fixed Points, Madison, Wis., 1979, Academic Press, New York ed.S.M. Robinson 1980, 139-181.

[22] Keller, H.B.: Accurate difference methods for nonlinear two point boundary value problems, SIAM J. Numer.Anal. 11, 305-320 (1974).

[23] Keller,H.B.: Numerical methods for two point boundary value problems, Blaisdell, Waltham, Mass. 1968 .

[24] Keller, H.B. and Peryra, V.: Difference methods and deffered corrections for ordinary boundary value problems, SIAM J. Num.Anal., 16, 241-259 (1979).

[25] Kellogg, R.B., Li, T.Y., and Yorke, J.: A constructive proof of the Brouwer fixed point theorem and computational results, SIAM J. Num.Anal. 4, 473-483 (1976).

[26] McCormick, S.F.: Multigrid methods;an alternate view, Lawrence Livermore Laboratory Reports, 1979.

[27] Peitgen, H.-O. and Prüfer, M.: The Leray Schauder continuation method is a constructive element in the numerical study of nonlinear eigenvalue and bifurcation problems, Funct.Diff. Equat. and Approx. of Fixed Pts., Springer L.N.730(1980)326-409

[28] Pereyra, V.L.: Iterated deferred corrections for non-linear operator equations, Numer.Math. 10, 316-323(1967).

[29] Pereyra, V., Proskurowsky, W.,and Widlund,O.: High order fast Laplace solvers for the Dirichlet problem on general regions, Math. Comp. 31, 1-16 (1977).

[30] Rheinboldt, W.: Solution field of nonlinear equations and continuation methods, Technical Report ICMA-79-04, March 1979.

[31] Richardson, L.F.: The approximate arithmetical solution by finite differences of physical problems involving differential equations, with applications to the stresses in a masonry dam, Philos. Trans.Roy. Soc. London Ser.A, 210, 307-357 (1910).

[32] Simpson, R.B.: Finite difference methods for mildly nonlinear eigenvalue problems, SIAM J. Num.Anal. 8, 190-211 (1971).

[33] Stetter, H.J.: Asymptotic expansions for the error of discretization algorithms for nonlinear functional equations, Numer. Math. 7, 18-31 (1965).

[34] Stetter, H.J.: Analysis of dicretization methods for ordinary differential equations, Springer-Verlag, Berlin - Heidelberg - New York, 1973.

[35] Stetter, H.J.: The defect correction principle and discretization methods, Numer.Math. 29, 425-443 (1978).

[36, Wasow, W.: Discrete approximations to elliptic differential equations, Z. Angew.Math. Phys. 6, 81-97 (1955).

[37] Watson, L.T.: An algorithm that is globally convergent with probability one for a class of nonlinear two-point boundary value problems, SIAM J. Num.Anal. (1979).

[38] Zadunaisky, P.E.: A method for the estimation of errors propagated in the numerical solution of a system of ordinary differential equations, in "The theory of orbits in the solar system and in stellar systems" Proc. of Intern. Astronomical Union, Symp. 25, Thessaloniki, Ed. G. Contopoulos, 1964.

[39] Zadunaisky, P.E.: On the accuracy in the numerical computation of orbits, in "Periodic Orbits, Stability and Resonances", Ed. G.E.O. Giacaglia, Dordrecht-Holland, 216-227 (1970).

[40] Zadunaisky, P.E.: On the estimation of errors propagated in the numerical integration of ordinary differential equations, Numer. Math. 27, 21-40 (1976).

A DUFFING EQUATION
WITH
MORE THAN 20 BRANCH POINTS

BY

K.-H. BECKER*

AND

R. SEYDEL**

*) Lehrstuhl A für Mechanik
Technische Universität München
Arcisstraße 21
D-8000 München 2

**) Institut für Mathematik
Technische Universität München
Arcisstraße 21
D-8000 München 2

A Duffing Equation with more than 20 Branch Points

K.-H. Becker
R. Seydel

The differential equation of a Duffing oscillator is presented which exhibits an interesting branching behaviour. Depending on the frequency of the excitation, there is a great variety of different types of solutions. Extensive numerical results are obtained by the means of classical numerical analysis.

1. Introduction

In this note, the special Duffing equation

$$\ddot{x} + \frac{1}{25}\dot{x} - \frac{1}{5}x + \frac{8}{15}x^3 = \frac{2}{5}\cos(\omega\bar{t}) \tag{1}$$

is considered. $\bar{t}$ is the physical time, $x(\bar{t})$ is the state variable, and ω is the parametric excitation which is considered as constant or varying "slowly", i.e. the case of stationary excitation will be handled. A Duffing equation with such coefficients has physical relevance: it is e.g. a simple model of the time dependence of a buckled beam undergoing forced lateral vibrations [8].

Equation (1) possesses a great number of qualitatively different kinds of periodic solutions $x(\bar{t})$ which depend on ω. In the following, some results of our numerical investigations will be given; the special attention is directed to the branching behaviour of the harmonic solutions. Beyond that there are aperiodic solutions (strange attractors) of this equation; they are treated in [12].

The above example seems to be valuable because of the very complex structure of the solutions; the authors would like to propose this example to become a standard example for testing continuation methods and methods for the handling of branching problems.

2. The boundary value problem

In order to be able to handle the problem comfortably, equation (1) will be suitably transformed. The periodic solutions satisfy the boundary conditions

$$(1') \qquad x(0) = x(T)\ , \quad \dot{x}(0) = \dot{x}(T)\ , \quad T = \frac{2\pi}{\omega}\ ,$$

where we confine us to the case of harmonic solutions; the other types of periodic solutions can be handled analogously. In version (1') the interval-length changes if the parameter ω is varied. This is avoided by the transformation to the unit interval $0 \le t \le 1$ by

$$(2) \qquad T\,t = \bar{t}\ , \quad y_1(t) = x(\bar{t})\ , \quad y_2 = \dot{x}\ .$$

For most values of ω there are several different solutions which have to be distinguished. This can be carried out by calculating the functional

$$(3) \qquad N(x) := ||x||^2 := \int_0^T x^2(\bar{t})\, d\bar{t}\ .$$

For the calculation of (3) two additional variables are introduced [10], which leads to the boundary value problem

$$(4) \qquad \begin{aligned} y_1' &= y_2\ 2\pi / y_3 \\ y_2' &= \left(-\frac{1}{25} y_2 + \frac{1}{5} y_1 - \frac{8}{15} y_1^3 + \frac{2}{5}\cos(2\pi t)\right) 2\pi / y_3 \\ y_3' &= 0 \\ y_4' &= y_1^2\ 2\pi / y_3 \ , \qquad \text{boundary conditions:} \\ 0 &= r_1 = y_1(0) - y_1(1) \\ 0 &= r_2 = y_2(0) - y_2(1) \\ 0 &= r_3 = y_4(0) \\ 0 &= r_4 = \begin{cases} y_4(1) - N\ , & \text{if } N \text{ is prescribed} \qquad (4a) \\ y_3(0) - \omega\ , & \text{if } \omega \text{ is prescribed} \qquad (4b) \end{cases} \end{aligned}$$

If one intends to calculate a solution of (1) with prescribed value of N, one has to solve the system (4a); the corresponding value of ω is given by $\omega = y_3$.

All the solutions of (1) which will be reported in Section 5 can be calculated by solving (4a) or (4b) repeatedly for various values of N or ω.

3. Symmetries; computation of the amplitude

The left-hand side of (1) is an odd function of x; the excitation on the right-hand side changes sign if the phase is shifted by T/2 . Thus one has the property:

Let $x(\bar{t})$ be solution of (1), (1') for $0 \le \bar{t} \le T$. Then

$$\tilde{x}(\bar{t}) := -x(\bar{t}+T/2) = \begin{cases} -x(\bar{t}+T/2) & \text{for } 0 \le \bar{t} \le T/2 \\ -x(\bar{t}-T/2) & \text{for } T/2 \le \bar{t} \le T \end{cases}$$

is also solution of (1), (1') for the same ω.
In the phase plane the graphs of x and $\tilde{x}$ are symmetric with respect to the origin.

Computation of the amplitude: In applications, one is more interested in the amplitude than in the integral variable N(x) in (3). The amplitude

$$A := \frac{1}{2}\{\max_{0 \le t \le 1} y_1(t) - \min_{0 \le t \le 1} y_1(t)\}$$

can be computed with high accuracy in the following way: During the integration of (4) the zeroes of the "switching function" $S(t)=y_2(t)$ are determined, see [1]. This yields the amplitude by the scheme

for t=0 : $a_{max} := a_{min} := y_1(0)$

for $0<t\le 1$ with S(t)=0 :

$$a_{max} := \max\{a_{max}, y_1(t)\}$$

$$a_{min} := \min\{a_{min}, y_1(t)\}$$

for t=1 : $A = 1/2\,(a_{max} - a_{min})$.

4. Computation of the branch points

Special solutions we are interested in are the branch points, often referred to as critical points. Following the method of [10], a branch point is calculated by solving (1) and its linearization simultaneously. Formally this can be formulated as follows: Solve the boundary value problem (4) where the fourth boundary condition is replaced by

(4c) r_4 : x has to satisfy the property: there exists a non-trivial solution of the linearization of (1).

Then the corresponding values of ω and N are determined by the system (4). - The numerical realization of (4c) requires two further variables; the resulting "branching system" (a system of 6 differential equations) is suitable for the computation of different types of branch points.

Before one solves the branching system one needs a reasonable initial guess; otherwise an efficient solution of such an enlarged system is not possible. Further, rough information on the location of the branch points is required. Both are supplied by the software in [11], for details see [10]. After these preparations an easy computation of the branch points is possible.

5. Numerical results

All the boundary value problems were solved by the multiple shooting algorithm BOUNDSOL of Bulirsch et al. [1, 13, 5, 6, 4]. The initial value problems were integrated by the extrapolation method DIFSY1 [3] (for the stepsize control see [9]) and by the Runge-Kutta-Fehlberg method RKF7 [7] with the usual stepsize control. The amplitudes were determined by means of the routine OPTSOL [1] which is a modification of BOUNDSOL for problems with switching functions. The computations were performed on the CYBER 175 of the Leibniz-Rechenzentrum der Bayerischen Akademie der Wissenschaften using single precision arithmetic (48-bits mantissa).

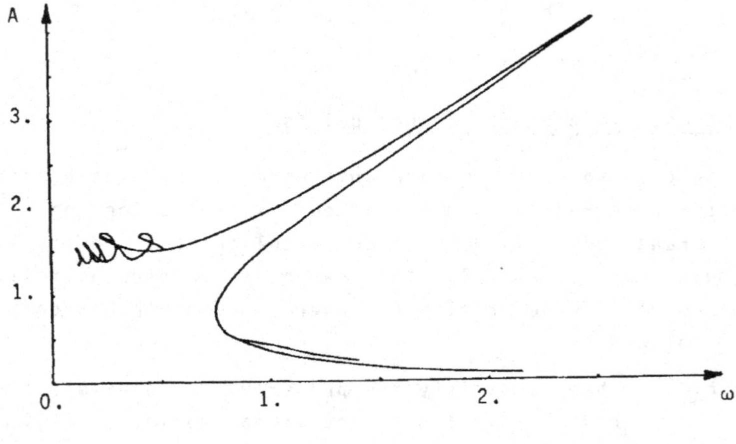

Fig.1 response curve of the Duffing oscillator (1)

A survey on the obtained solutions is given by the response curve in Fig.1. The "main branch" (whose graph is smooth in the diagram Fig.1) is characterized by $x=\tilde{x}$. This main branch shows for $\omega>0.6$ the familiar response of a Duffing oscillator with hard spring. Beside these expected solutions there are "secondary branches". The solutions of the secondary branches can be characterized by $x\neq\tilde{x}$.

The branches emanating at the branch point with $\omega=0.84...$ are caused by the additional changes of the sign of the restoring force at the stable equilibrium of the unforced oscillation $x=\pm(3/8)^{1/2}$. For increasing ω the zeroes of $x(t)$ disappear, see the phase plane Fig.2. In Fig.2 and in all the following phase planes the horizontal axes are the x (y_1)-axes, the vertical axes are the $\dot{x}$ (y_2)-axes; both axes are scaled by 0.1 .

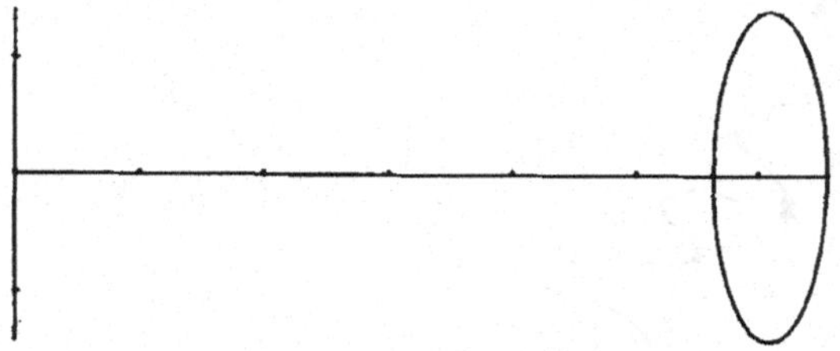

Fig.2 phase plane of a secondary solution for $\omega=3$.

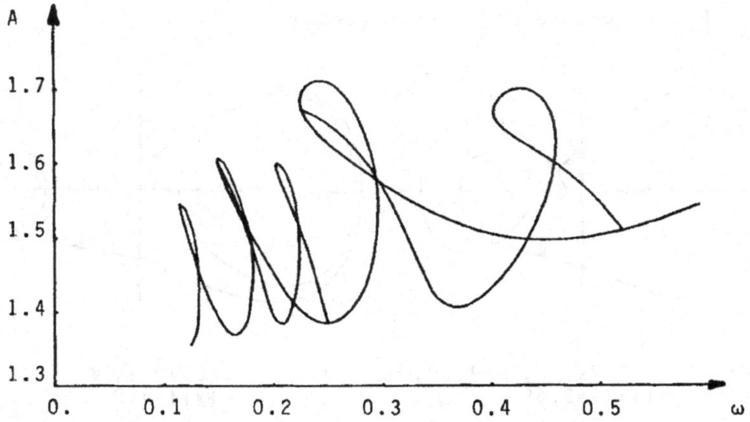

Fig.3 detail of the response diagram

As can be seen in Fig.1, for parameter values $\omega<0.5$, there are many different solutions; the variation of the amplitude A with ω is presented in Fig.3. The same solutions are given in Fig.4 in the ω-$||x||$-diagram which is easier to survey; in Fig.4 the branch points are marked. In Fig.5 and Fig.6 two examples of secondary solutions are given. As is shown in the examples of Fig.5, Fig.6, Fig.7, Fig.8, and Fig.9, the solutions for "small" values of ω have two centers. For decreasing ω the solutions oscillate stronger; each "loop" (Fig.3, Fig.4) attaches a further oscillation. The gain in oscillations makes the numerical computation of the solutions harder.

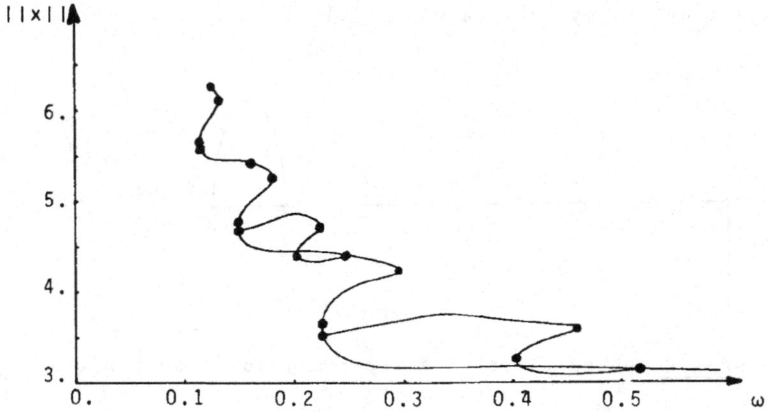

Fig.4 branching diagram (detail)

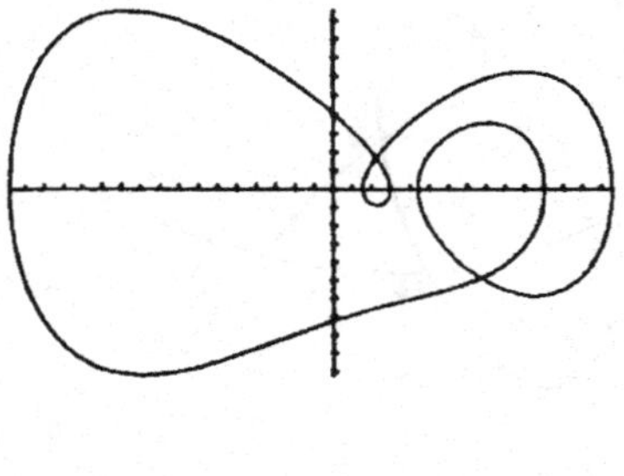

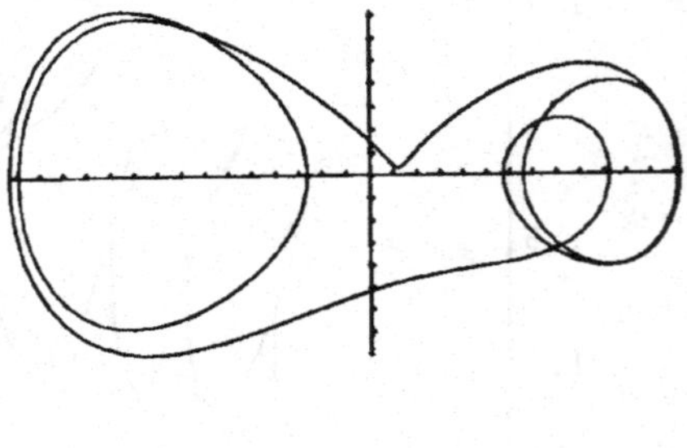

Fig.5: $\omega=0.2995$, $A=1.57$, $||x||=3.69$

Fig.6: $\omega=0.1921$, $A=1.43$, $||x||=4.84$

Fig.5, Fig.6 phase planes of two secondary solutions

During the continuation along the branches (by solving (4a) or (4b)) the branch points were detected and calculated (cf. Sect.4). The results are given in the table; three of the corresponding solutions are presented in Fig.7, Fig.8, and Fig.9. The third column of the table yields the information on the type of the branch points; two emanating branches refer to a "turning point", four emanating branches refer to a "secondary bifurcation point".

ω	A (amplitude)	number of emerging branches
2.502052	4.110751	2
0.842728	0.494041	4
0.748053	0.798077	2
0.520572	1.509612	4
0.460134	1.627274	2
0.402191	1.669646	2
0.296267	1.538703	2
0.250065	1.385417	4
0.225884	1.673456	4
0.224722	1.684075	2
0.224279	1.487556	2
0.202391	1.596173	2
0.181368	1.456409	2
0.165875	1.369795	4
0.149098	1.600829	4
0.148825	1.604388	2
0.132017	1.422928	2
0.124410	1.355929	4
0.114505	1.542689	4
0.114355	1.544986	2

<u>Table</u> 20 branch points of the Duffing equation (1)

Remark on the accuracy in computing the secondary bifurcation points: About one half of the number of digits of the mantissa can be obtained very efficiently. If the demands for accuracy are increased too much, the convergence begins to become slow because the branching system gets singular at a secondary bifurcation point. However, the computation of branch points in higher accuracy seems to be beyond the demands in applications.

We stopped the calculation after having calculated 20 branch points. Further reduction of the parameter ω does not seem to be worth wile as one has to expect an analogous branching behaviour for smaller values of ω.

<u>Acknowledgement</u>. The authors appreciate the influence and the interest of Prof.Dr.R. Bulirsch.

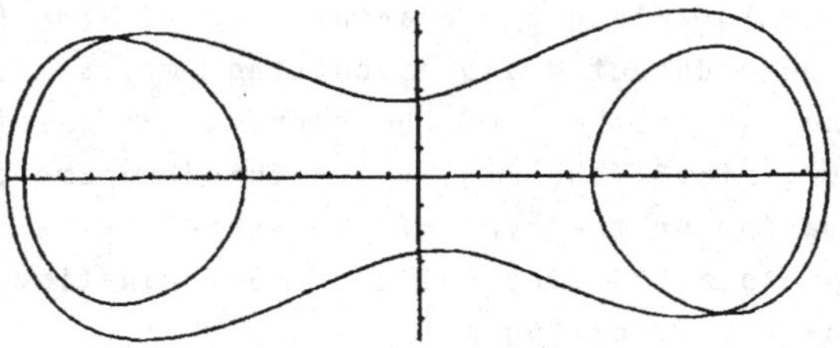

Fig.7:
ω=0.2501

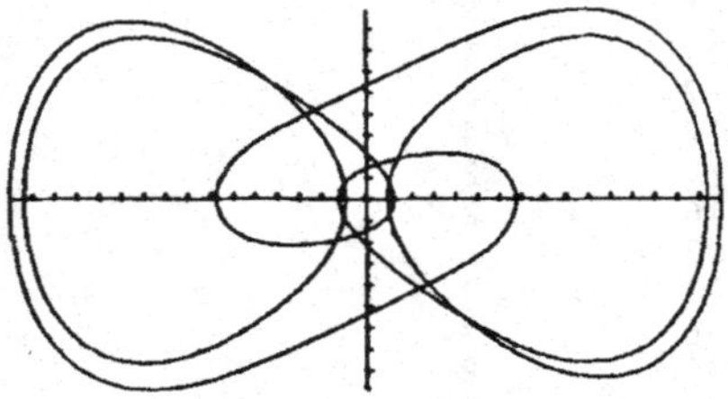

Fig.8:
ω=0.1488

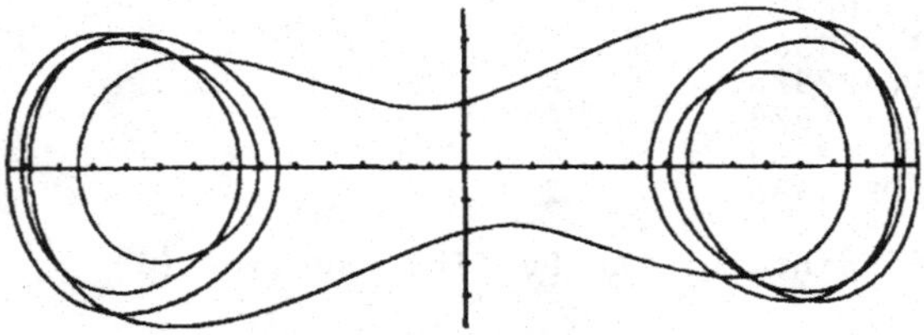

Fig.9:
ω=0.1244

Fig.7, Fig.8, Fig.9 phase planes of three branch points, compare the table

6. References

[1] R. Bulirsch: Die Mehrzielmethode zur numerischen Lösung von nichtlinearen Randwertproblemen und Aufgaben der optimalen Steuerung. Report der Carl-Cranz-Gesellschaft (1971)

[2] R. Bulirsch, W. Oettli, J. Stoer (ed.): Optimization and optimal control. Proceedings of a conference held at Oberwolfach, 1974. Lecture Notes, Vol. 477, Springer, Berlin-Heidelberg- New York (1975)

[3] R. Bulirsch, J. Stoer: Numerical treatment of ordinary differential equations by extrapolation methods. Numer. Math. 8 (1966), 1-13

[4] R. Bulirsch, J. Stoer, P. Deuflhard: Numerical solution of nonlinear two-point boundary value problems I. to appear in Numer. Math., Handbook Series Approximation

[5] P. Deuflhard: A modified Newton method for the solution of ill-conditioned systems of nonlinear equations with application to multiple shooting. Numer. Math. 22 (1974) 289-315

[6] P. Deuflhard: A relaxation strategy for the modified Newton method. in [2]

[7] E. Fehlberg: Klassische Runge-Kutta Formeln fünfter und siebenter Ordnung mit Schrittweitenkontrolle. Computing 4 (1969), 93-106

[8] P. Holmes: A nonlinear oscillator with a strange attractor. Phil. Trans. Roy. Soc. London A. 292 (1979), 419-448

[9] H.G. Hussels: Schrittweitensteuerung bei der Integration gewöhnlicher Differentialgleichungen mit Extrapolationsverfahren. Universität zu Köln, Mathem. Institut, Diplomarbeit (1973)

[10] R. Seydel: Numerical computation of branch points in ordinary differential equations. Numer. Math. 32 (1979), 51-68

[11] R. Seydel: Programme zur numerischen Behandlung von Verzweigungsproblemen bei nichtlinearen Gleichungen und Differentialgleichungen. ISNM 54 (1980), 163-175

[12] R. Seydel: The strange attractors of a Duffing equation - dependence on the exciting frequency. Technische Universität München, Institut für Mathematik, Bericht M8019 (1980)

[13] J. Stoer, R. Bulirsch: Introduction to numerical analysis. Springer, Berlin- Heidelberg- New York (1980)

K.-H. Becker,
Lehrstuhl A für Mechanik
R. Seydel,
Institut für Mathematik

Technische Universität München
Arcisstraße 21
Postfach 202420
D-8000 München 2

EINSCHLIESSUNGSSÄTZE FÜR FIXPUNKTE

BY

L. COLLATZ

Institut für Angewandte Mathematik
der Universität Hamburg
Bundesstraße 55
2000 Hamburg 13

Einschließungssätze für Fixpunkte

L. Collatz, Hamburg

Inhalt In diesem Überblick wird auf die Möglichkeit hingewiesen, Fixpunkte in angebbare Schranken einzuschließen, falls Monotoniesätze gelten; die leichte Handhabung der Monotoniemethoden unter Benutzung von Algorithmen für Approximation und Optimierung wird an einem einfachen Beispiel einer nichtlinearen Integralgleichung aus der Aerodynamik vorgeführt.

Summary It is shown in this survey that in certain cases it is possible to give inclusion theorems for fixed points if monotonicity principles are applicable. A simple example of a nonlinear integral equation occuring in aerodynamics illustrates in detail how easy it is to apply the monotonicity methods by using the algorithms of approximation and optimization.

Es sind zahlreiche numerische Verfahren zur Berechnung von Fixpunkten entwickelt worden. Die meisten dieser Verfahren benutzen mehr oder weniger direkt Diskretisierungen, indem z.B. Differenzenverfahren oder finite Elementmethoden angewendet werden oder daß man mit Energiemethoden, Variationsungleichungen, Ritz-Galerkin-schen Verfahren u.a. arbeitet und dabei die auftretenden Integrale näherungsweise mit Hilfe von Quadraturmethoden durch endliche Summen ersetzt. Bei diesen genannten Methoden ist es oft schwer oder nur mit erheblichem Mehraufwand an Rechnung möglich, untere und obere Schranken für den Fixpunkt aufzustellen, d.h. eine Garantie dafür zu geben, wieviele der vom Computer gelieferten Dezimalen richtig sind. Wenn aber Monotoniesätze gültig sind, besteht im Prinzip die Möglichkeit, zu den gewünschten Einschließungen des Fixpunktes zu kommen und bei nicht zu komplizierten Fällen erhält man durch Anwendung der üblichen Algorithmen für Approximation (Meinardus [67]) und Optimierung oft mit sehr geringem Rechenaufwand und guter Genauigkeit Schranken für die Lösung. Auch bei freien Randwertaufgaben waren diese Methoden erfolgreich (Hoffmann [78], Glashoff-Werner [79], Collatz [80]).

I. Verschiedene Arten von Monotonie

Es sind im wesentlichen 2 Arten von monotonem Verhalten, die in einer Vielzahl durchgerechneter Fälle erfolgreich zur Aufstellung von Fehlerschranken geführt haben:

1. Operatoren monotoner Art: Es seien R und R^* lineare halbgeordnete Banachräume und es werden Gleichungen $Tu = \theta$ betrachtet, wobei T ein gegebener linearer oder nichtlinearer Operator ist, der Elemente f eines in R gelegenen konvexen Definitionsbereiches D in den Raum R^* abbildet. u ist ein gesuchtes Element in D und θ das Nullelement in R^*. Wenn aus $Tv \leq Tw$ für alle $v,w \in D$ stets $v \leq w$ folgt, heißt T ein "Operator monotoner Art" (Collatz [52],[68], Redheffer [67], Walter [70], Bredendiek [76] u.a.). Falls eine Lösung u mit $Tu = \theta$ existiert, bestimmen 2 Elemente

$$v,w \text{ mit } Tv \leq \theta, \quad Tw \geq \theta$$

ein Einschließungsintervall für die Lösung

$$(1.1) \qquad v \leq u \leq w \text{ .}$$

Für diese Methode gibt es in der Literatur zahlreiche Beispiele (vergl. etwa Collatz |68|,|81|).

2. Monoton zerlegbare Operatoren: Hier wird die Fixpunktgleichung $Tu = u$ betrachtet, wobei jetzt der (lineare oder nichtlineare) Operator T den Definitionsbereich $D \subset R$ wieder in R abbilden möge. Der Operator T heißt synton, wenn

$$(1.2) \qquad \text{aus } v \leq w \text{ für alle } v,w \in D \text{ stets } Tv \leq Tw \text{ folgt,}$$

er heißt antiton, wenn

$$(1.3) \qquad \text{aus } v \leq w \text{ für alle } v,w \in D \text{ stets } Tv \geq Tw \text{ folgt,}$$

er heißt monoton zerlegbar, wenn T als Summe eines syntonen Operators

T_1 und eines antitonen Operators T_2 darstellbar ist (J. Schröder |62|,|8o|, Bohl |74|):

(1.4) $\quad T = T_1 + T_2$, symbolisch $T_1\nearrow$, $T_2\searrow$.

In diesem Falle kann man iterativ zwei Folgen v_n, w_n von Elementen aufstellen:

$$(1.5)\quad \begin{cases} v_{n+1} = T_1 v_n + T_2 w_n & (n = 0,1,2,\ldots) \\ w_{n+1} = T_1 w_n + T_2 v_n \end{cases}$$

wobei man von 2 Startelementen $v_o, w_o \in D$ ausgeht und voraussetzt, daß auch v_1, w_1 zu D gehören und daß die Anfangsbedingungen

$$(1.6)\quad v_o \leq v_1 \leq w_1 \leq w_o$$

erfüllt sind. Ist der Operator T überdies kompakt, so liefert der Schaudersche Fixpunktsatz (Schauder |3o|, Collatz |68|, Schröder |8o|) die Existenz von mindestens einem Fixpunkt u der Gleichung $u = Tu$ im Interwall

$$(1.7)\quad v_1 \leq u \leq w_1 \ .$$

Es gilt dann auch

$$(1.8)\quad v_n \leq u \leq w_n \quad (n = o,1,2,\ldots).$$

Mit dieser Methode sind zahlreiche nichtlineare Randwertaufgaben bei gewöhnlichen und partiellen Differentialgleichungen numerisch behandelt worden, und auch nichtlineare Integralgleichungen vom Hammersteinschen Typ.

(1.9) $u = Tu$ mit $Ts(x) = \int_B K(x,t)\, \phi(s(t))dt$

mit Vektoren

$$x = (x_1,\dots,x_n) \in B \subset R^n,\ t = (t_1,\dots,t_n) \in B\ ,$$

K als gegebener reeller stetiger Funktion auf B x B, B gegebener meßbarer Bereich im R^n, $\phi(z)$ eine gegebene (im allgemeinen nichtlineare) Funktion einer reellen Veränderlichen von beschränkter Schwankung. Die im (1.9) beschriebenen Integraloperatoren sind monoton zerlegbar und mit der beschriebenen Theorie im Prinzip erfaßbar.

II. Laminare Grenzschichtgleichung.

Das folgende Beispiel soll die bequeme Anwendbarkeit der beschriebenen Theorie im Falle monotonen Verhaltens zeigen. Die klassische Anfangswertaufgabe der Theorie der laminaren Grenzschicht führt auf eine im Intervall $[0,\infty)$ gesuchte Funktion y(x)

(2.1) $y'''(x) + y''(x)\ y(x) = 0,\ y(0) = 0,\ y'(0) = a,\ y''(0) = 1.$

Die Aufgabe geht mit der Transformation

(2.2) $u(x) = y''(x),\ \frac{u'}{u} + y = 0$

über in die nichtlineare Integralgleichung

(2.3) $-\ln u(x) = T^*u(x)$ mit $T^*z(x) = \int_0^x \frac{1}{2}(x-s)^2 z(s)ds + \frac{1}{2}ax^2$

oder

(2.4) $u(x) = \exp\left[-\int_0^x \frac{1}{2}(x-s)^2\, u(s)ds - \frac{1}{2}ax^2\right],$

d.h.

(2.5) $u(x) = Tu(x)$ mit $Tz(x) = \exp\,(-T^*z(x))$

Für die Funktion

$$g(x) = -\ln\, u(x)$$

liegt dann eine nichtlineare Hammersteinsche Integralgleichung vom Typ (1.9) vor. Der Operator T^* ist synton und T ist antiton. Es ist daher in (1.4)

$$T = T_2,\ T_1 = 0,$$

und die Iterationsvorschrift (1.5) lautet

$$(2.6) \qquad v_{n+1} = Tw_n,\quad w_{n+1} = Tv_n.$$

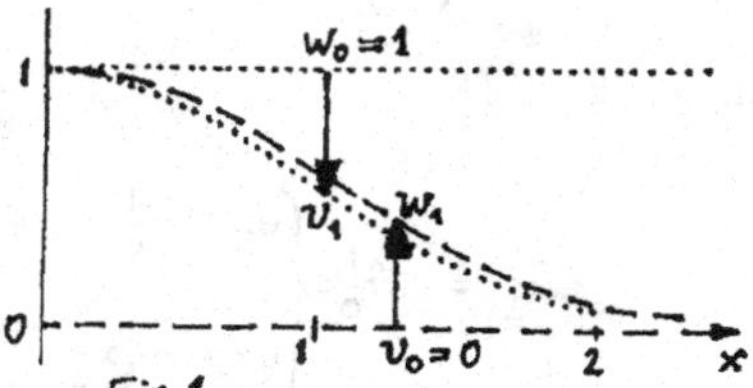

Fig.1

Als einfachster Ansatz werde

$$v_o(x) \equiv 0$$

versucht, dann erhält man ohne Rechnung sofort nach (2.3) (2.4)

$$T^*v_o = \tfrac{1}{2}ax^2,\ w_1 = Tv_o = \exp\,(-\tfrac{1}{2}ax^2).$$

Es ist $0 \leq w_1 \leq 1$, und es werde daher versucht, ob $w_o = 1$ eine obere Schranke ist; eine ganz kurze Rechnung mit $z(x) \equiv 1$ liefert bei (2.3) (2.4)

$$T^*w_o = \tfrac{1}{2}ax^2 + \tfrac{1}{6}x^3,\ v_1 = Tw_o = \exp\,(-\tfrac{1}{2}ax^2 - \tfrac{1}{6}x^3).$$

Da die Anfangsbedingungen (1.6) mit

$$v_o \leq v_1 \leq w_1 \leq w_o$$

offensichtlich erfüllt sind, Fig. 1, hat man ohne Computer mit einem Minimum von Rechenaufwand die Existenz einer Lösung u(x) und eine Einschließung für u(x) erhalten

$$(2.7) \qquad v_1 = \exp(-\tfrac{1}{2}ax^2 - \tfrac{1}{6}x^3) \leq u(x) \leq w_1 = \exp(-\tfrac{1}{2}ax^2).$$

Durch diese Schranken wird das qualitative Verhalten der Lösung $u(x)$ bereits beschrieben; man kann aber die Schranken auch leicht verbessern: offenbar wird bei der unteren Schranke $v_0 = 0$ sehr viel "verschenkt"; es werde daher $v_0 = 0$ durch einen "Spline" $\hat{v}_0$ ersetzt:

$$(2.8) \qquad \hat{v}_o = \begin{Bmatrix} 1-(\frac{x}{x_0})^2 & \text{für } 0 \leq x \leq x_0 \\ 0 & \text{für } x_0 \leq x \end{Bmatrix}, \text{Fig. 2.}$$

Dann wird

$$\hat{w}_1 = T\hat{v}_o = \begin{Bmatrix} \exp(-\frac{1}{6}x^3 + \frac{1}{60x_0^2}x^5 - \frac{a}{2}x^2) & \text{für } 0 \leq x \leq x_0 \\ \exp(-\frac{1}{3}x_0 x^2 + \frac{1}{4}x_0^2 x - \frac{1}{15}x_0^3 - \frac{a}{2}x^2) & \text{für } x_0 \leq x \end{Bmatrix};$$

x_0 muß so klein gewählt werden, daß $\hat{v}_0 \leq v_1$ ist; dann gilt die Abschätzung

$$v_1(x) \leq u(x) \leq \hat{w}_1(x).$$

Für

$$a = 1 \text{ und } x_0 = 1.71072$$

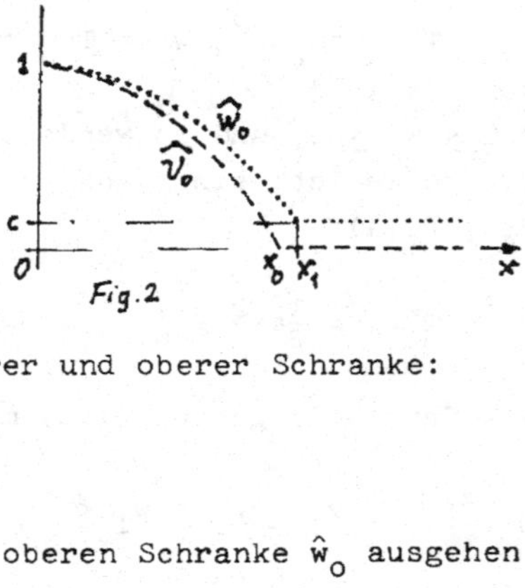

Fig. 2

beträgt die Differenz zwischen unterer und oberer Schranke:

$$\hat{w}_1(x) - v_1(x) \leq 0.0195.$$

Man kann nun auch von einer besseren oberen Schranke $\hat{w}_0$ ausgehen und auch $\hat{w}_0$ als Spline wählen, Fig. 2

$$(2.9) \qquad \hat{w}_o = \begin{cases} 1 - \frac{1-c}{x_1^2}x^2 & \text{für } 0 \le x \le x_1 \\ c & \text{für } x_1 \le x \end{cases}.$$

Man bekommt dann in $\hat{v}_1 = T\hat{w}_o$ eine bessere untere Schranke für u(x) und kann damit wieder in (2.8) einen günstigeren Wert für x_o aufstellen. Wählt man in (2.8) bei v_o den Wert x_o = 1.716 324 und in (2.9) bei $\hat{w}_o$ die Parameter x_1 = 1.805 566 und c = 0.184982, so unterscheiden sich die mit diesen Funktionen berechneten untere Schranke $\hat{v}_1$ und die obere Schranke $\hat{w}_1$ nur um höchstens 0.0054: es ist also

$$(2.1o) \qquad \hat{v}_1(x) \le u(x) \le \hat{w}_1(x) \text{ mit } \hat{w}_1(x) - \hat{v}_1(x) \le 0.0054.$$

In diesem Beispiel ist die beschriebene Monotoniemethode wohl den meisten der sonst üblichen numerischen Verfahren überlegen; aber man darf aus einzelnen Beispielen keine allgemeinen Schlüsse über den Wert einer Methode ziehen; es gibt bei der numerischen Behandlung nichtlinearer Differentialgleichungen keine für alle Fälle "beste" Methode, es gibt Fälle, in denen z.B. die Methode der finiten Elemente oder andere numerische Verfahren überlegen sind.

Ich danke Herrn Dipl.-Math. Uwe Grothkopf für numerische Rechnungen auf dem Computer.

Literatur

Bohl, E. [74] Monotonie, Lösbarkeit und Numerik bei Operatorgleichungen. Springer, 1974, 255 S.

Bredendiek, E. - L. Collatz [76] Simultan Approximation bei Randwertaufgaben, Internat.Ser.Num.Math. 30 (1976), 147-174.

Collatz, L. [52] Aufgaben monotoner Art, Arch.Math. 3 (1952) 366-376.

Collatz, L. [68] Funktional Analysis und Numerische Mathematik, Springer 1968, 371 S.

Collatz, L. [80] Monotonicity and Free Boundary Value Problems Lect. Notes in Math. Vol. 773, 1980, 33-45.

Collatz, L. [81] Anwendungen von Monotoniesätzen zur Einschließung der Lösungen von Gleichungen, Überblicke der Mathematik 1981, erscheint demnächst.

Glashoff, K. - B. Werner [79] Inverse Monotonicity of Monotone L-Operators with Applications..., J. Math. Anal. Appl. 72 (1979) 89-1o5.

Grothkopf, U. [81] Anwendungen der nichtlinearen Optimierung auf Randwertaufgaben bei partiellen Differentialgleichungen, erscheint demnächst in Intern.Ser.Numer.Mathematics (Vortrag Clausthal 198o).

Hoffmann, K.H. [78] Monotonie bei nichtlinearen Stefan Problemen. Internat.Ser.Num.Math. Vol. 39 (1978) 162-19o.

Meinardus, G. [67] Approximation of functions, Theory and numerical methods, Springer 1967, 198p.

Redheffer, R.M. [67] Differentialungleichungen unter schwachen Voraussetzungen, Abhandl.Math.Sem.Univ.Hamburg, 31 (1967) 33-5o.

Schauder, J. [30] Der Fixpunktsatz in Funktionenräumen, Stud.Math.2 (1930) 171-182.

Schröder, J. [62] Invers-monotone Operatoren, Arch.Rat.Mech.Anal.1o (1962), 276-295.

Schröder, J. [80] Operator Inequalities, Acad.Press 1980, 367 S.

Walter, W. [70] Differential and Integral Inequalities, Springer 1970, 352 S.

Prof. Dr. L. Collatz
Institut für Angewandte Mathematik
der Universität Hamburg
Bundesstraße 55
2ooo Hamburg 13

A NUMERICALLY STABLE UPDATE
FOR
SIMPLICIAL ALGORITHMS
BY

K. GEORG*

Institut für Angewandte Mathematik
Universität Bonn
Wegelerstraße 6
D-5300 Bonn

---

*) Partially supported by Deutsche Forschungsgemeinschaft through SFB 72, Bonn

Abstract .

In simplicial algorithms , a linear system of equations is solved at each step . Similar to the pivoting steps of linear programming , this can be done by numerically stable techniques [11] . In the following short note , we point out that an even stabler method may be used by looking at the underdetermined linear system involved . The computational cost is less expensive than might be expected since , by the special structure of the labeling , some calculations can be avoided .

Introduction .

(1) We consider a simplicial algorithm acting on an N-dimensional pseudomanifold T where $N>1$. On the set T^0 of vertices , a labeling $L : T^0 \to \mathbb{R}^N$ is given , and we assume that the N-th coordinate of L is always 1 . This means that we want to trace zeroes of some piecewise affine map $H : T \to \mathbb{R}^{N-1}$. For simplicity , we consider a pseudomanifold T without boundary [9,p.148] .

(2) Let us briefly recall the main features of a simplicial algorithm . For general references , see [7,10,4,3,1] . If $\sigma = [v_1,\dots,v_k,\dots,v_{N+1}]$ is an N-simplex of T , since T has no boundary there exists exactly one vertex $\overline{v_k} \neq v_k$ such that also $\overline{\sigma} = [v_1,\dots,\overline{v_k},\dots,v_{N+1}]$ is an

N-simplex of T . We say that v_k is pivoted into $\overline{v_k}$ with respect to σ_k . In simplicial algorithms , a pseudo-manifold is not stored simplex by simplex but via its pivoting rules . If $\rho = [v_1,\dots,v_m]$ is an $(m-1)$-simplex in T , its labeling matrix (of dimension (N,m)) is denoted by $L(\rho) = L(v_1,\dots,v_m) = (L(v_1),\dots,L(v_m))$. An $(N-1)$-simplex $\tau = [v_1,\dots,v_N]$ is completely labeled iff the inverse $L(\tau)^{-1}$ exists and has lexicographically positive rows . The zero-tracing is obtained by the following crucial fact of linear programming : An N-simplex contains exactly two or none completely labeled $(N-1)$-faces .

(3) We sketch the procedure of a simplicial algorithm .

(3a) Start . We assume that the following is given : an N-simplex $\sigma = [v_1,\dots,v_{N+1}]$ and an index i such that the $(N-1)$-face $\tau_i = [v_1,\dots,\hat{v}_i,\dots,v_{N+1}]$ is completely labeled . Here and in the following the symbol ^ indicates omission .

(3b) Pivoting step . Pivot v_i ! We consider the vertices $v_1,\dots,v_{N+1}$ as variables in the sense of usual programming languages . Hence , the above command means : replace the vertex stored in v_i by the corresponding pivoted vertex .

(3c) Linear equation step . Find $k \neq i$ such that $\tau_k = [v_1,\dots,\hat{v}_k,\dots,v_{N+1}]$ is completely labeled .

(3d) End of cycle . Let $i = k$ and go to (3b) .

Stopping rules are added according to the user's intentions .

The underdetermined linear system .

(4) Similar to a technique described in [1,p.61] we calculate the linear equation step (3c) by using an underdetermined system of linear equations (4a-b) . We do not claim that the following lemma is new . Its proof is an easy exercise , see also [5] :

Lemma . Let $\sigma = [v_1,\dots,v_{N+1}]$ be an N-simplex which contains the two different completely labeled (N-1)-faces $\tau_i = [v_1,\dots,\hat{v}_i,\dots,v_{N+1}]$ and $\tau_k = [v_1,\dots,\hat{v}_k,\dots,v_{N+1}]$, $i \neq k$. The idea is that the first face is known and the second has to be found . Let $b = (\beta_1,\dots,\beta_{N+1})^T \in \mathbb{R}^{N+1}$ be a non trivial solution of the homogeneous equation

(4a) $$L(\sigma)b = 0$$

and let $c = (\gamma_1,\dots,\gamma_{N+1})^T \in \mathbb{R}^{N+1}$ be a solution of the inhomogeneous equation

(4b) $$L(\sigma)c = e_1 \quad ,$$

where here and in the following e_m denotes the m-th unit vector the length of which will be clear from the context . Note that (4b) is a linear system of N equations in N+1 unknowns . Then the following is true :

(4c) $$\beta_i\beta_k < 0$$, hence , without loss of generality , we assume $\beta_i < 0 < \beta_k$ by choosing the right sign for b .

(4d) $$\gamma_k/\beta_k = \inf\{\gamma_m/\beta_m : \beta_m > 0\} \quad .$$

(5) For numerical purposes, it is sufficient to use condition (4d) in order to determine the linear equation step (3c), i.e. the case of degeneracy where in (4d) more than one index assumes the infimum may be neglected for numerical calculations. This is a well-known fact in linear programming and holds here for similar reasons. Hence, the above algorithm (3) is specified by giving an explicit description of how to solve the linear equations (4a-b). Since the current system differs from the previous one in just one column, this is normally done by a rank-one updating, hence the computational work for solving (4a-b) is of the order $O(N^2)$.

(6) Usually, in order to obtain a unique solution in (4b), one considers the additional condition $e_i^T c = \gamma_i = 0$ or, equivalently, the square matrix $L(\tau_i)$, i.e. a system of N equations in N unknowns:

(6a) $L(\tau_i)\,\bar{c} = e_1$, where $\bar{c} = (\gamma_1,\dots,\hat{\gamma}_i,\dots,\gamma_{N+1})^T$

Since $L(\tau_k)$ is obtained from $L(\tau_i)$ by a rank-one update, the inverse of this matrix or, if more numerical stability is wished, a triangular decomposition (e.g. QR) is updated with $O(N^2)$ arithmetic operations in each pivoting step and used in order to solve (4a-b), cf. [11].

<u>The Moore-Penrose inverse</u>.

(7) In this paper, we consider the case that such a square

labeling matrix $L(\tau_i)$ may have a bad condition whereas the condition of (4b) may still be reasonable if the solution c with minimal norm $\|c\|_2$ is taken . This naturally leads to solving (4b) by means of the Moore - Penrose inverse :

(7a) $$c = L(\sigma)^+ e_1 \quad .$$

We introduce some features of this inverse , see e.g. [2] for a detailed study .

(8) Let A be an $(N,N+1)$-matrix with maximal rank N . Let us denote by $t(A) \in \mathbb{R}^{N+1}$ a normalized element from the null space of A which is uniquely defined by

(8a) $$A\, t(A) = 0 \quad ,$$

(8b) $$\| t(A) \|_2 = 1 \quad ,$$

(8c) $$\det \begin{pmatrix} A \\ t(A)^T \end{pmatrix} > 0$$

The last property (orientation) is not important here and used only in order to obtain uniqueness . The following lemma gives some well-known facts on the Moore- Penrose inverse :

<u>Lemma</u> . If B is an $(N+1,N)$-matrix then B coincides with the Moore-Penrose inverse $A^+ = A^T(A\,A^T)^{-1}$ iff $A\,B = Id$ and $t(A)^T B = 0$. Furthermore ,

(8d) $$A^+ A = (Id - t(A)\, t(A)^T) \quad .$$

(9) In the following , we propose two different updating procedures for solving (4a-b) by means of the Moore-Penrose inverse $L(\sigma)^+$ of $L(\sigma)$. We emphasize that (4a) is

solved by $t(L(\sigma))$ and (4b) is solved by the first column of $L(\sigma)^+$. The first updating procedure keeps track of $t(L(\sigma))$ and $L(\sigma)^+$ whereas the second more stable updating keeps track of a QR - decomposition of $L(\sigma)$ which yields the solutions $t(L(\sigma))$ and $L(\sigma)^+ e_1$ without additional computational effort .

Of course , independently of the above considerations , the linear systems should be equilibrated , e.g. the labeling L may be normalized in such a way that $\| L(v) \|_2 = \sqrt{N}$ for $v \in T^o$ and $L(v) \neq 0$. Clearly , this normalization of L (or a similar one) does not effect the definition (2) of complete labeling .

Updating the Moore - Penrose inverse .

(10) Let the N-simplices $\sigma = [v_1,\ldots,v_i,\ldots,v_{N+1}]$ and $\bar{\sigma} = [v_1,\ldots,\bar{v}_i,\ldots,v_{N+1}]$ have the common completely labeled (N-1)-face $\tau = [v_1,\ldots,\hat{v}_i,\ldots,v_N]$, i.e. let us consider the pivoting step $v_i \rightarrow \bar{v}_i$. The corresponding labeling matrices are linked by a rank-one update

(10a) $$L(\bar{\sigma}) = L(\sigma) + u\, v^T$$

where $v = e_i$ (= i-th unit vector) and

(10b) $$u = L(\bar{v}_i) - L(v_i) \qquad \text{or}$$

(10c) $$u = L(\bar{v}_i) - L(\sigma) e_i \quad .$$

We note that the update (10a+c) is selfcorrecting whereas (10a+b) is not . This means : if $L(\sigma)$ is approximated very coarsely by a matrix and the update (10a+c) is subsequently used on all vertices v_i, then the resulting matrix is correct.

(11) In this first procedure, we assume that only $L(\sigma)^+$ and $t(L(\sigma))$ are carried along, i.e. at the beginning of a cycle in algorithm (3) we assume only $L(\sigma)^+$ and $t(L(\sigma))$ to be known. In order to perform one cycle of (3) we therefore have to show how to obtain $t(L(\bar{\sigma}))$ and $L(\bar{\sigma})^+$ from these data. This is done in the following lemma. To simplify our writing, we use the short notation

(11a) $L = L(\sigma)$, $t = t(L)$

$\bar{L} = L(\bar{\sigma})$, $\bar{t} = t(\bar{L})$, $v = e_i$.

<u>Lemma</u>. Let

(11b) $d = t - (1 + v^T L^+ u)^{-1} (v^T t) L^+ u$ and

(11c) $D = L^+ - (1 + v^T L^+ u)^{-1} L^+ u v^T L^+$, where

(11d) $Lu = L^+(L(\bar{v}_i)) - v + (t^T v) t$.

Then

(11e) $\bar{t} = \varepsilon d / \|d\|_2$, $\varepsilon \in \{+1, -1\}$, and

(11f) $\bar{L}^+ = D - \bar{t}\, \bar{t}^T D$.

Though this is of no interest here, it can be shown that $\varepsilon = \text{sign}\,(1 + v^T L^+ u)$.

<u>Proof</u>. It is immediately calculated that $\bar{L}d = 0$, $\bar{L} D = \text{Id}$ and hence $\bar{L} E = 0$ with $E = D - \bar{t}\, t^T D$. Since also $\bar{t}^T E = 0$, assertions (11e-f) follow from Lemma (8) while (11d) follows from (8d). ■

We emphasize that formulas (11b-f) are self-correcting in the sense of (10) . Note also that (11c) is similar to the Sherman - Morrison [8] formula .

Updating a QR - decomposition .

(12) A numerically stabler way of using update (10a) is by updating a corresponding QR-decomposition . Let us again use the short notation (11a) . We now suppose that an orthoganal (N+1,N+1)-matrix Q and a triangular (N,N+1)-matrix R of type $\begin{pmatrix} 0 & x & x & x & x \\ 0 & 0 & x & x & x \\ 0 & 0 & 0 & x & x \\ 0 & 0 & 0 & 0 & x \end{pmatrix}$ are given such that

(12a) $L\,Q = R$.

It follows that

(12b) $t(L) = \varepsilon\, Q\, e_1$, $\varepsilon \in \{0,1\}$, and

(12c) $L^+ e_1 = (e_1^T R e_2)^{-1} Q e_2$.

The second equation is obtained by multiplying (12a) from the left with L^+ , applying the resulting matrices on e_2 , and using (8d) , (12b) and the orthogonality of Q .

(13) Hence , given a QR-decomposition as described above , the first two columns of Q solve (4a-b) without any additional computational effort . Let us now describe how to obtain a similar decomposition $\bar{Q}$, $\bar{R}$ for $\bar{L}$ using

the update (10a) . This is done in the spirit of [6] , i.e. by using (general) Givens transformations : let us say an $(N+1,N+1)$-matrix P is a Givens transformation of type (m) , $m \in \{1,\dots,N\}$, if P is orthogonal and $Pe_n = e_n$ for $n \neq m,m+1$, i.e. P acts on the coordinates m and $m+1$ as orthogonal transformation and on the other coordinates as identity .

(14) Multiplying (10a) from the right by Q , we obtain

(14a) $$\bar{L}\,Q = R + u\,v^T Q \quad .$$

It is easily seen that one can choose Givens transformations P_n of type (n) such that

(14b) $$v^T Q\,P_1 \dots P_N = \varepsilon\,e_{N+1}^T \quad , \quad \varepsilon \in \{+1,-1\} \quad ,$$

and since $R\,P_1 \dots P_N + \varepsilon u e_{N+1}^T$ is of the Hessenberg form

$$\begin{pmatrix} x & x & x & x & x \\ 0 & x & x & x & x \\ 0 & 0 & x & x & x \\ 0 & 0 & 0 & x & x \end{pmatrix}$$

it is easy to choose Givens transformations Q_n of type (n) such that

(14c) $$\bar{R} = (R\,P_1 \dots P_n + \varepsilon u e_{N+1})\,Q_N \dots Q_1$$

is of the desired form . Hence

(14d) $$\bar{Q} = Q\,P_1 \dots P_N\,Q_N \dots Q_1 \quad .$$

In order to obtain a self-correcting procedure , we suggest to calculate u by (10c) and use $L = R\,Q^T$, i.e.

(14e) $$u = L(\bar{v}_i) - R\,Q^T e_i$$

Hence , the procedure consists in carrying only the matrices Q and R along . They are updated according to (14b-e) .

Finally , we note that [6] provides recursion formulas in order to obtain more efficient calculations of the operations $.P_1 \ldots P_N$ respectively $.Q_N \ldots Q_1$.

Author's address :
Inst. f. Angew. Math.
Wegelerstr. 6 D - 53 Bonn

References .

[1] ALLGOWER,E. and GEORG,K. : Simplicial and continuation methods for approximating fixed points and solutions to systems of equations . SIAM Review 22 (1980) 28-85.

[2] BEN-ISRAEL,A. and GREVILLE,T.N.E. : Generalized inverses: theory and applications . Wiley-Interscience publ.(1974).

[3] EAVES,B.C. : A short course in solving equations with PL homotopies . SIAM-AMS Proceedings 9 (1976) 73-143.

[4] EAVES,B.C. and SCARF,H. : The solution of systems of piecewise linear equations . Mathematics of Operations Research 1 (1976) 1-27.

[5] GEORG,K. : Algoritmi simpliciali come realizzazione numerica del grado di Brouwer . In : A survey on the theoretical and numerical trends in nonlinear analysis,I . Gius.Laterza e Figli , Bari (1979) 69-120.

[6] GILL,P.E. , GOLUB,G.H. , MURRAY,W. and SAUNDERS,M.A. : Methods for modifying matrix factorizations . Mathematics of Computation 28 (1974) 505-535.

[7] SCARF,H.E. with HANSEN,T. : Computation of economics equilibria . Yale Univ. Press, New Haven (1973).

[8] SHERMAN,J. and MORRISON,W.J. : Adjustment of an inverse matrix corresponding to changes in the elements of a given column or a given row of the original matrix . Ann. Math. Statist. 20 (1949) p.621.

[9] SPANIER,E.H. : Algebraic topology . McGraw-Hill (1966).

[10] TODD,M.J. :The computation of fixed points and applications. Lecture Notes in Economics and Mathematical Systems 124, Springer-Verlag (1976).

[11] TODD,M.J. : Numerical stability and sparsity in piecewise linear algorithms . To appear in the proceedings of a symposium on analysis and computation of fixed points , S.M.Robinson (ed.), Academic Press.

NUMERICAL INTEGRATION OF THE DAVIDENKO EQUATION

BY

K. GEORG*

Institut für Angewandte Mathematik
Universität Bonn
Wegelerstraße 6
D-5300 Bonn

---

*) Partially supported by the Deutsche Forschungsgemeinschaft through SFB 72, Bonn

Abstract .

Given a solution curve c(s) in the kernel $H^{-1}(0)$ of a smooth map $H : \mathbb{R}^{N+1} \to \mathbb{R}^N$, we consider a differential equation such that c(s) is an asymptotically stable solution . The equation may be viewed as a continuous version of Haselgrove's [17] predictor - corrector method and is a modification of Davidenko's [12] equation . In order to numerically trace c(s) , this modified equation may be integrated by some standard IVP - code [40] .

A curve - tracing algorithm is then discussed which makes one predictor step along the kernel of the Jacobian DH and one subsequent corrector (Newton) step perpendicular to this kernel . Instead of using the exact Jacobian , we update an approximate Jacobian in the sense of Broyden [9] . The algorithm differs somewhat from the recently described methods [15,20] in that we emphasize on "safe" curve - following . A simple and robust step - size control is given which may be improved in particular for less "nasty" problems .

Finally , it is discussed how such derivative - free curve - tracing methods may be used to deal with bifurcation points caused by an index jump in the sense of Crandall - Rabinowitz [11] . Instead of using a local perturbation [15] in the sense of Jürgens - Peitgen - Saupe [18] , a technique more closely related to Sard's theorem [37] is proposed . This had the advantage that sparseness of DH is not destroyed near a bifurcation point , and hence the given method may be applied to large eigenvalue problems arising from discretizations of differential equations .

The following numerical examples are discussed :

1. A homotopy method for solving a difficult fixed point test problem [49] .

2. A bifurcation problem for highly symmetric periodic solutions of a differential delay equation [18,28] .

3. A secondary bifurcation problem for periodic solutions of a differential delay equation , where a highly symmetric solution bifurcates into a solution with less symmetries [18,28] .

Some ideas are only roughly sketched and will be appropriately discussed elsewhere [16] . The numerical calculations were performed on a Hewlett Packard 85 and are illustrated by the standard plots which have a rather coarse grid .

Intoduction .

(1) We consider a (sufficiently) smooth map $H : \mathbb{R}^{N+1} \to \mathbb{R}^N$ such that zero is a regular value of H , i.e. we assume that the Jacobian $DH(x)$ has maximal rank N for all zeros $x \in H^{-1}(0)$. As is known [1,26] , the above assumptions imply that $H^{-1}(0)$ is a 1-dimensional smooth manifold , and each component of $H^{-1}(0)$ is a smooth curve diffeomorphic either to the real line $\mathbb{R}$ or to the circle S^1 .

Our aim here is to discuss how a curve in $H^{-1}(0)$ may be traced numerically . Such techniques are important in embedding (homotopy , continuation) methods , see for example the extensive bibliography in [3,14,22,25,33,35,39,47] , but also for calculating solution branches of nonlinear eigenvalue problems [18,19,21,23,28,34] . Let us describe some cases one may want to deal with :

(1a) Approximation of a single point b on the curve $c(s)$. this point b may be given as a zero or a minimal point of a functional on $H^{-1}(0)$. Here it is not important to exactly trace the whole curve but $c(s)$ is only considered a means to approximate b and hence should be traversed fast , i.e. just safely and without high precision . This is the typical situation of embedding methods .

(1b) Approximation of the whole solution curve $c(s)$. A user may be interested to numerically study the qualitative and quantitative behavior of a solution of a nonlinear problem in its dependence on an additional parameter . Here no point of the curve is a priori more important than the other .

(1c) Bifurcation . A particular case of (1a) is a bifurcation point b . This problem has to be treated separately since the system considered becomes degenerate in such points . The problem may be to precisely approximate b or to numerically trace the solution curve branching away from c(s) at b .

Notations .

(2) We introduce two definitions which will be repeatedly used in the sequel . If A is an (N,N+1) - matrix with maximal rank N then the one - dimensional kernel $A^{-1}(0)$ is spaned by a normalized element $t(A) \in \mathbb{R}^{N+1}$ which is uniquely defined by the following three conditions :

(2a) $$A\, t(A) = 0 ,$$

(2b) $$\| t(A) \|_2 = 1 ,$$

(2c) $$\det \begin{pmatrix} A \\ t(A)^T \end{pmatrix} > 0 \quad \text{(orientation)} .$$

We will also consider the Moore - Penrose inverse

(2d) $$A^+ = A^T (A\, A^T)^{-1} ,$$

which has the following properties :

(2e) $$A\, A^+ = Id ,$$

(2f) $$A^+ A = Id - t(A)\, t(A)^T ,$$

see e.g. [6] for a detailed study . In the following , we

will use these notations for the case that the (N,N+1) - matrix A coincides with or approximates the Jacobian DH .

The Davidenko equation .

(3) Consider a solution curve $c(s)$ in $H^{-1}(0)$ and suppose for the moment that $c(s)$ is parametrized according to arc length s , which is always possible under the assumptions in (1) . Then , by differentiating the equation $H(c(s)) = 0$, it is immediately seen that the tangent $\dot{c}(s)$ of $c(s)$ lies in the kernel of the Jacobian $DH(c(s))$, and orienting $c(s)$ appropriately we see that $c(s)$ may be regarded as a solution of the following differential equation going back to Davidenko [12] :

(3a) $$\dot{x} = t(DH(x)) \quad .$$

In order to numerically trace a curve in $H^{-1}(0)$ starting at a point $x_0 \in H^{-1}(0)$ some authors , e.g. [49,50-52] , propose to use an IVP - code [40] on (3a) with the initial value

(3b) $$x(0) = x_0 \in H^{-1}(0) \quad .$$

We now want to show that this is not an efficient way to deal with this type of numerical curve - tracing since no advantage is taken of the fact that (3a-b) is solved implicitely by $H(x) = 0$.

(4) Indeed , this will become clear from the following stability consideration . Let $x(s)$ be a solution of (3a) for s in

some neighborhood $I \subset \mathbb{R}$ of zero . This means in particular that $DH(x(s))$ has maximal rank N for $s \in I$. Now ,since $H(x(s))^{\cdot} = DH(x(s))\ t(DH(x(s))) = 0$, it follows that $H(x(s))$ is constant for $s \in I$. Measuring the "distance" from $x \in \mathbb{R}^{N+1}$ to $H^{-1}(0)$ by $\|H(x)\|$, it is seen that every solution of (3a) runs "parallel" to $H^{-1}(0)$ or , in other words , that the solution curve $c(s)$ of (3a-b) is stable but not asymptotically stable . Hence , an IVP - code applied to (3a) also runs "parallel" to $H^{-1}(0)$, which means that discretization errors are not "damped away" , and in order to safely follow the solution curve $c(s)$ one has to use a small step length for high precision integration . Thus the computational work is considerable and the method is inferior to the ones described below .

Haselgrove's corrector .

(5) Probably Haselgrove [17] was the first to propose an integration method for (3a) with a corrector - step particularly adapted to the special situation , namely a Newton step . It is well known [5] that the Moore - Penrose inverse can be used to obtain a Newton method for underdetermined systems of nonlinear equations , and Haselgrove's corrector may be formulated in this way :

(5a) $$x_{n+1} = x_n - DH(x_n)^{+}H(x_n) \quad .$$

Starting with a point x_o sufficiently near $H^{-1}(0)$, it can be shown [5] that (5a) converges to a point $x^* \in H^{-1}(0)$, and it is not difficult to prove quadratic convergence. A continuous version of (5a) is given by the differential equation

(5b) $$\dot{x} = -DH(x)^+H(x) \quad ,$$

i.e. (5a) is obtained from (5b) by Euler steps of length one, and it is shown in [43] that the flow defined by (5b) approaches the solution curve $c(s)$ of (3a-b) in a perpendicular way. Heuristically speaking, one may hence say that the corrector steps (5a) single out the point on $c(s)$ which is "nearest".

<u>The modified Davidenko equation</u>.

(6) Let us consider the simplest predictor - corrector method in the sense of Haselgrove where the predictor is just a tangential step of length $\delta > 0$ and the corrector is one Newton step:

(6a) $$x_{n+1} = x_n + \delta \cdot t(DH(x_n)) - DH(x_n)^+H(x_n) \quad .$$

Similar to considerations in (5), this formula may be regarded as one Euler step of length one for the differential equation

(6b) $$\dot{x} = \delta \cdot t(DH(x)) - DH(x)^+H(x) \quad ,$$

which we consider a modified version of Davidenko's equation

(6c) $$\dot{x} = \delta \cdot t(DH(x))$$

corresponding to a parametrization of (3a) according to

$\delta\cdot$arclength . We mention that the modification (6b) of Davidenko's equation has been motivated by a continuous version of Rosen's [36] gradient projection method given by Tanabe [42] .

(7) If $x(s)$ is a solution to (6b) for s in some neighborhood $I \subset \mathbb{R}$ of zero , by

$H(x(s))^{\cdot} = DH(x(s))\ [\delta\cdot t(DH(x(s))) - DH(x(s))^{+}H(x(s))] = H(x(s))$

it follows immediately that

$$\text{(7a)} \qquad H(x(s)) = e^{-s}H(x(0)) \quad , s \in I \quad .$$

Considering the initial value (3b) , it is clear that a curve $c(s)$ through $c(0) = x_o$ is a solution to (6b) iff it is a solution to (6c) . In (4) we have seen that such a solution curve $c(s)$ has no satisfactory stability properties with respect to (6c) ; now , from (7a) , it follows that $c(s)$ is exponentially stable for (6b) . Hence , if one applies an IVP - code on (6b) , the asymptotic behavior (7a) tells us that discretization errors are "damped away" , and consequently the step size for integrating (6b) may be chosen much coarser than the step size for integrating (6c) .

(8) If one does not want to invest much time for programming a special code for following a curve $c(s)$ in $H^{-1}(0)$, it now seems reasonable to apply a standard IVP - subroutine on (6b) . One modification , though , should be made in order

to take best advantage of the asymptotic behavior (7a) : the step length of the code should always be one , and step size control should be monitored by the additional parameter $\delta > 0$ in (6b) . This ensures that any integration step is accompanied by a full Newton step in the sense of Haselgrove .

(9) A preliminary test [29] has been made for a nasty test problem , see (16) below , and four different integration methods proposed in [42] :

(9a) Fourth - order multistep method : Crane - Klopfenstein predictor + Adams - Moulton corrector ,

(9b) Fourth - order Runge - Kutta method due to Ralston ,

(9c) Improved Euler method ,

(9d) Euler method , cf. (6a) .

It turned out that the simple Euler method (9d) performed best if one just wanted to safely follow the curve , whereas the multistep method (9a) performed best if a higher precision approximation of the curve was wished . A simple step size control was used . Details will appear elsewhere .

Updating methods .

(10) Let us consider now the case that an exact Jacobian DH is not available for computational purposes . This may result from a mapping H given only by a subroutine , or it may be

too complicated to programm DH , etc... Hence , an approximate Jacobian DH has to be updated while integrating along c(s) . Such a technique probably has been first considered by Broyden [10] for the case of a more classical embedding [12] , i.e. when the parameter s of a solution curve c(s) in $H^{-1}(0)$ is taken to be the last coordinate of $x = (\bar{x},s)$, $\bar{x} \in \mathbb{R}^N$, $s \in \mathbb{R}$, and hence e.g. (6a) reduces to the Newton iteration

$$\text{(10a)} \quad \begin{cases} s_{n+1} = s_n + \delta \quad , \\ \bar{x}_{n+1} = \bar{x}_n + D_{\bar{x}}H(\bar{x}_n,s_n)^{-1}\, H(\bar{x}_n,s_n) \quad , \end{cases}$$

see also [14] . Branin - Hoo [8] give updating procedures for the global Newton method [7,41] some thoughts but report no numerical experience and seem to have abandoned the idea . Schmidt [38] uses updating methods for approximating (3a) and then applies a standard IVP - code . This is not advisable since it adds to an unfavorable stability property (4) a coarse approximation of the tangent t(DH(x)) . Furthermore , cf. (12) below , we will see that updating the Jacobian DH along c(s) leads to bad approximations of DH perpendicular to c(s) .

In [15,20] several corrector steps (5a) with corresponding updates are used in order to get a good approximation of a point on the solution curve c(s) . However , if one wants to safely follow the curve , it is "robuster" to make always one predictor-step and one subsequent corrector-step with updatings . Performing a very small predictor-step then means that one essentially performs one additional corrector-step . But in this way , the tangent $t(DH(x))$ is better approximated (where this is needed) and consequently one obtains a better imitation of the flow (5b) which has the "robust" property of running perpendicularly towards the solution curve $c(s)$. Let us now describe this method in more detail .

Broyden update .

(11) For simplicity , we use Broyden's [9] update which may be defined in the following way [13] : Let the $(N,N+1)$-matrix A be an approximation of DH , and assume that the evaluation of H is given on two points $x , \bar{x} \in \mathbb{R}^{N+1}$. The (Broyden-)update of A on $x , \bar{x}$ is the matrix $\bar{A}$ which minimizes the distance $\|\bar{A} - A\|_F$ in the Frobenius norm $\|\cdot\|_F$ subject to the constraint $\bar{A}(\bar{x} - x) = H(\bar{x}) - H(x)$. This leads immediately to Broyden's formula

(11a) $$\bar{A} = A + \|\bar{x} - x\|_2^{-2} \; (\, (H(\bar{x}) - H(x) - A(\bar{x} - x)) \; (\bar{x} - x)^T .$$

The above formulation has the advantage that one sees immediately how the update has to be modified for the case that DH is sparse , namely by adding this sparseness structure as additional constraints [13,44-46] . The resulting formula for $\bar{A}$ is not more difficult than (11a) . Thus , we emphasize that our considerations are applicable to large sparse systems of nonlinear equations as given e.g. by discretizations of nonlinear boundary value problems .

(12) It is important to discuss how the above update changes the kernel of A in the two cases "predictor" and "corrector" .

Predictor - update : If $\bar{x} - x = \delta \cdot t(A)$, $\delta > 0$, then

$$\text{(12a)} \quad r \cdot t(\bar{A}) = t(A) - A^+(H(\bar{x}) - H(x)) \quad , \quad r > 0 \quad ,$$

and

$$\text{(12b)} \quad \bar{A}^+ = (Id - t(\bar{A})\ t(\bar{A})^T)\ A^+ \quad .$$

Hence , updating along the kernel of A causes essentially a rotation by less than 90° .

Corrector - update : If $\bar{x} - x = -A^+H(x)$ then

$$\text{(12c)} \quad t(\bar{A}) = t(A) \cdot \text{sign}\ \alpha$$

and

$$\text{(12d)} \quad \bar{A}^+ = A^+ - \alpha^{-1} A^+ H(\bar{x})\ d^T A^+$$

where

$$\text{(12e)} \quad \alpha = 1 + d^T A^+ H(\bar{x})$$

and

$$\text{(12f)} \quad d = (\bar{x} - x) / \|\bar{x} - x\|_2^2 \quad .$$

Here , the null space of A is unchanged though the orientation (12c) may change sign , and the update essentially takes only place in the hyperplane perpendicular to $t(A)$. The above formulas are easily deduced and will be discussed in more detail elsewhere [16] .

Curve tracing algorithm .

(13) We are now ready to formulate a simple algorithm which is only intended to illustrate the above ideas . Hence , we are aware of the fact that many improvements (particularly for the step size control) should be made in order to get a useful routine .

Given a point x approximately in $H^{-1}(0)$ and an $(N,N+1)$ - matrix A approximating $DH(x)$, the idea is to make a predictor step $y = x + \delta \cdot t(A)$ and a subsequent corrector step $z = y - A^{+}H(y)$. We then perform some tests in order to ensure that we are safely following the curve $c(s)$. If a test is negative , we reduce δ and try again . If all tests are positive , we accept z as a new point approximately in $H^{-1}(0)$ and increase δ .

(14) Algorithm .

(14a) Given :

$\delta_p > 0$: a minimal predictor step size

$\delta_c > 0$: a maximal corrector step length

$\varepsilon > 0$: a maximal error for the norm of H

$\kappa > 0$: a maximal contraction number

x : a point approximately in $H^{-1}(0)$

A : an $(N,N+1)$ - matrix approximating $DH(x)$

(14b) Start :

Let $\delta = \delta_p$

(14c) Predictor - corrector step :

Let $y = x + \delta \cdot t(A)$ and $z = y - A^+H(y)$

(14d) Test :

Is $\| H(y) \| < \varepsilon$ and $\| A^+H(y) \| < \delta_c$ and $\| H(z) \| / \| H(y) \| < \kappa$?

1. If yes , let $\delta = 2 \cdot \delta$ and goto (14e) .

2. If no and $\delta = \delta_p$, goto (14e) .

3. If no and $\delta > \delta_p$, let $\delta = \delta_p$ and goto (14c) .

(14e) Update :

Perform updates for A on x,y and y,z ,

let x = z and goto (14c) .

A stopping criterion is introduced according to the user's intentions .

(15) Let us make some comments . Our idea here is to consider a coarse approximation by making the step size just so small that the curve is "safely" followed . Thus , a predictor step of higher order is difficult to obtain , since the generated points are bad approximations of the curve $c(s)$ and the $t(A)$ are bad approximations of the tangents $\dot{c}(s)$. What one may try is to approximate the generated points by a curve in a smoothering way (e.g. least square) and use this approximation for a predictor step . The idea will be investigated elsewhere .

In (14d) , the values $\|H(y)\|$ and $\|A^+H(y)\|$ are two different ways of measuring how far the point y is away from the curve $c(s)$, whereas the third value $\|H(z)\|/\|H(y)\|$ is a measure for the contraction number obtained by the corrector step $y \to z$. This last value gives a very good idea of how "safely" we are approximating the curve .

The step size control given here just chooses between "$\delta = \delta_p$" and "double δ" . This is , in fact , a very crude control which can be easily improved . However , as will be seen in our numerical example (16) , the algorithm is already surprisingly efficient on a "nasty" problem which has much in common with a stiff differential equation . The simple form of our step size control has the advantage that a user may add whatever he wants to the list of tests in (14d) . Nevertheless , a good control should adjust the predictor step size δ in dependence on the values currently observed in (14d) , see e.g. [14] .

The fact that a minimal step of length $\delta = \delta_p$ is always performed , prevents the algorithm from "breaking down" and , indeed , often maneuvers it through "edges" of $c(s)$, i.e. through points where the tangent of the (still continuous) solution curve $c(s)$ jumps . This will be of interest in the context of piecewise smooth curve following [2] and has to be investigated further . The fact that $t(A)$ is always chosen in an oriented way (2c) prevents the algorithm from turning and running back on the curve when encountering a difficult situation .

Usually , however , a minimal step length $\delta = \delta_p$ just ensures the performance of an additional corrector step , and the predictor update (12a-b) rotates the corrector step into the hyperplane of the flow (5b) which is desirable since this flow tends perpendicularly towards the curve $c(s)$. At the same time , A is updated and a better approximation of the Jacobian is obtained . Hence , a minimal step $\delta = \delta_p$ may be regarded as a moment in which the algorithm "recovers" before going on .

There are two ways of performing the updates (14e) and calculating $t(A)$ and $A^+H(y)$. One way is roughly described in (12) . The other possibility is to update a triangular decomposition (e.g. QR) of A^T . The vectors $t(A)$ and $A^+H(y)$ are then easily obtained from Q and R . In both cases , one cycle of algorithm (14) costs two evaluations of H and $O(N^2)$ arithmetic operations . Details will appear elsewhere [16] .

<u>Numerical example</u> .

(16) To illustrate the above algorithm , let us consider the following numerical test example [3,15,20,49] : Define

$F : \mathbb{R}^N \to \mathbb{R}^N$ by $e_i^T F(x) = \exp(\cos(i \cdot e^T x))$ where e_i is the i-th unit vector in $\mathbb{R}^N$ and $e = \Sigma e_i$, $i = 1, \ldots, N$. A homotopy $H : \mathbb{R}^N \times \mathbb{R} \to \mathbb{R}^N$ is performed by $H(x,\lambda) = x - \lambda F(x)$. Since F is bounded , Brouwer's theorem immediately implies that F has at least one fixed point x^* , and such a point is characterized by $H(x^*,1) = 0$.

We follow a curve $c(s)$ in $H^{-1}(0)$, beginning at $(0,0)$ in positive λ-direction , and stop after encountering the level $\lambda = 1$. Figure (16a) below illustrates the performance of an algorithm very similar to the one described in (14) .

Figure (16a)

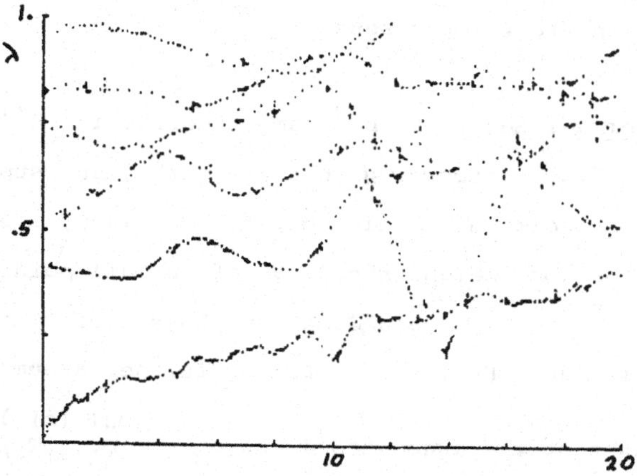

arc length modulo 20

As arc length we used the sum of the predictor step lengths which is not very precise . The parameters for monitoring the algorithm were chosen just to ensure a safe curve following , and no attempt was made to obtain higher precision . 3407 evaluations of H were performed for dimension $N = 10$ before the algorithm was stopped at level $\lambda = 1$.

Bifurcation points .

(17) Let us now discuss how a derivative - free algorithm of the above type may be used to recognize bifurcation points b and either jump over such points and follow the old curve or follow the new curve branching off at b . We consider here only bifurcation points which are induced by an index jump [24,32] in the following restricted sense :

(17a) Lemma . Let $c(s)$ be a smooth curve in $H^{-1}(0)$ parametrized , say , according to arc length , and suppose that $b = c(0)$ is an isolated singular point of $H^{-1}(0)$, i.e. we assume that for some neighborhood V of b all points $x \in V \cap H^{-1}(0)$, $x \neq b$ have a Jacobian $DH(x)$ of maximal rank N while rank $DH(b) < N$. Let us further assume that the determinant of the augmented Jacobian $\begin{pmatrix} DH(c(s)) \\ \dot{c}(s)^T \end{pmatrix}$, cf. (2c) , changes sign at $s = 0$. Then $b = c(0)$ is a bifurcation point .

The above lemma is easily proved by using a local degree argument [24,32] . Similar situations are considered by Crandall - Rabinowitz [11] . A numerical approach related to their considerations has been given by Keller [21,23] and Rheinboldt [34] . Both authors have to calculate first the bifurcation point $b = c(0)$ rather precisely and make explicit use of the Jacobian $DH(b)$ there .

(18) Derivative - free treatment of bifurcation problems arising from an index jump can be given very generally by simplicial methods [28,18,19] . As has been shown by Peitgen and Prüfer [28,30,31] , these algorithms always branch off . Our aim here is to imitate this desirable behavior .

In [15] we were able to obtain rather satisfactory results by using a local perturbation due to Jürgens - Peitgen - Saupe [18] . However , this perturbation destroys sparseness structure of the Jacobian DH and hence is not adequate for large sparse systems arising from discretizations of boundary value problems . Instead , we will consider here a more natural perturbation which will respect sparseness structure . Indeed , our perturbation will be directly motivated by Sard's [37] lemma : For almost all $d \in \mathbb{R}^N$ the map $x \to H(x) - d$ has zero as regular value .

(19) Under the assumptions of lemma (17a) we consider the two

flows of (6b resp. c) whenever they are defined . Figure (19a) illustrates the situation . Black lines indicate the flow corresponding to the Davidenko equation (6c) and the dotted line corresponds to the modified Davidenko equation (6b) .

Figure (19a)

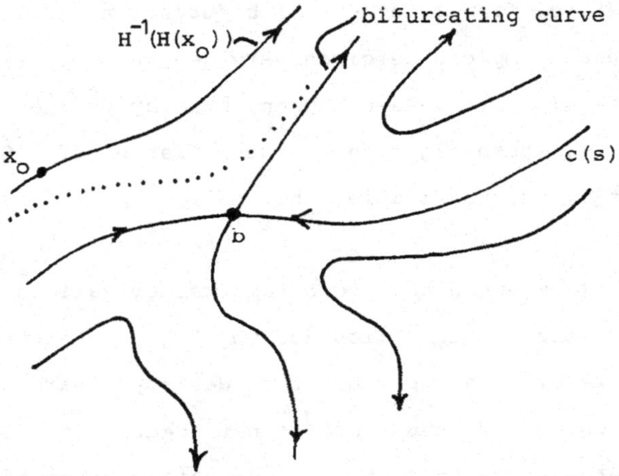

It is clear from orientation considerations that the flows "branch off" at b .

<u>Bifurcating algorithm</u> .

(20) If we follow the curve c(s) of lemma (17a) by using

algorithm (14) , the bifurcation point b causes a change of sign of the tangent t(A) in the corrector update (12c) , provided the step size is small with respect to the curvature of c(s) , i.e. the ratio "corrector step length / predictor step length" is sufficiently small , and provided the bifurcating curves are transversal to c(s) at b [11] . This is clear from the change of sign in lemma (17a) since the orientation t(A) has been defined accordingly in (2c) , and various numerical tests , see also [15] , have confirmed the above observation .

Hence , a bifurcation point of type (17a) can easily be detected by algorithm (14) , and a simple change of sign will enable us to continue following c(s) beyond b , now using the "opposite" orientation . A more difficult task is to numerically trace a bifurcating curve . This will be discussed in the following .

(21) The idea is to follow the dotted line in figure (19a) , i.e. to follow the flow corresponding to the modified Davidenko equation (6b) starting in a point x_0 away from $H^{-1}(0)$. In order to do this and avoid a flip flopping due to the change of sign in the corrector - update (12c) we essentially use the following devices :

(21a) We perform only damped corrector (Newton) steps .

(21b) When the algorithm gets too close to $H^{-1}(0)$ due to the asymptotic behavior (7a) we deliberately go away from $H^{-1}(0)$ by using a corrector step in the opposite direction..

As in (14) , the algorithm sketched below is only intended to illustrate the above ideas and far from being a good routine .

(22) <u>Algorithm</u> .

(22a) Given :

$\delta_p > 0$: a minimal predictor step size

$\delta_c > 0$: a maximal relative change of H

$\varepsilon > 0$: an ideal value for the error in the norm of H

$\kappa > 0$: a maximal contraction number

x : a point such that $||H(x)|| \sim \varepsilon$

A : an $(N,N+1)$ - matrix approximating $DH(x)$

(22b) Start :

Let $\delta = \delta_p$

(22c) Predictor - corrector step :

Let $y = x + \delta.t(A)$ and $z = y - \sigma.A^+H(y)$ where $|\sigma| = .2$ and $\text{sign}(\sigma) = \text{sign}(||H(x)|| - \varepsilon)$

(22d) Test :

Is $||H(y)|| < 4\varepsilon$ and $||H(y) - H(x)|| / ||H(x)|| < \delta_c$ and $||H(z)|| / ||H(y)|| < \kappa$?

1. If yes , let $\delta = 2\delta$ and go to (22e) .
2. If no and $\delta = \delta_p$, go to (22e) .
3. If no and $\delta > \delta_p$, let $\delta = \delta_p$ and go to (22c) .

(22e) Update :

Perform updates for A on x,y and y,z , let $x = z$ and go to (22c) .

(23) We give some comments . Algorithm (22) is similar to the one given in (14) , and hence many of the comments given in (15) apply equally here . The main differences are the damped corrector steps $z = y - \sigma . A^{+}H(y)$ with occasional $\sigma < 0$ and the tests in (22d) . The absolute value .2 of σ has proven to work well , but is not motivated theoretically . Probably , as the predictor step length , it should be adjusted according to the data currently observed .

The factor 4 in the first test is arbitrary and not motivated at all . The second test , i.e. $||H(y) - H(x)|| / ||H(x)|| < \delta_c$, checks for a drastic change in the value of H and has proven to be useful in order to prevent the algorithm from jumping over a bifurcation point .

Numerical examples .

(24) We consider a differential delay equation

(24a) $\dot{x}(t) = -\lambda f(x(t-1))$, $t > 1$

with a continuous function

(24b) $x : [0,1] \to \mathbb{R}$

as initial value . We are interested in initial values such that the resulting solution of (24a) is periodic . Solution branches of periodic solutions , and in particular bifurcation , have been studied with great success by using simplicial curve - tracing [28,30,31,18,19] . To simplify the discussion , we make the following quite strong assumption on f which will be verified by our examples (25b) and (26b) :

(24c) $f : \mathbb{R} \to \mathbb{R}$ is continuous , bounded , odd , differentiable at $x = 0$ with $Df(0) = 1$, and $f(x).x > 0$ for $x \in \mathbb{R}$, $x \neq 0$.

Obviously , $x(.) \equiv 0$ is a periodic solution of (24a-b) for all λ , and it can be shown [27] that $\lambda = \pi/_2$ is a bifurcation point of this trivial axis of solutions .

Bifurcation from the trivial axis .

(25) Given the assumptions (24c) , it is easily seen [4,18] that high symmetry solutions of period 4 can be obtained by the following simpler integral equation on $E = C[0,1]$:

(25a) $$x(t) + \lambda \int_{1-t}^{1} f(x(\xi))d\xi = 0 \quad .$$

In our numerical example we took the nonlinearity

$$\text{(25b)} \quad f(x) = \begin{cases} \tan(x) & \text{for} \quad |x|<1 \,, \\ \tan(\mathrm{sign}(x)) & \text{for} \quad |x|\geq 1 \,, \end{cases}$$

which causes bifurcation to the left [48] . We discretized $x \in E$ on N points and integrated by Simpson's rule . The figures (25c - e) below illustrate the performance of algorithm (22) on this example . For a given set of parameters in test (22d) , the numerical work (measured in evaluations of H , e.g. 115,146,92) seems to be rather indipendent of the discretization N . This promising observation should be tested on further examples and is currently investigated .

(25c) Figure .

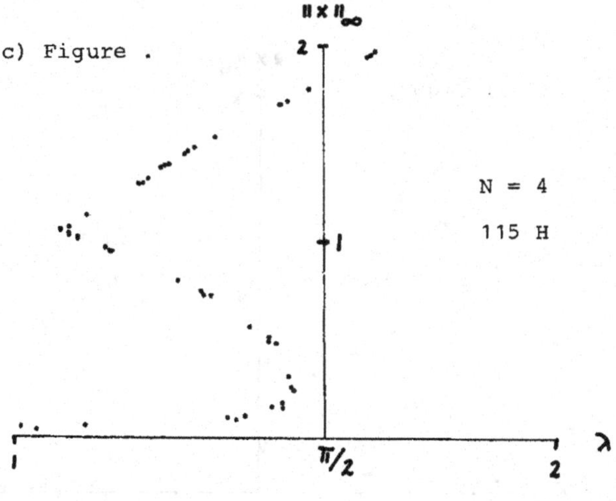

(25d) Figure

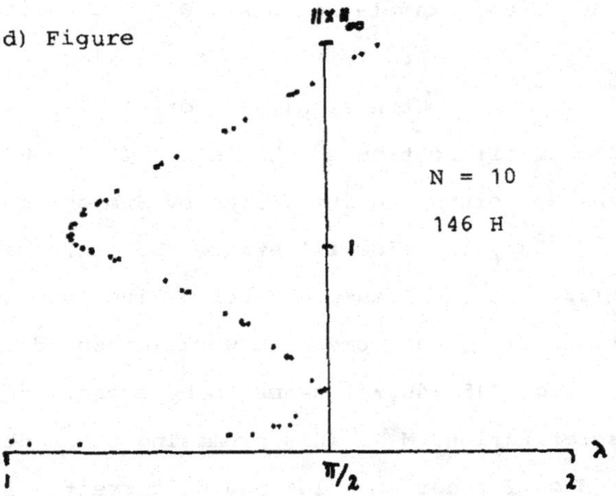

(25e) Figure

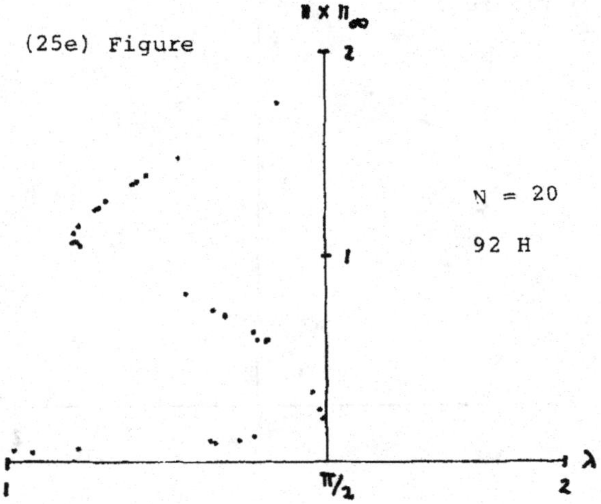

Secondary bifurcation .

(26) Let us now reduce E to the space $E = \mathrm{Var}[0,1]$ of functions $x : [0,1] \to \mathbb{R}$ having bounded variation , and define the variation $|x| \in E$ by $|x|(t) = \int_0^t |\dot{x}(\xi)|\,d\xi$ (note that x is differentiable for almost all $\xi \in [0,1]$) . Following [28,30,31,18,19] we introduce a "fixed point" operator $S_\lambda : E \to E$ $(\lambda > 0)$ essentially due to Nussbaum [27] : Given $x \in E$ we consider the solution y_λ of (24a) corresponding to the initial values $y_\lambda(t) = |x|(t)$, $0 \leq t \leq 1$. If y_λ has no zero for $t \geq 1$ we set $S_\lambda x = 0$. If y_λ has a zero for $t \geq 1$, it has a smallest such zero $t_o \geq 1$, and we set $(S_\lambda x)(t) = -y_\lambda(t_o + t)$, $0 \leq t \leq 1$. Fixed points of the type

$$(26a)_p \qquad S_\lambda^p x = x \quad , \quad x \in E \quad , \quad p \in \mathbb{N}$$

are seen to be periodic solutions of (24a - b) .

Clearly , if $q,p \in \mathbb{N}$ such that $p/q \in \mathbb{N}$, a fixed point of type $(26a)_q$ is also of type $(26a)_p$. Roughly speaking , periodic solutions corresponding to fixed points of type $(26a)_p$ which are not already of type $(26a)_q$ for $q < p$, have less and less symmetry structure for increasing p . Jürgens - Peitgen - Saupe [18] give a numerical study of secondary bifurcation points $(\lambda_o, x_o) \in \mathbb{R} \times E$, $\lambda_o > 0$, $x_o \neq 0$, in the following sense : (λ_o, x_o) lies on a solution branch of $(26a)_q$ and on an additional solution branch of $(26a)_p$ not belonging to $(26a)_q$ for two integers $0 < q < p$.

We calculate an example for the interesting [18] non-linearity

$$(26b) \qquad f(x) = \frac{x}{1 + x^8} \quad .$$

Again , we discretized $x \in E$ on N points and calculated the operator S_λ by integrating with Simpson's rule and determining the zero point t_o by interpolation with polynomials of degree 2 . Figure (26c) below illustrates the performance of algorithm (22) on the two equations $(26a)_1$ and $(26a)_2$. The observed secondary bifurcation coincides with results of [18] .

(26c) Figure .

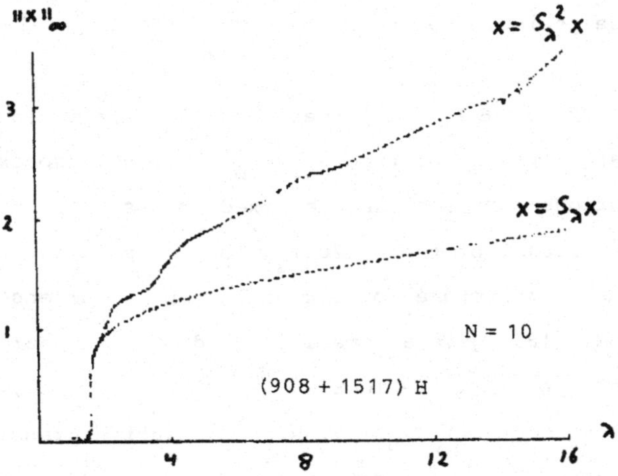

Final remarks .

(27) If one wants to follow solution branches over a larger region without knowing a priori whether bifurcation points will be encountered or not , it seems reasonable to combine two algorithms of type (12) and (22) interactively . The first algorithm follows a curve and detects a bifurcation point , cf. (20) . A user may then decide whether to continue on the old branch by changing the orientation and using (12) or whether to trace a bifurcating branch by using (22) . If the bifurcating branch is safely followed , one may switch back to (12) which is a faster procedure and gives better approximations of the curve .

We emphasize again that the algorithms described in (12) and (22) are not more that illustrations of some numerical ideas . Modifications are presently tested and hopefully will result in providing a useful routine for derivative - free numerical curve tracing allowing bifurcation .

References .

[1] ABRAHAM,R. and ROBBIN,J. : Transversal mappings and flows . W.A.Benjamin (1967).

[2] ALEXANDER,J. , KELLOGG,R.B. , LI,T.Y. and YORKE,J.A. : Piecewise smooth continuation . Preprint, University of Maryland (1979).

[3] ALLGOWER,E. and GEORG,K. : Simplicial and continuation methods for approximating fixed points and solutions to systems of equations . SIAM Review 22 (1980) 28-85.

[4] ANGELSTORF,N. : Global branching and multiplicity results for periodic solutions of functional differential equations. In : Functional Differential Equations and Approximation of Fixed Points , H.O.Peitgen , H.O.Walther (eds) , Springer Lecture Notes in Math. 730 (1979) 32-45.

[5] BEN-ISRAEL,A. : A modified Newton - Raphson method for the solution of systems of equations . Israel J. Math. 3 (1965) 94-98.

[6] BEN-ISRAEL,A. and GREVILLE,T.N.E. : Generalized inverses : theory and applications . Wiley - Interscience publ. (1974).

[7] BRANIN,JR.,F.H. : Widely convergent method for finding multiple solutions of simultaneous nonlinear equations . IBM J. Res. Develop. 16 (1972) 504-522.

[8] BRANIN,JR.,F.H. and HOO,S.K. : A method for finding multiple extrema of a function of n variables . Numerical Methods for Nonlinear Optimization , F.Lootsma, ed., Academic Press (1972) 231-237.

[9] BROYDEN,C.G. : A new method of solving nonlinear simultaneous equations . The Computer Journal 12 (1969) 94-99.

[10] BROYDEN,C.G. : Quasi-Newton , or modification methods . Numerical Solution of Systems of Nonlinear Equations, G.Byrne and C.Hall (eds), Academic Press (1973) 241-280.

[11] CRANDALL,M.G. and RABINOWITZ,P.H. : Bifurcation from simple eigenvalues . J. Functional Analysis 8 (1971) 321-340.

[12] DAVIDENKO,D. : On a new method of numerical solution of systems of nonlinear equations . Doklady Akad. Nauk SSSR (N.S.) 88 (1953) 601-602.

[13] DENNIS JR.,J.E. and SCHNABEL,R.B. : Least change secant updates for Quasi - Newton methods . SIAM Review 21 (1979) 443-459.

[14] DEUFLHARD,P. : A stepsize control for continuation methods and its special application to multiple shooting techniques . Numer. Math. 33 (1979) 115-146.

[15] GEORG,K. : On tracing an implicitly defined curve by Quasi - Newton steps and calculating bifurcation by local perturbation . To appear in : SIAM Journal of Scientific and Statistical Computing .

[16] GEORG,K. : Zur numerischen Lösung nichtlinearer Gleichungssysteme mit simplizialen und kontinuierlichen Methoden . Unfinished manuscript.

[17] HASELGROVE,C.B. : The solution of non-linear equations and of differential equations with two-point boundary conditions. Computing J. 4 (1961) 255-259.

[18] JÜRGENS,H. , PEITGEN,H.-O. and SAUPE,D. : Topological perturbations in the numerical study of nonlinear eigenvalue and bifurcation problems . in : Proceedings Symposium on Analysis and Computation of Fixed Points , S.M.Robinson (ed.), Academic Press, 1980, 139-181.

[19] JÜRGENS,H. and SAUPE,D. : Methoden der simplizialen Topologie zur numerischen Behandlung von nichtlinearen Eigenwert- und Verzweigungsproblemen . Diplomarbeit, Bremen (1979).

[20] KEARFOTT,R.B. : A derivative-free arc continuation method and a bifurcation technique . Preprint.

[21] KELLER,H.B. : Numerical solution of bifurcation and nonlinear eigenvalue problems . Applications of Bifurcation Theory , P.H.Rabinowitz (ed.), Academic Press (1977) 359-384.

[22] KELLER,H.B. : Global homotopies and Newton methods . Numerical Analysis , Academic Press (1978) 73-94.

[23] KELLER,H.B. : Constructive methods for bifurcation and nonlinear eigenvalue problems . Lect. Notes Math. 704 (1979) 241-251.

[24] KRASNOSEL'SKII,M.A. : Topological methods in the theory of nonlinear integral equations . Pergamon Press (1964).

[25] MENZEL,R. and SCHWETLICK,H. : Zur Lösung parameterabhängiger nichtlinearer Gleichungen mit singulären Jacobi-Matrizen . Numer. Math. 30 (1978) 65-79.

[26] MILNOR,J.W. : Topology from the differentiable viewpoint . University Press of Virginia (1969).

[27] NUSSBAUM,R.D. : A global bifurcation theorem with applications to functional differential equations . J. Func. Anal. 19 (1975) 319-338.

[28] PEITGEN,H.-O. and PRÜFER,M. : The Leray - Schauder continuation method is a constructive element in the numerical study of nonlinear eigenvalue and bifurcation problems . In : Proceedings Functional Differential Equations and Approximation of Fixed Points , H.O.Peitgen and H.O.Walther (eds), Springer Lecture Notes in Math. 730 (1979) 326-409.

[29] POTTHOFF,M. : Diplom thesis , in preparation.

[30] PRÜFER,M. : Calculating global bifurcation . In : Continuation Methods , H.J.Wacker (ed.) Academic Press (1978) 187-213.

[31] PRÜFER,M. : Simpliziale Topologie und globale Verzweigung . Dissertation , Bonn (1978).

[32] RABINOWITZ,P.H. : Some global results for nonlinear eigenvalue problems . J. Functional Analysis 7 (1971) 487-513.

[33] RHEINBOLDT,W.C. : Methods for solving systems of nonlinear equations . Regional conference series in applied mathematics 14 , SIAM (1974).

[34] RHEINBOLDT,W.C. : Numerical methods for a class of finite dimensional bifurcation problems . SIAM J. Numer. Anal. 15 (1978) 1-11.

[35] RHEINBOLDT,W.C. : Solution fields of nonlinear equations and continuation methods . SIAM J. Numer. Anal. 17 (1980) 221-237.

[36] ROSEN,J.B. : The gradient projection method for nonlinear programming . Part II : Nonlinear constraints . Journal of Society of Industrial and Applied Mathematics 9 (1961) 514-532.

[37] SARD,A. : The measure of the critical values of differential maps . Bull. Amer. Math. Soc. 48 (1942) 883-890.

[38] SCHMIDT,C. : Approximating differential equations that describe homotopy paths . Preprint No 7931 , Univ. of Santa Clara (1979).

[39] SCHWETLICK,H. : Numerische Lösung nichtlinearer Gleichungen . VEB Deutscher Verlag der Wissenschaften (1979).

[40] SHAMPINE,L.F. and GORDON,M.K. : Computer solution of ordinary differential equations : The initial value problem . W.H.Freeman and Company (1975).

[41] SMALE,S. : A convergent process of price adjustment and global Newton methods . Journal of Mathematical Economics 3 (1976) 1-14.

[42] TANABE,K. : A geometric method in nonlinear programming . Preprint STAN-CS-77-643, Stanford University (1977).

[43] TANABE,K. : Continuous Newton-Raphson method for solving an underdetermined system of nonlinear equations . Nonlinear Analysis, Theory, Methods and Applications 3 (1979) 495-503.

[44] TOINT,PH.L. : On sparse and symmetric matrix updating subject to a linear equation . Mathematics of Computation 31 (1977) 954-961.

[45] TOINT,PH.L. : Some numerical results using a sparse matrix updating formula in unconstrained optimization . Mathematics of Computation 32 (1978) 839-851.

[46] TOINT,PH.L. : On the superlinear convergence of an algorithm for solving a sparse minimization problem . SIAM J. Numer. Anal. 16 (1979) 1036-1045.

[47] WACKER,H.J. : A summary of the developments on imbedding methods . Continuation Methods, H.J.Wacker (ed), Academic Press (1978) 1-35.

[48] WALTHER,H.O. : A theorem on the amplitudes of periodic solutions of delay equations with applications to bifurcation. J. Diff. Eq. 29 (1978) 396-404.

[49] WATSON,L.T. : A globally convergent algorithm for computing fixed points of C^2 maps . Appl. Math. Comp. 5 (1979) 297-311.

[50] WATSON,L.T. : An algorithm that is globally convergent with probability one for a class of nonlinear two-point boundary value problems . SIAM J. Numer. Anal. 16 (1979) 394-401.

[51] WATSON,L.T. and FENNER,D. : Algorithm 555 : Chow-Yorke algorithm for fixed points or zeros of C^2 maps. ACM Transactions on Mathematical Software, Vol. 6 (1980) 252-259.

[52] WATSON,L.T. , LI,T.Y. and WANG,C.Y. : The elliptic porous slider – a homotopy method . J. Appl. Mech. 45 (1978) 435-436.

Author's address :

Institut für Angewandte Mathematik
Wegelerstr. 6
D-5300 Bonn

FIXPUNKTPRINZIPIEN
UND
FREIE RANDWERTAUFGABEN

VON

K.-H. HOFFMANN

Institut für Mathematik III
Freie Universität Berlin
D-1000 Berlin 33

FIXPUNKTPRINZIPIEN UND FREIE RANDWERTAUFGABEN

K.-H. Hoffmann

Institut für Mathematik III
Freie Universität Berlin
1000 Berlin 33

Summary: It is the aim of this paper to give some insight how fixed point principles work to develop results in pure analytical as well as in numerical respect on the field of free boundary problems for partial differential equations. In the beginning a series of examples is presented where free boundaries become involved in all three classical types of partial differential equations elliptic, hyperbolic and parabolic. Later on equations of parabolic type only are studied in detail. It is shown how Schauder's fixed point theorem can be applied to prove existence in melting problems as well as in a model describing the mixture of different fluids. Numerical experiments confirm that these methods can also be useful to obtain practical results.

1. Beispiele

Während gewöhnlich bei Differentialgleichungsproblem die Aufgabe darin besteht in einem a priori bekanntem Gebiet des $\mathbb{R}^n$ unter geeigneten Randbedingungen Lösungen in einem genauer festzulegenden Sinn zu bestimmen, deren Eindeutigkeit und Regularität zu untersuchen, so kommt als Teil der Aufgabe bei freien Randwertproblemen hinzu, daß man das Lösungsgebiet gleichzeitig mitbestimmen muß und auch dessen Regularitätseigenschaften wissen möchte. Probleme dieser Art treten in natürlicher Weise bei allen drei Typen partieller Differentialgleichungen auf. Ohne Anspruch auf Vollständigkeit zu erheben, geben wir hier einige charakteristische Beispiele an, von denen wir nur die parabolischen Typs detaillierter behandeln und bei den anderen auf die Literatur verweisen.

1.1 Elliptische Differentialgleichungen

Ein in der Literatur häufig behandeltes Modellproblem ist die mathematische Beschreibung der stationären Strömung einer inkompressiblen Flüssigkeit durch ein homogenes poröses Medium, das zwei Reservoirs verschiedener Höhe trennt (vgl. z.B. C. BAIOCCHI u.a. [1], C. BAIOCCHI u.a. [2]). Die Abbildung 1 zeigt im Querschnitt eine grobe Vereinfachung dieses Dammproblems.

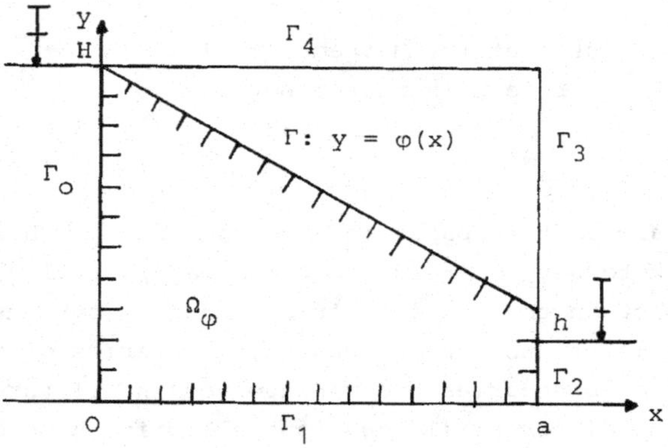

Abbildung 1

Die Aufgabe besteht darin, Funktionen $\varphi: [o,a] \to \mathbb{R}$ und $u \in C(\bar{\Omega}_\varphi)$ zu finden, die den folgenden Bedingungen genügen:

i. $\varphi(o) = H \;,\quad \varphi(a) > h$

ii. u ist harmonisch in Ω_φ , $u(o,y) = H$ für alle $o \le y \le H$

iii. $$u(a,y) = \begin{cases} h & \text{für } 0 \le y \le h \;, \\ y & \text{für } h \le y \le \varphi(a) \;, \end{cases}$$

$$\frac{\partial}{\partial y} u(x,o) = o \quad \text{für alle} \quad o < x < a$$

iv. $u(x,y) = y$ und $\frac{\partial}{\partial n} u(x,y) = o$ für alle $o < x < a$ und $y = \varphi(x)$.

Die Funktion φ beschreibt die Sickergrenze und somit den freien Rand des Gebietes Ω_φ, während u als Verteilung des Wasserdruckes interpretiert werden kann. Probleme dieser Art wurden von C. BAIOCCHI und anderen intensiv bezüglich Existenz, Eindeutigkeit und Regularität sowie der numerischen Lösbarkeit untersucht, wobei meist der Zugang über Variationsungleichungen eingeschlagen wurde.

1.2 Hyperbolische Differentialgleichungen

Im Zusammenhang mit hyperbolischen Differentialgleichungen sind freie

Randwertprobleme schon sehr früh behandelt worden. Stoßwellenprobleme und Fragestellungen bei der Untersuchung von Wasserwellen gehören bereits zur klassischen Mathematik. Es soll daher hier ein weniger bekanntes Problem aus der Gasdynamik aufgegriffen werden, das von C.D. HILL [9] untersucht und von J.R. CANNON, K.-H. HOFFMANN [5] weiter verfolgt wurde. Die Bewegung eines Kolbens in einem mit Gas gefüllten Rohr, rechts und links des Kolbens unterschiedliche Zustände des Gases vorausgesetzt, kann unter einigen vereinfachenden Annahmen, die hier nicht weiter diskutiert werden, durch das nachfolgende freie Randwertproblem beschrieben werden. Zu den Bezeichnungen vergleiche man Abbildung 2.

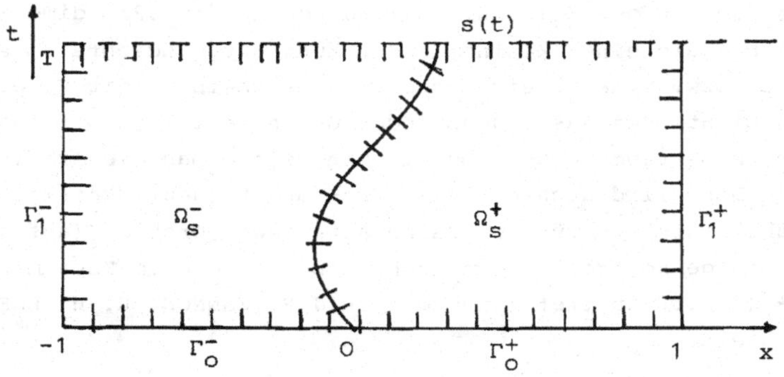

Abbildung 2

Es wird ein Quintupel von Funktionen $(s, \sigma_1, \eta_1, \sigma_2, \eta_2)$ gesucht, welches dem System von linearen hyperbolischen Differentialgleichungen mit Anfangs-Randbedingungen (1.2.1), (1.2.2) und ihrer Koppelung (1.2.3) über ein Anfangswertproblem für s genügt.

$$(1.2.1)\quad \begin{cases} \frac{\partial}{\partial t}\sigma_1 + c\,\frac{\partial}{\partial x}\eta_1 = 0 & \\ \frac{\partial}{\partial t}\eta_1 + c\,\frac{\partial}{\partial x}\sigma_1 = 0 & \text{in } \Omega_s^- \,, \\ \sigma_1 = f_1 \,,\quad \eta_1 = g_1 & \text{auf } \Gamma_o^- \,, \\ \eta_1 = h_1 & \text{auf } \Gamma_1^- \,. \end{cases}$$

$$(1.2.2)\quad \begin{cases} \dfrac{\partial}{\partial t}\sigma_2 + c\,\dfrac{\partial}{\partial x}\eta_2 = 0 & \\ \dfrac{\partial}{\partial t}\eta_2 + c\,\dfrac{\partial}{\partial x}\sigma_2 = 0 & \text{in } \Omega_s^+ \,, \\ \sigma_2 = f_2 \,,\quad \eta_2 = g_2 & \text{auf } \Gamma_o^+ \,, \\ \eta_2 = h_2 & \text{auf } \Gamma_1^+ \,. \end{cases}$$

$$(1.2.3)\quad d\frac{d^2}{dt^2}\, s(t) = c^2 \varsigma_o (\sigma_1 - \sigma_2)(s(t),t) \,, \quad 0 < t \leq T \,,$$

$$s(0) = 0 \,,\ \dot{s}(0) = b.$$

Hierbei sind b, c, ς_o bekannte Konstanten und f_i, g_i, h_i $(i=1,2)$ vorgegebene Rand- bzw. Anfangsfunktionen. Die gesuchte freie Randfunktion s beschreibt den Weg des Kolbens, während σ_i, η_i $(i=1,2)$ dimensionslose Größen sind, die eine (kleine) Schwankung der Gasdichte um eine mittlere Dichte bzw. eine Abweichung der Gasgeschwindigkeit um eine zur mittleren Dichte des Gases gehörenden Geschwindigkeit c angeben. Für die genauere Begründung des Modells vergleiche man die Arbeit von C.D. HILL [9]. Dort wird auch ein Existenz- und Eindeutigkeitsbeweis, sowie ein Stabilitätssatz für das reine Anfangswertproblem (auf ganz R) gegeben. Einige kontrolltheoretische Aspekte, die im Zusammenhang mit (1.2.1) - (1.2.3) auftreten wurden von J.R. CANNON, K.-H. HOFFMANN [5] behandelt.

1.3 Parabolische Differentialgleichungen

Physikalische Vorgänge der Wärmeleitung, die die Betrachtung von Phasenänderungen einschließen, wurden 1889 von J. STEFAN [16] erstmals systematisch untersucht und sind unter dem Namen Stefanproblem klassische Beispiele für freie Randwertaufgaben parabolischer Differentialgleichungen. Das nachfolgende Modell ist das auf ein Zweiphasenproblem verallgemeinerte Eis - Wasser System, wie es von W.T. KYNER [15] behandelt wurde. Hierzu vergleiche man auch die Arbeiten [10] und [11]. In den Abschnitten 2 und 3 werden wir dieses Modell analytisch und numerisch weiter untersuchen.

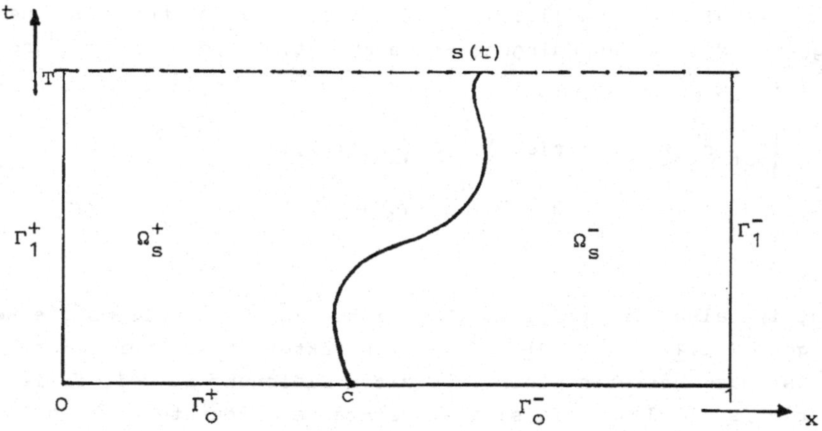

Abbildung 3

Die Gebiete Ω_s^+ bzw. Ω_s^- repräsentieren die (räumlich eindimensionalen) warmen bzw. kalten Zonen des Eis-Wasser-Systems. Ohne im Augenblick die Voraussetzungen und Bezeichnungen zu spezifizieren (das geschieht im Abschnitt 2) geben wir die Differentialgleichungen für die Temperaturverteilungen u in der warmen und v in der kalten Phase an.

$$(1.3.1)\quad \begin{cases} \frac{\partial}{\partial x}(p(x,t,u)\,\frac{\partial}{\partial x}\,u) = k_1\frac{\partial}{\partial t}\,u & \text{in } \Omega_s^+\,, \\ u = \varphi & \text{auf } \Gamma_o^+\,, \\ p_o(t)\frac{\partial}{\partial x}\,u = -\,a(t) & \text{auf } \Gamma_1^+\,, \\ u(s(t),t) = 0 & \text{für } 0 \le t \le T. \end{cases}$$

$$(1.3.2)\quad \begin{cases} \frac{\partial}{\partial x}\,(q(x,t,v)\frac{\partial}{\partial x}\,v) = k_2\,\frac{\partial}{\partial t}\,v & \text{in } \Omega_s^-\,, \\ v = \psi & \text{auf } \Gamma_o^-\,, \\ q_o(t)\frac{\partial}{\partial x}\,v = -\,d(t) & \text{auf } \Gamma_1^-\,, \\ v(s(t),t) = 0 & \text{für } 0 \le t \le T. \end{cases}$$

Beide Systemgleichungen (1.3.1) und (1.3.2) sind über eine Energiebedingung - der Stefanbedingung - verknüpft, die den freien Rand s festlegt:

$$(1.3.3)\quad \begin{cases} \frac{d}{dt}\, s(t) = b - p(s(t),t,0)\frac{\partial}{\partial x}u(s(t),t) + \\ \qquad\qquad + q(s(t),t,0)\frac{\partial}{\partial x}v(s(t),t) , \\ s(0) = c. \end{cases}$$

Gesucht ist also ein Tripel von Funktionen (s,u,v), welches die Beziehungen (1.3.1) - (1.3.3) erfüllt. Charakteristisch für solche Aufgaben ist das Auftreten einer expliziten Gleichung für die freie Randkurve s (1.3.3), die sich als einfache Fixpunktgleichung formulieren lassen wird. Ähnliche Methoden lassen sich noch Anwenden auf Probleme, die das Ineinanderfließen von Flüssigkeiten beschreiben (Muskat's Modell), obgleich hier die Herleitung der benötigten a priori-Abschätzungen wesentlich komplizierter ist (vgl. L.C. EVANS [7]). Muskat's Modell läßt sich mit den Bezeichnungen von Abbildung 3 durch die nachfolgenden Gleichungen beschreiben, wobei u bzw. v die Geschwindigkeiten der beiden Flüssigkeiten und s ihre Kontaktgrenze repräsentieren.

$$(1.3.4)\quad \begin{cases} \alpha\frac{\partial^2}{\partial x^2}\, u = \frac{\partial}{\partial t}\, u & \text{in } \Omega_s^+ , \\ u = \varphi & \text{auf } \Gamma_0^+ , \\ u = f_1 & \text{auf } \Gamma_1^+ . \end{cases}$$

$$(1.3.5)\quad \begin{cases} \beta\frac{\partial^2}{\partial x^2}\, v = \frac{\partial}{\partial t}\, v & \text{in } \Omega_s^- , \\ v = \psi & \text{auf } \Gamma_0^- , \\ v = f_2 & \text{auf } \Gamma_1^- . \end{cases}$$

Es kommen zwei "Interface"-Bedingungen und die Bewegungsgleichung für den freien Rand s hinzu.

$$(1.3.6)\quad \begin{cases} u(s(t),t) = v(s(t),t), \\ \alpha\frac{\partial}{\partial x}\, u(s(t),t) = \beta\frac{\partial}{\partial x}\, v(s(t),t). \end{cases} \qquad 0 \le t \le T ,$$

(1.3.7) $\frac{d}{dt}s(t) = -\alpha\frac{\partial}{\partial x}u(s(t),t)$, $0 \le t \le T$,

$s(0) = c.$

Auch dieses Modell wird in den folgenden Abschnitten nochmals aufgegriffen werden.
Das "Dammbeispiel" in Abschnitt 1.1 ausgenommen wurden hier alle anderen Beispiele als räumlich eindimensionale Probleme formuliert. Selbstverständlich ist es sinnvoll und auch unmittelbar einsichtig, eine Formulierung in mehr als einer Raumdimension anzugeben. Der analytische wie numerische Mehraufwand, solche Probleme zu lösen, ist allerdings beträchtlich (vgl. A. FRIEDMANN [8], Y. ICHIKAWA, N. KIKUCHI [12]). Diese Arbeit wird sich im wesentlichen auf die Darstellung der Situation im räumlich eindimensionalen Fall beschränken.

2. Fixpunktmethoden bei Problemen vom "Stefan-Typ"

Die in 1.3 beschriebenen Probleme werden wieder aufgegriffen und detaillierter behandelt.

2.1 Das Schmelzproblem

Es sei $D := \{(x,t,z) | 0 < x < 1, \; 0 < t < \infty, \; -\infty < z\}$ und $p,q \in C^2(D)$ mit in D beschränkten partiellen Ableitungen bis zur 2. Ordnung. Die Funktionen a und d seien zweimal stetig differenzierbar und die Ableitungen beschränkt. Von p bzw. q setzen wir voraus, daß für eine Konstante γ gilt:

$0 < \gamma \le p(x,t,z) = p_0(t) +$
+ Ausdrücke, die bei $z=0$ verschwinden

$0 < \gamma \le q(x,t,z) = q_0(t) +$
+ Ausdrücke, die bei $z=0$ verschwinden.

Ferner mögen $p_z \le 0$, $p_x \ge 0$, $q_z \ge 0$, $q_x \le 0$ und die Abschätzungen

$$0 < \underline{a} \le a(t)/p_0(t) \; , \quad 0 < \underline{d} \le d(t)/q_0(t)$$

gelten. Die Konstanten k_1, k_2 seien positiv und $b \ge 0$, $0 < c < 1$.

Da u die warme und v die kalte Phase darstellen sollen, wird entsprechend von den Anfangsdaten $0 \le \varphi(x)$ und $0 \ge \psi(x)$ gefordert.

Die weiteren an φ und ψ zu stellenden Bedingungen bezüglich Glattheit und Wachstumsverhalten sind analog zu den Voraussetzungen in [11] und sind dort nachzulesen. Weiter werden die gleichen Voraussetzungen an die Größenordnung der Daten gestellt wie in der Arbeit [11].

Die Stefanbedingung (1.3.3) läßt sich in bekannter Weise als Integralgleichung formulieren, indem man die Differentialgleichungen (1.3.1) bzw. (1.3.2) über Ω_s^+ bzw. Ω_s^- integriert und die Rand- und Anfangsbedingungen berücksichtigt:

$$0 = \int_0^t \int_0^{s(\tau)} \{p(x,\tau,u)u_x)_x - k_1 u_t\}dxd\tau = \int_0^t p(s(\tau),\tau,0)u_x(s(\tau),\tau)d\tau$$

$$- \int_0^t p_0(\tau)u_x(0,\tau)d\tau - k_1\int_0^t \{[\int_0^{s(\tau)} u(x,\tau)dx]_\tau - \dot{s}(\tau)u(s(\tau),\tau)\}d\tau =$$

$$= \int_0^t p(s(\tau),\tau,0)u_x(s(\tau),\tau)d\tau + \int_0^t a(\tau)d\tau - k_1\int_0^{s(t)} u(x),t)dx$$

$$+ k_1\int_0^c \varphi(x)dx.$$

Eine analoge Beziehung erhält man für die kalte Phase v. Kombiniert man beide Relationen und beachtet die Stefanbedingung (1.3.3), so ergibt sich für s eine Fixpunktgleichung der folgenden Form:

$$(2.1.1) \qquad s(t) = r(t) + (Rs)(t) \;,$$

wobei zur Abkürzung

$$r(t) := c + bt + \int_0^t a(\tau)d\tau - \int_0^t d(\tau)d\tau +$$

$$+ k_1 \int_0^c \varphi(x)dx + k_2 \int_c^1 \varphi(x)dx \qquad \text{und}$$

$$(Rs)(t) := - k_1 \int_0^{s(t)} u(x,t)dx - k_2 \int_{s(t)}^1 v(x,t)dx \qquad \text{gesetzt wurde.}$$

Es ist nun das Ziel, in diesem Abschnitt zu zeigen, daß der Operator $R^* := r + R$ in einem geeigneten Banach-Raum einen Fixpunkt besitzt. Da die ausführlichen Beweise zu den nachfolgenden Aussagen teilweise sehr technisch und umfangreich sind, beschränken wir uns auf die Darstellung der wesentlichen Beweisideen. Als geeigneter Banach-Raum erweist sich der Raum $C[0,T_g]$ versehen mit der sup-Norm, wobei T_g bedeutet, daß man sich auf kleine Zeiten beschränken muß, die vor allem durch die Daten des Problems bestimmt sind. Man vergleiche hierzu [11], wo die Verhältnisse, allerdings für ein vereinfachtes Modell dargestellt, ähnlich sind.

LEMMA 2.1: R ist auf einer geeigneten Teilmenge $D \subset C[0,T_\varsigma]$ stetig.

Zum Beweis sei $\underline{s} := \min(s_1,s_2)$ und $\overline{s} := \max(s_1,s_2)$, wobei die Supremumsbildungen punktweise zu verstehen sind. Wegen der Ungleichung

$$|(Rs_1 - Rs_2)(t)| \le k_1\Big(\int_0^{\underline{s}(t)} |u_2 - u_1|dx + \int_{\underline{s}(t)}^{\overline{s}(t)} |u_2 - u_1|dx\Big) +$$

$$+ k_2\Big(\int_{\overline{s}(t)}^{1} |v_2 - v_1|dx + \int_{\underline{s}(t)}^{\underline{s}(t)} |v_2 - v_1|dx\Big)$$

müssen die unter den Integralen stehenden Differenzen in Abhängigkeit von $s_i\,(i=1,2)$ abgeschätzt werden.

LEMMA 2.2: Es seien $s_1 s_2 \in C^1[0,T_\delta]$ mit $s_1(0) = s_2(0) = C$, und $0 < \delta \le s_i(t) \le 1-\delta$ für $0 \le t \le T_\delta$ und $i=1,2$. Ferner seien (u_1,v_1) bzw. (u_2,v_2) die zu s_1 bzw. s_2 gehörenden Lösungen von (1.3.1),(1.3.2). Dann gilt:

Es gibt Konstanten $A,B > 0$, die nicht von den Rändern s_i (i=1,2) abhängen, so daß die folgenden Abschätzungen gelten:

(i) $\forall 0 \le x \le \underline{s}(t)$: $|(u_1-u_2)(x,t)| \le A(1-e^{-S\|s_1-s_2\|_t})$,

(ii) $\forall \overline{s}(t) \le x \le 1$: $|(v_1-v_2)(x,t)| \le B(1-e^{-S\|s_1-s_2\|_t})$.

Der Beweis geschieht durch Konstruktion von Vergleichsfunktionen und Anwendungen des Maximumprinzips für parabolische Differentialgleichungen. Hier werden die eingangs erwähnten Wachstumsbeschränkungen für die Datenfunktionen benötigt. Mit Lemma 2.2 wäre der Stetigkeitsbeweis für R zu führen, wenn die Größe S unabhängig von s_i (i=1,2) wäre. Sie hängt allerdings von $\dot{s}_i$ (i=1,2) explizit ab. Also ist es nötig, a priori-Schranken für $\dot{s}$ herzuleiten und das bedeutet a priori-Schranken für $u_x(s(t),t)$ bzw. $v_x(s(t),t)$ wegen (1.3.3).

Wieder mit Hilfe der Majorantenkonstruktion läßt sich zeigen:

Lemma 2.3: Es existieren Konstanten $a_i,\varepsilon_i,\eta_i > 0$ (i=1,2), die nur von den Daten des Problems abhängen, mit:

$$|u_x(s(t),t)| \le \frac{\varepsilon_1 a_1 (\|\dot{s}\|_t + \eta_1) k_1}{(1-\exp(-(\|\dot{s}\|_t+\eta_1)\delta k_1/\gamma))\gamma},$$

$$|v_x(s(t),t)| \le \frac{\varepsilon_2 (a_2+1)(\|\dot{s}\|_t + \eta_2) k_2}{(1-\exp(-(\|\dot{s}\|_t+\eta_2)\delta k_2/\gamma))\gamma} \quad \text{für} \quad 0 \le t \le T_\delta .$$

Nach Einsetzen dieser Abschätzungen in die Stefan-Bedingungen und kurzer Rechnung erhält man a priori-Schranken für $\|\dot{s}\|_{T_\delta}$, wobei auf der rechten Seite der Ungleichung bekannte Konstanten stehen, die nur noch von den Daten abhängen (vgl. [11]).

<u>KOROLLAR 2.4</u>: Es gilt die a priori-Abschätzung

$$\|\dot{s}\|_{T_\delta} \le \frac{b(1-\exp(-\underline{\eta}\xi)+\overline{\eta}\Gamma}{1-\exp(-\eta\xi)-\Gamma} =: \Lambda .$$

Das Definitionsgebiet für den Operator R wird durch
$D := \{s \in C^1[0,T_\delta] \mid s(0) = c, \ \|\dot{s}\|_{T_\delta} \le \Lambda, \ \delta \le s(t) \le 1-\delta\}$ festgelegt.
Durch einfaches Nachrechnen bzw. Anwenden des Satzes von ARZELA-ASCOLI unter Berücksichtigung von Korollar 2.4 zeigt man:

<u>LEMMA 2.5</u>: Es gelten:

(i) $R^*(D) \subset D$,

(ii) D ist präkompakt.

Damit läßt sich SCHAUDER'S Fixpunktsatz anwenden und nachweisen, daß die Gleichung (2.1.1) eine Lösung besitzt, die den gesuchten freien Rand s des Problems darstellt. Durch eine einfache Zusatzüberlegung sieht man ein, daß die Banach-Iteration (2.1.2)

$$s_{n+1} = R^* s_n$$

in jedem Schritt Einschließungen der Lösung liefert.

<u>LEMMA 2.6</u>: R ist antiton.

<u>Beweis</u>: Seien $s_1, s_2 \in D$ mit $s_1 \le s_2$ im Sinne der punktweisen Ordnung. Dann folgt:

$$(R(s_2-s_1))(t) = -k_1 \int_0^{s_1(t)} (u_2-u_1)dx - k_1 \int_{s_1(t)}^{s_2(t)} u_2 dx + $$

$$+ k_2 \int_{s_2(t)}^{1} (v_2-v_1)dx + k_2 \int_{s_1(t)}^{s_2(t)} v_1 dx \quad ,$$

und, Anwendung des Maximumprinzips auf die Funktionen $u := u_2-u_1$ bzw. $v := v_2-v_1$, liefert $u \geq 0$ bzw. $v \leq 0$. Damit gilt $Rs_2 \leq Rs_1$, womit die Antitonie von R gezeigt ist.

Geeignete Startnäherungen für die Iteration (2.1.2) vorausgesetzt, läßt sich die mit Lemma 2.6 bewiesene Eigenschaft von R numerisch gut verifizieren. Das wird im nächsten Abschnitt demonstriert werden.

2.2 Das Mischproblem

Für das in Abschnitt (1.3) beschriebene "Muskat" Modell bewies L.C. EVANS [7] mit Hilfe des Schauder'schen Fixpunktsatzes das folgende Existenzresultat.

Satz 2.7 (L.C. EVANS [7]): Es sei $0 < c < 1$ und $\alpha, \beta > 0$. Ferner gelte für die Anfangsdaten

$$w_0 := \begin{cases} \varphi & \text{auf } \Gamma_0^- \\ \psi & \text{auf } \Gamma_0^+ \end{cases}$$

$w_0 \in W_2^1(0,1)$ und die Randdaten $f_1, f_2 \in W_{3/2}^1(0,1)$ mit den Verträglichkeitsbedingungen $w_0(0) = f_1(0)$, $w_0(1) = f_2(0)$.

Dann folgt:

Es gibt eine Endzeit T, $0 < T < 1$, und Funktionen s auf $[0,T]$ bzw. w auf $[0,1] \times [0,T]$ mit den Eigenschaften:

(2.2.1) $s \in W_3^1(0,T)$, $s(0) = c$, $0 < s(t) < 1$,

(2.2.2) w ist gleichmäßig Hölder-stetig in x mit Exponent $\varkappa$ und in t mit Exponent $\varkappa/2$, $0 < \varkappa < 1$,

(2.2.3) w nimmt das Anfangsdatum w_0 und die Randdaten f_1, f_2 an,

(2.2.4) w ist in Ω_s^+ bzw. Ω_s^- beliebig oft differenzierbar und Lösung der Differentialgleichung (1.3.4) bzw. (1.3.5),

(2.2.5) $w(\cdot,t) \in W_2^1(0,1)$ für alle $t \in [0,T]$ und

$w(\cdot,t) \in W_2^2(0,s(t)) \cap W_2^2(s(t),1)$ für f.a. $t \in [0,T]$,

(2.2.6) $\lim\limits_{\xi \to s(t)-0} w_x(\xi,t)$, $\lim\limits_{\xi \to s(t)+0} w_x(\xi,t)$ existieren für f.a.

$t \in [0,T]$ und genügen den "Interface"-Bedingungen (1.3.6),

(2.2.7) $\dot{s}$ existiert für f.a. $t \in [0,T]$ und erfüllt (1.3.7).

Zum Beweis wendet man den Schauder'schen Fixpunktsatz auf die Abbildung

$$H: D \subset L^3(0,T) \to L^3[0,T] \quad \text{mit} \quad H := H_2 \circ H_1 ,$$

$$(H_1 r)(t) := c + \int_0^t r(\tau)d\tau =: s(t) \quad \text{und}$$

$$(H_2 s)(t) := - \alpha u_x(s(t),t) \quad .$$ an.

Zur Auswertung des Operators H hat man neben einer Integration also das "Interface"-Problem (1.3.4) - (1.3.6) zu lösen. Die Teilmenge D muß wieder geeignet konstruiert werden. Bei der numerischen Lösung des Problems liegt es nahe, die Iteration

$$(2.2.8) \quad s_n(t) := c + \int_0^t r_n(\tau)d\tau \ , \quad r_{n+1}(t) = (H_2 s_n)(t)$$

zu versuchen. Bei den numerischen Tests zeigt sich wiederum ein monotones Verhalten der Iterierten. Ein Beweis hierfür ist jedoch nicht gelungen.

2.3 Bemerkung zu mehrdimensionalen Problemen

Bei Problemen in mehr als einer Raumdimension ist es häufig möglich, durch Übergang zum "freezing index" (siehe z.B. G. DUVAUT [6] eine Variationsungleichung herzuleiten, die man z.B. durch Anwendung der Methode der Finiten Elemente der numerischen Behandlung zugänglich machen kann. Die resultierende endlichdimensionale Variationsungleichung kann durch SOR-Verfahren oder Fixpunktalgorithmen numerisch gelöst werden. Eine Gegenüberstellung dieser Verfahren findet man in den Arbeiten von Y. ICHIKAWA, N. KIKUCHI [12], [13].

3. Numerische Resultate

Neben den hier präsentierten Resultaten, die alle mit Hilfe der Banach-Fixpunktiteration erzielt wurden, ist auch die Anwendung des Newton-Verfahrens möglich (vgl. [3], [4], [14]). Man hat dabei eine höhere Konvergenzgeschwindigkeit, die theoretische Absicherung dieser Algorithmen ist allerdings noch nicht vollständig geklärt.

3.1 Beispiel zum Schmelzproblem

Es seien $p(x,t,z) := 1 + z^2$, $q(x,t,z) := 1 + \frac{1}{10} z^2$,

$a(t) := \exp(t+0.5)$, $d(t) := \exp(t-0.5)$.

Ferner setzen wir die Anfangsfunktionen

$$\varphi(x) := \exp(0.5-x) - 1 ,$$

$$\psi(x) := \exp(0.5-x) - 1 .$$

Es wurde mit den Konstanten $k_1 = k_2 = 1$, $c = 0.5$ und $b = 0$ gerechnet.

Dieses Beispiel verletzt einige der im Existenzbeweis benötigten Voraussetzungen (Beschränktheitsvoraussetzungen und $p_z \leq 0$), trotzdem sind die numerischen Resultate befriedigend. Die zulässige Endzeit wäre $T_C = 0.2$, das Verfahren läuft in der Praxis jedoch für weit größere Endzeiten. Numerisch wurden die Randwertprobleme in jedem Schritt mit dem Crank-Nicolson Verfahren diskretisiert und das auftretende nichtlineare Gleichungssystem mit dem Newton-Verfahren auf 7 Stellen genau gelöst. Die exakte Lösung des freien Randwertproblems ist in diesem Beispiel unbekannt. Wir unterscheiden bei der numerischen Durchführung eine globale und eine lokale Version. Bei der globalen Version wird die Iteration für das gesamte Raster auf der Zeitachse gleichzeitig durchgeführt, während in der lokalen Version auf jeder Zeitschicht neu gestartet wird. Die Genauigkeit ist dabei etwa gleich, während das lokale Verfahren erheblich weniger Rechenzeit benötigt.

3.1.1 Die lokale Version

Diskretisierung: $\Delta t := 0.002$, $\Delta x := 0.0066$.

Tabelle 1 zeigt das Resultat der numerischen Rechnung.

t	s(t)
0.032	0.487 455
0.066	0.480 076
0.100	0.475 450
0.134	0.471 855
0.168	0.468 400
0.200	0.465 032

Tabelle 1

3.1.2 Die globale Version

Diskretisierung : $\Delta t := 0.0017$, $\Delta x := 0.0083$,

Startfunktion : $s_0(t) := 0.52 + 2.25\ t$.

Tabelle 2 zeigt das Resultat der numerischen Rechnung, wobei mit ESi die errechnete i-te Iterierte von s bezeichnet wurde, und die Tabelle so angeordnet ist, daß die Iterierten mit ungeradem Index von oben und die mit geradem Index von unten zu lesen sind. So ergibt sich der optimale Einschluß der Lösung in der Mitte. In Abbildung 4 sind die Iterierten skizziert.

T	0.033	0.067	0.100	0.133	0.167	0.200
ES 1:	0.446 935	0.383 217	0.296 652	0.180 923	0.022 851	-0.205 677
ES 3:	0.482 519	0.458 928	0.405 592	0.292 815	0.123 592	-0.072 804
ES 5:	0.486 699	0.477 006	0.458 203	0.396 484	0.232 905	0.030 640
ES 7:	0.487 076	0.479 680	0.472 465	0.449 729	0.349 275	0.121 531
ES 9:	0.487 105	0.479 975	0.475 096	0.467 101	0.424 355	0.225 252
ES11:	0.487 108	0.480 002	0.475 464	0.471 159	0.456 036	0.340 524
ES13:	0.487 108	0.480 004	0.475 506	0.471 892	0.465 797	0.415 892
ES15:	0.487 108	0.480 004	0.475 510	0.471 999	0.468 127	0.449 785
ES14:	0.487 108	0.480 004	0.475 512	0.472 056	0.469 932	0.495 650
ES12:	0.487 108	0.480 004	0.475 525	0.472 344	0.474 940	0.561 205
ES10:	0.487 109	0.480 012	0.475 651	0.474 130	0.494 445	0.700 938
ES 8:	0.487 117	0.480 103	0.476 669	0.483 078	0.555 015	0.851 179
ES 6:	0.487 224	0.481 019	0.483 152	0.517 996	0.686 209	0.910 229
ES 4:	0.488 509	0.488 407	0.514 178	0.614 165	0.826 554	0.922 656
ES 2:	0.501 559	0.531 407	0.611 177	0.754 200	0.882 798	0.923 943
ES 0:	0.595 000	0.670 000	0.745 000	0.820 000	0.895 000	0.970 000

Tabelle 2

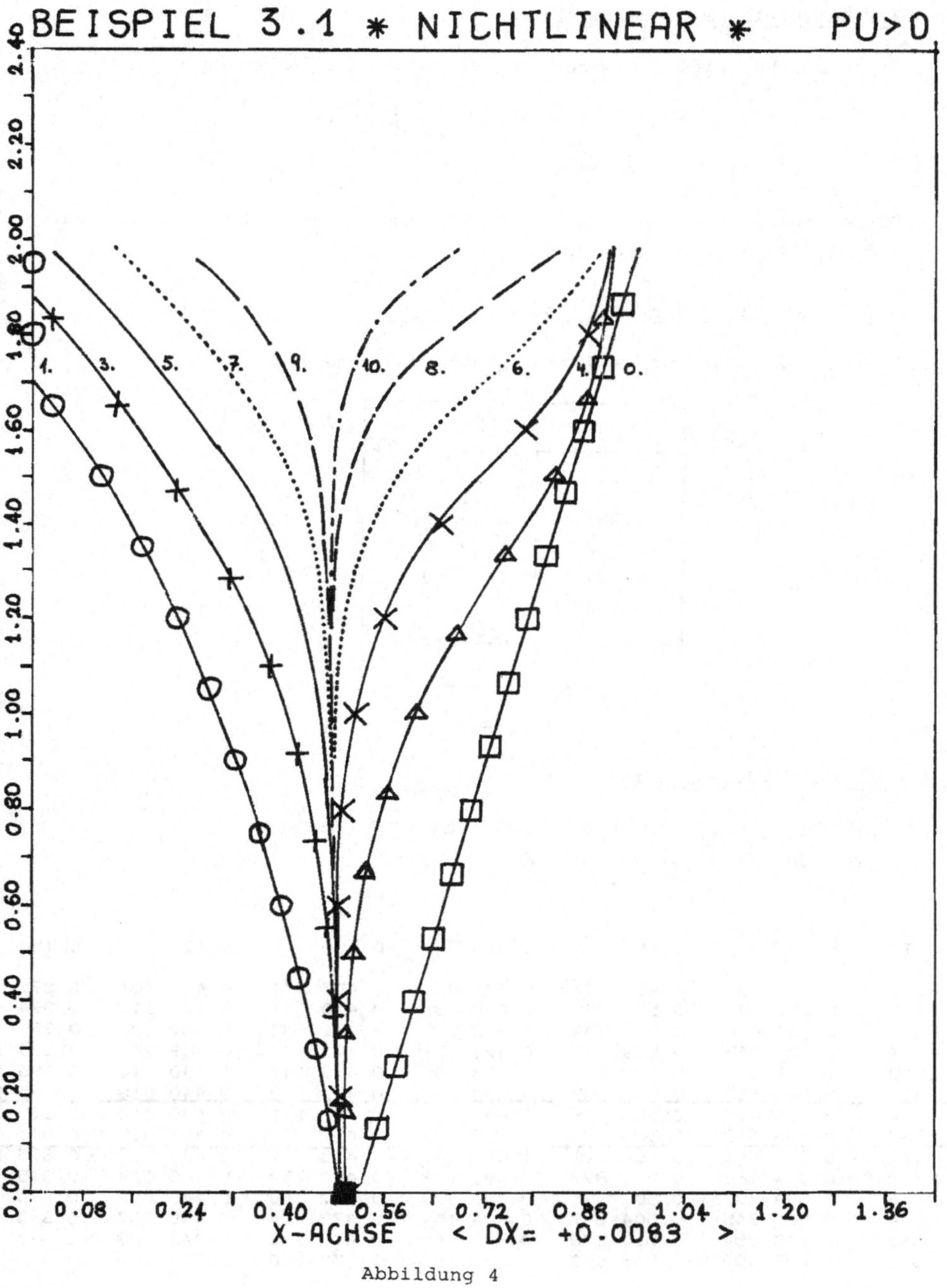
BEISPIEL 3.1 * NICHTLINEAR * PU>0
* 10^-1
0.00
0.20
0.40
0.60
0.80
1.00
1.20
1.40
1.60
1.80
2.00
2.20
2.40
1.
3.
5.
7.
9.
10.
8.
6.
4.
0.
0
0.08
0.24
0.40
0.56
0.72
0.88
1.04
1.20
1.36
X-ACHSE < DX= +0.0083 >

Abbildung 4

3.2 Beispiel zum Schmelzproblem

Gegenüber Beispiel 3.1 werden nur p,q und a abgeändert:

$$p(x,t,z) := 1 - \frac{1}{10} z^2 \quad , \quad q(x,t,z) := 1 + \frac{1}{10} z^2 \quad ,$$

$$a(t) := - \exp(t+0.5) \quad .$$

Dieses Beispiel verletzt die Voraussetzung $a > 0$. Es wurde bis zur Endzeit $T_\delta := 0.3$ gerechnet.

3.2.1 Die lokale Version

Diskretisierung: $\Delta t := 0.0030$, $\Delta x := 0.0066$.

t	s(t)
0.042	0.504 506
0.087	0.483 580
0.132	0.428 085
0:177	0.327 974
0.222	0.158 759
0.267	0.015 793

Tabelle 3

3.2.2 Die globale Version

Diskretisierung : $\Delta t := 0.0013$, $\Delta x := 0.0083$,

Startfunktion : $s_0(t) = 0.52 + 0.30\ t$.

T	0. 025	0. 050	0. 075	0. 100	0. 125	0. 150
ES 1:	0. 492 051	0. 489 347	0. 476 946	0. 455 574	0. 426 701	0. 391 603
ES 3:	0. 502 640	0. 501 275	0. 490 685	0. 470 141	0. 438 613	0. 394 194
ES 5:	0. 504 450	0. 503 554	0. 492 869	0. 471 882	0. 439 833	0. 395 005
ES 7:	0. 504 645	0. 503 854	0. 493 171	0. 472 122	0. 439 998	0. 395 112
ES 9:	0. 504 662	0. 503 885	0. 493 204	0. 472 148	0. 440 016	0. 395 123
ES11:	0. 504 663	0. 503 887	0. 493 207	0. 472 150	0. 440 018	0. 395 124
ES13:	0. 504 663	0. 503 888	0. 493 207	0. 472 151	0. 440 018	0. 395 124
ES12:	0. 504 663	0. 503 888	0. 493 207	0. 472 151	0. 440 018	0. 395 124
ES10:	0. 504 664	0. 503 888	0. 493 208	0. 472 151	0. 440 018	0. 395 125
ES 8:	0. 504 668	0. 503 897	0. 493 218	0. 472 159	0. 440 024	0. 395 128
ES 6:	0. 504 727	0. 503 995	0. 493 320	0. 472 240	0. 440 079	0. 395 163
ES 4:	0. 505 345	0. 504 857	0. 494 159	0. 472 904	0. 440 537	0. 395 456
ES 2:	0. 510 084	0. 510 278	0. 499 409	0. 477 178	0. 443 293	0. 397 405
ES 0:	0. 527 500	0. 535 000	0. 542 500	0. 550 000	0. 557 500	0. 565 000

Tabelle 4

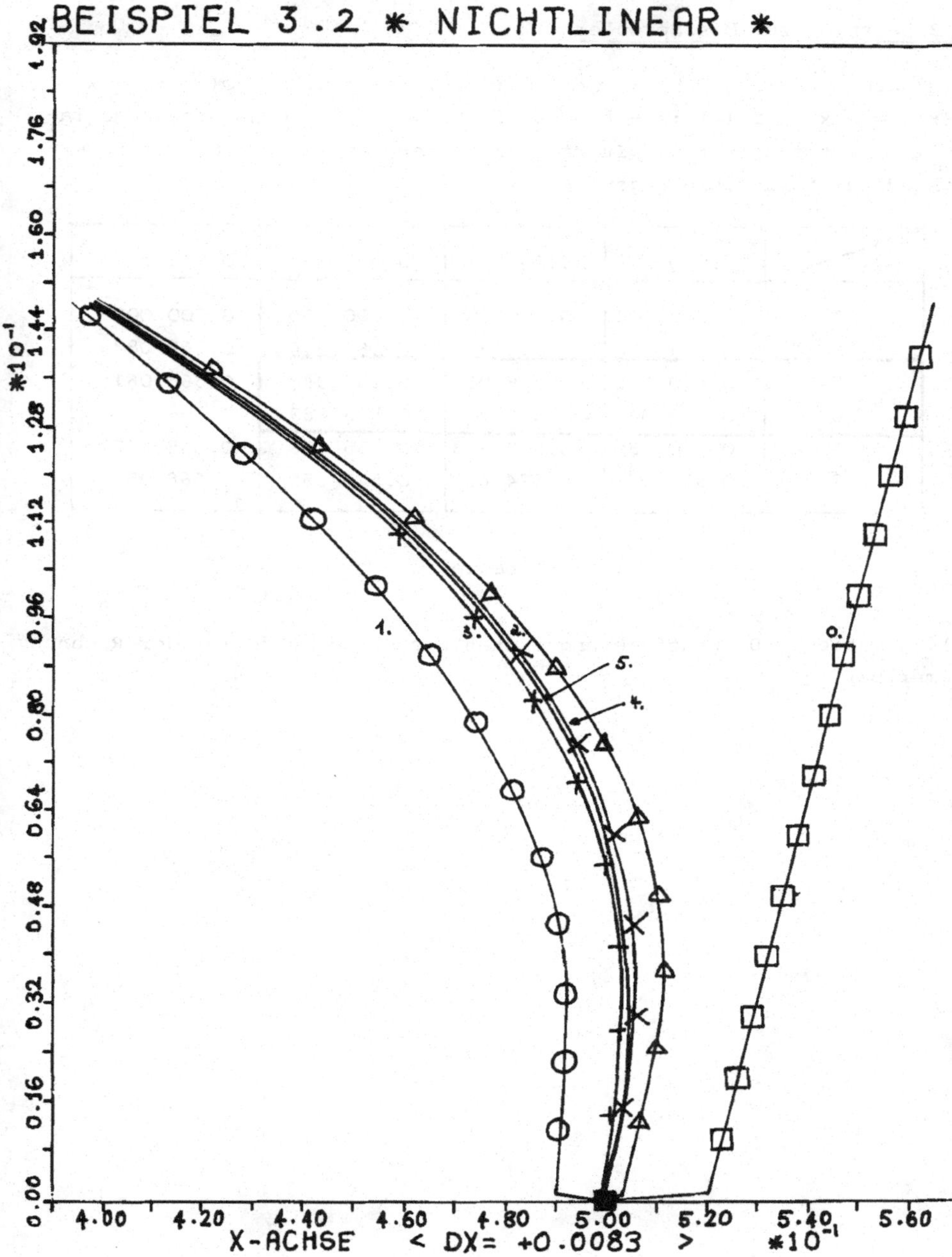
BEISPIEL 3.2 * NICHTLINEAR *
*10⁻¹
0.00
0.16
0.32
0.48
0.64
0.80
0.96
1.12
1.28
1.44
1.60
1.76
1.92
4.00
4.20
4.40
4.60
4.80
5.00
5.20
5.40
5.60
X-ACHSE < DX= +0.0083 > *10⁻¹
1.
2.
3.
4.
5.
0.

Abbildung 5

3.3 Beispiel zum Mischproblem

Als Daten wurden gewählt: $\alpha := \beta := 1$, $c := 0.5$, $\varphi(x) := -x$, $\psi(x) := -x$, $f_1(t) := -t$ und $f_2(t) := t$. Die nachfolgende Tabelle zeigt den Wert der Iterierten an der Stelle $t = 0.2$ für verschiedene Diskretisierungen Δt.

It. \ Δt	0.050	$0.03\overline{3}$	0.025	$0.01\overline{6}$
1	0.500 000	0.500 000	0.500 000	0.500 000
3	0.569 476	0.566 278	0.564 946	0.565 083
5	0.570 042	0.566 935	0.565 388	0.565 061*
6	0.570 044	0.566 937	0.565 389	--
4	0.570 080	0.568 978	0.565 408	0.565 062*
2	0.577 071	0.574 689	0.572 255	0.568 955

Tabelle 5

*) Bei dieser Diskretisierung machen sich Rundungsfehler der Rechnung bemerkbar.

4. Literatur

[1] BAIOCCHI, C., V. COMINCIOLI, E. MAGENES, G.A. POZZI: Free boundary problems in the theory of fluid flow through porous media. Existence and uniqueness theorems. Ann. Math. Pura Appl. 97 (1973), 1-82.

[2] BAIOCCHI, C., V. COMINCIOLI, L. GUERRI, G. VOLPI: Free boundary Problems in the theory of fluid flow through porous media. A numerical approach. Calculo 10, 1 (1973).

[3] BAUMEISTER, J., K.-H. HOFFMANN, P. JOCHUM: Numerical solution of a parabolic free boundary problem via Newton's method. J. Inst. Maths. Applics 25 (1980), 99-109.

[4] BRAESS, D.: Private Mitteilung (1981).

[5] CANNON, J.R., K.-H. HOFFMANN: Optimale Kontrolle eines freien Randes in der Gasdynamik. Preprint Nr. 110/80, FU-Berlin.

[6] DUVAUT, G.: Rêsolution d'un problême de Stefan (fusion d'un bloc de glace a zêro degrê). C.R. Acad. Sci. Paris 276 (1973), 1461-1463.

[7] EVANS, L.C.: A free boundary problem: The flow of two immiscible fluids in a one-dimensional porous medium: I. Ind. Univ. Math. J. 26 (1977), 915-931.

[8] FRIEDMANN, A.: The Stefan problem in several space variables. Amer. Math. Soc. Trans. 133 (1968), 51-87.

[9] HILL, C.D.: A hyperbolic free boundary problem. J. Math. Anal. Appl. 31 (1970), 117-129.

[10] HOFFMANN, K.-H.: Monotonie bei nichtlinearen Stefan-Problemen. ISNM 39 (1978), 162-190.

[11] HOFFMANN, K.-H.: Monotonie bei Zweiphasen-Stefan-Problemen. Numer. Funct. Anal. Optim. 1 (1979), 79-112.

[12] ICHIKAWA, Y., N. KIKUCHI: A one-phase multidimensional Stefan problem by the method of variational inequalities. Internat. J. Numer. Methods Engrg. 14 (1979), 1197-1220.

[13] ICHIKAWA, Y., N. KIKUCHI: ibi dem

[14] KRÜGER, H.: Zum Newtonverfahren für ein Stefanproblem. Erscheint in INSM (1981).

[15] KYNER, W.T.: An existence and uniqueness theorem for a nonlinear Stefan problem. J. of Math. and Mech. 8 (1959), 483-498.

[16] STEFAN, J.: Über einige Probleme der Theorie der Wärmeleitung. S.B. Wien, Akad. Mat. Naturw. 98, 173-484.

A DERIVATIVE-FREE ARC CONTINUATION METHOD AND A BIFURCATION TECHNIQUE

BY

R.B. KEARFOTT

Department of Mathematics
University of Southwestern Louisiana
Lafayette, Louisiana 70504
USA

A DERIVATIVE-FREE ARC CONTINUATION METHOD AND A BIFURCATION TECHNIQUE

by

Ralph Baker Kearfott

ABSTRACT

Algorithms and comparison results for a derivative-free predictor-corrector method for following arcs of $H(x,t) = \theta$, where $H : R^n \times [0,1] \to R^n$ is smooth, are given. The method uses a least-change secant update for H', adaptive controlled predictor stepsize, and Powell's indexing procedure to preserve linear independence in the updates. Considerable savings in numbers of theoretical function calls are observed over high order methods requiring explicit H'. The framework of a promising technique for handling general bifurcation problems is presented.

key words: arc continuation, quasi-Newton methods, least change secant updates, Brouwer degree, numerical computation, nonlinear algebraic systems, Powell's method.

1. Introduction

An approach to the numerical analysis of nonlinear systems in R^n is to study arcs of $H(x,t) = \theta$, where $H : R^n \times [0,1] \to R^n$ is smooth (cf. e.g., [2], pp. 39-48). This technique is used, for example, to study nonlinear eigenvalue problems and to solve algebraic systems for which Newton's method is not globally convergent. The original such methods, referred to as the "Davidenko" or "continuous Newton" methods, have been improved. Present solution techniques involve integrating the initial value problem:

$$(1) \qquad H'(y(s))y'(s) = \theta, \qquad \| y'(s) \| = 1, \qquad y(0) = y^0$$

where H' is the n by $n+1$ Jacobi matrix of H and $y'(s)$ is the componentwise derivative of $y = (x,t) \in R^{n+1}$ relative to arclength s.

The integration may be effected by high-order methods [16] or by "predictor-corrector" techniques (e.g., [13], [4], [5], [14]). In the latter, a direction $b^0 \in R^{n+1}$ is found to approximately satisfy $H'(y^0)b^0 = \theta$, and the predicted value $y(\delta_0)$ is set to:

(2) $$z^0 = y^0 + \delta_0 b^0 .$$

Corrections to z^0 are made by applying Newton's method to the system:

(3) $$G(z) = \begin{pmatrix} H(z) \\ (b^0)^t(z - z^0) \end{pmatrix} = \theta .$$

Note that this corrects z^0 in a hyperplane perpendicular to the step b^0 (cf. [4]). The entire process is repeated to obtain sequences y^i, b^i, z^i, and δ_i. The length δ_{i+1} can be chosen according to the angle between b^{i-1} and b^i ([14]), but in any case must be such that both the algorithm functions efficiently and the corrector iterations are stable.

Disadvantages of such arc continuation methods include the necessity of computing H' several times per predictor-corrector step. Also, new derivative-free techniques to handle multiple bifurcations (i. e., to pass points y, $H(y) = \theta$ where the null space of $H'(y)$ is of dimension greater than 1) are desirable.

Simplicial methods (cf. [2]) are derivative-free and have been applied to nonlinear bifurcation (see [8] and references therein). However, difficulties remain in the interplay between the triangulation, mesh, tracing of bifurcation branches, and the proximity of approximate solution arcs to true ones.

The purpose of this paper is to present a derivative-free arc continuation algorithm modelled on the predictor-corrector approach. In Section 2 the arc continuation algorithm is presented and explained. In Section 3 we give some numerical comparisons. In Section 4 a derivative-free method for bifurcation problems is presented.

It should be mentioned that Kurt Georg has independently developed similar derivative-free path-following algorithms, to appear in the SIAM Journal on Scientific and Statistical Computing [7] and in these proceedings [6]. Georg also gives a method of handling odd-order bifurcation points. Specific techniques from those methods and the method given below may be merged to effect improvements.

2. The Algorithm

The algorithm follows the general pattern outlined above and in [14].

The techniques herein may be applied to a variety of settings. For explanatory purposes, however, we assume H is of the form:

(4) $$H(z) = H(x,t) = t\,f(x) + (1 - t)\,g(x)$$

where $z = (x,t) \in R^n \times R$, $f : R^n \to R^n$, and $g : R^n \to R^n$. We also assume that a root x^0; $g(x^0) = \theta$ is known; the object is to find roots of F by following (possibly bifurcating) arcs of $H(z) = \theta$ from $(x^0, 0)$ to the $t = 1$ hyperplane.

The main modifications of the general scheme ([14] and above) are: (1) use of a least-change secant update to H' [3] instead of complete evaluation of H'; (2) use of Powell's indexing ([15], pp. 133-138) to assure accuracy in H'; and (3) special choice and adjustment of the stepsize to assure stability and accurate H'. The least-change update is given in Algorithm 2.1, Step 8, and is documented in [3], while the Powell indexing procedure is given in Algorithm 2.2 (infra) and is documented in [15].

Several parameters in Algorithm 2.1 are chosen to control the inner iteration and stepsize. These include the initial predictor stepsize δ_0, the maximum allowable predictor step δ_{max}, the criterion c_i for doubling the stepsize, and the criterion c_d for halving the stepsize. As in [4], $0 < c_d < c_i < 1$, where $\delta_{i+1} \leftarrow 2\delta_i$ if $b^i \cdot b^{i-1} > c_i$, but $\delta_{i+1} \leftarrow \delta_i/2$ if $b^i \cdot b^{i-1} < c_d$. (In all tests, the algorithm functioned well with $c_d = .95$ and $c_i = .99$.)

Additional parameters include the predictor function magnitude tolerance ε_δ, the maximum number of inner iterations N_i, the singular matrix indicator ε_{mat}, the inner iteration convergence criterion ε_y, the $t = 1$ convergence criterion ε_t, the relative stepsize for finite differences Δ, and the eigenvalue criterion ε_e. Upon taking a predictor step: $z \leftarrow y + \delta b$ (Steps 5-6 of Algorithm 2.1), δ is halved and z is revised if:

(5) $$\|H(z)\|(n+1)^{1/2} / \|H'\|_F > \varepsilon_\delta ,$$

where $\|\cdot\|_F$ is the Frobenius norm. If the number of inner iterations between successive predictor steps exceeds $N_i + n + 1$, H' is reinitialized using finite differences, δ is halved, and the initial predictor step is revised. In solving (3) during the inner iteration, it is necessary to solve $G'X = -G$ repeatedly; if, after normalizing G', a maximum Gaussian elimination pivot element of magnitude less than ε_{mat} is found in partial pivoting, G' is considered singular. In that case, the technique in Section 4 can be used to continue. The inner iteration is terminated when $\|H\|(n+1)^{1/2} / \|H'\|_F < \varepsilon_y$, whereas outer iteration is terminated when $\| y_{n+1} - 1 \| = |t - 1| < \varepsilon_t$. The additional parameter ε_e is used to determine when eigenvalues of $H'^t H'$ are approximately equal to 0; this will be explained in Section 4. M_z is an estimate of the maximum magnitude of any point on the arc.

2.1 Algorithm

0. Input the dimension n, the function $H : R^{n+1} \to R^n$, δ_0, δ_{max}, c_i, c_d, ε_δ, N_i, ε_{mat}, ε_y, ε_t, Δ, ε_e, and M_z.
1. Set $y \leftarrow y^0$; compute $H \leftarrow H(y)$ and compute $H' \leftarrow H'(y)$ using finite differences with stepsize $\| y \| \Delta$; initialize the Powell vectors: $d^i \leftarrow e^i$, $i = 1, \ldots, n+1$, where e^i is the i-th coordinate vector in R^{n+1}, and $\omega_i \leftarrow 1$, $i = 1, \ldots, n+1$; set flag $f_1 \leftarrow 1$. ($f_1 = 1$ indicates H' has just been initialized.)
2. Initialize b so that $H'b = \theta$. (cf. [4].)
3. (Initialization of the stepsize and counter for outer iterations) $\delta \leftarrow \delta_0$, nit $\leftarrow$ 1.
4. (Initialization of the counter for inner iterations) ninit $\leftarrow$ 1.
5. (Take predictor step) $s \leftarrow \delta b$; $z^0 \leftarrow y + s$; $z \leftarrow z^0$; $H_{old} \leftarrow H$.
6. $H \leftarrow H(z)$.
7. (Halve predictor step if new H value is too large in magnitude)
   If $\| H \| (n + 1)^{1/2} / \| H' \|_F > \varepsilon_\delta$, do the following:
   (a) $H \leftarrow H_{old}$.
   (b) $\delta \leftarrow \delta / 2$.
   (c) Return to Step 4.
   Otherwise, continue.
8. Make a "least change" (Broyden) update to H':
   $H' \leftarrow H' + (H - H_{old} - H's)s^t / \| s \|^2$.
9. (Powell's checking procedure applied to the predictor step)
   (a) Compute the inner products: $a_i \leftarrow (s / \| s \|) \circ d^i$, $i = 1, \ldots, n+1$.
   (b) If $\omega_1 \geq 2(n + 1)$ and $2 |a_1| < 1$ do (i)-(vi); otherwise, continue to Step 9(c). (Here, a special correction update is made to H' if necessary.)
      (i) $s \leftarrow \Delta d^1$; $z \leftarrow z + s$.
      (ii) $H_{old} \leftarrow H$; $H \leftarrow H(z)$.
      (iii) $H' \leftarrow H' + (H - H_{old} - H's)s^t / \| s \|^2$.
      (iv) $a_i \leftarrow (s / \| s \|) \circ d^i$, $i = 1, \ldots, n+1$.

(v) $H \leftarrow H_{old}$; $z \leftarrow z - s$.

(vi) Proceed to Step 21.

Steps 10-23 involve refinement of the predictor step (i. e., inner iteration).

(c) Update the Powell indices ω_i and d^i, $i = 1, \ldots, n+1$.

10. (Reinitialization if maximum number of inner iterations has been exceeded) If $\text{ninit} > N_i + n + 1$, do Step 11; otherwise, continue to Step 12.

11. (Reinitialization)

(a) $\delta \leftarrow \delta/2$.

(b) $H \leftarrow H(y)$.

(c) Initialize the Powell indices and Powell vectors: $\omega_i \leftarrow 1$, $d^i \leftarrow e^i$, $i = 1, \ldots, n+1$.

(d) Set $H' \leftarrow H'(y)$, where $H'(y)$ is computed using finite differences.

(e) Set the initialization indicator flag : $f_1 \leftarrow 1$.

(f) Return to Step 4.

12. (Storage of convergence criteria) $\tau_1 \leftarrow |s_{n+1}|$; $\tau_2 \leftarrow \|H\|(n+1)^{1/2}/\|H'\|_F$.

13. Compute a new corrector step: $s \leftarrow -(G')^{-1}G$, where

$$G = \begin{pmatrix} H \\ b^t(z - z^0) \end{pmatrix} \text{ and } G' = \begin{pmatrix} H' \\ b^t \end{pmatrix}.$$

14. If a singular G' is detected in Step 13 (cf. the explanation of ε_{mat} above) then do the following:

(a) (Reinitialize if the singularity is possibly due to the update process) If $f_1 = 0$, go to Step 11; otherwise, continue to (b).

(b) Compute and store direction vectors b, y, δ, and nit via Algorithm 4. 1.

(c) Retrieve a direction vector b, y, δ and nit via Algorithm 4. 2.

(d) $\text{nit} \leftarrow \text{nit} + 1$.

(e) Go to Step 4.

Otherwise, continue.

15. (Execute Powell's special step if independence is not maintained) If $\omega_1 \geq 2(n+1)$ and $2|\alpha_1| < 1$, go to step 9(b); otherwise, continue to Step 16.

16. (Termination of inner iteration if convergence has been achieved) If $\tau_2 < \varepsilon_y$ and $\tau_1 < |z_{n+1}|/10$ or $\text{ninit} = 1$, go to Step 24;

otherwise, continue. (Convergence in the t variable is tested separately in case of poor scaling.)

17. (Reinitialize if the ratio of magnitudes of the corrector step to the predictor step is too large) If $\|s\|/\delta > 2(1 - c_d^2)/c_d$, go to Step 11; otherwise, continue.

18. $\alpha_i \leftarrow (s/\|s\|) \circ d^i$, $i+1, \ldots, n+1$.

19. (Take the corrector step) $z \leftarrow z + s$; $H_{old} \leftarrow H$; $H \leftarrow H(z)$.

20. (Make a Broyden update as in Step 8 and reset initialization indicator flag)

    (a) $H' \leftarrow H' + (H - H_{old} - H's)s^t/\|s\|^2$.

    (b) $f_1 \leftarrow 0$.

21. Update the Powell indices ω_i and the Powell vectors d^i, $i = 1, \ldots, n+1$ via Algorithm 2.2.

22. (Advance counter for inner iterations) ninit $\leftarrow$ ninit + 1.

23. (Do another inner iteration) Return to Step 10.

24. (Reset initialization flag in case no inner iterations were necessary) $f_1 \leftarrow 0$.

    The remaining steps consider possible reasons for ending the outer iterations.

25. (Divergence or a return to the $t = 0$ hyperplane) If $z_{n+1} \leq 0$ or if $z \geq M_z$, then do the following:

    (a) Print an appropriate message.

    (b) If there are no more bifurcation branches to be considered, then stop; otherwise, continue to step (c).

    (c) Retrieve a vector b, y, δ, and nit via Algorithm 4.2.

    (d) nit $\leftarrow$ nit + 1; go to Step 4.

26. (Termination if the $t = 1$ hyperplane has been successfully reached) If $|z_{n+1} - 1| < \varepsilon_t$ do the following:

    (a) Store z.

    (b) If there are no more bifurcation branches to be considered, then stop; otherwise, continue to Step (c).

    (c) Retrieve a vector b, y, δ, and nit via Algorithm 4.2.

    (d) nit $\leftarrow$ nit + 1; go to Step 4.

27. (Interpolation if the $t = 1$ hyperplane has been passed) If $z_{n+1} > 1$, do the following:

    (a) Find the point $q = (q_1, q_2, \ldots, q_{n+1})$ on the line connecting y and

z such that $q_{n+1} = 1$.

(b) $y \leftarrow q$; $s \leftarrow y - z$; $z^0 \leftarrow y$; $z \leftarrow z^0$; $b \leftarrow e^{n+1}$; $H \leftarrow H(y)$.

(c) (Refinement of the interpolation at the $t = 1$ hyperplane by inner iteration) Go to Step 8.

In the remaining steps, δ is adjusted and a new b is computed for further outer iteration.

28. $b_{old} \leftarrow b$; $b \leftarrow (z - y)/\| z - y \|$; $y \leftarrow z$.
29. (Adjusting the stepsize according to the angle between the previous and present direction)

    (a) If $b \bullet b_{old} > c_i$, set $\delta \leftarrow 2\delta$.

    (b) If $\delta > \delta_{max}$, set $\delta \leftarrow \delta_{max}$.

    (c) If $b \bullet b_{old} < c_d$, set $\delta \leftarrow \delta/2$.
30. (Increment counter for number of outer iterations) nit $\leftarrow$ nit + 1.
31. (Do another outer iteration) Go to Step 4.

The following algorithm, given and explained in [15], pp. 133-138, is for keeping track of the directions in which generalized secant updates are made. It is repeated here for convenience.

Algorithm 2.2 (Powell's indexing)

0. Input the vectors s and d^i, $i = 1, \ldots, n+1$ and the scalars ω_i and α_i, $i = 1, \ldots, n+1$ from Algorithm 2.1.
1. Set m equal to the smallest k: $1 \le k \le n + 1$ such that $\sum_{i=1}^{k} \alpha_i^2 \ge 1/4$.
2. $\omega_j \leftarrow \omega_j + 1$ for $j = 1$ to $m - 1$.
3. $\omega_j \leftarrow \omega_{j+1} + 1$ for $j = m$ to n.
4. $\omega_{n+1} \leftarrow 1$.
5. $d^1_{new} \leftarrow d^m$; $d^i_{new} \leftarrow d^{i-1}$ for $i = 2$ to m; $d^i \leftarrow d^i_{new}$, for $i = 1$ to m.
6. $\alpha_{1,new} \leftarrow \alpha_m$; $\alpha_{i,new} \leftarrow \alpha_{i-1}$, $i = 2$ to m; $\alpha_i \leftarrow \alpha_{i,new}$, $i = 1$ to m.
7. Set $r \leftarrow \alpha_1^2$ and $\sigma_i \leftarrow 0$, for $i = 1$ to $n + 1$.
8. Repeat the following in sequence for $i = 2$ to $n + 1$:

    (a) denom $\leftarrow (r(r + \alpha_i^2))^{1/2}$.

    (b) For $j = 1$ to $n + 1$:

    (i) $\sigma_j \leftarrow \sigma_j + \alpha_{i-1} d_j^{i-1}$

    (ii) $d_j^{i-1} \leftarrow (r d_j^i - \alpha_i \sigma_j)/\text{denom}$.

    (c) $r \leftarrow r + \alpha_i^2$.

9. $d^{n+1} \leftarrow s/\|s\|$.

10. Return to Algorithm 2.1.

3. Numerical Results

Introduction of the Broyden update with special Powell steps, the choice of predictor direction, and special control of predictor and corrector step lengths have these ends: (1) provision of an arc-continuation algorithm applicable where derivatives are difficult to obtain; (2) provision of a more efficient algorithm; and (3) provision of a more reliable algorithm. To test the achievement of these ends, we have made comparisons on the following four problems used by Watson in [12]:

(1) $f_k(x) = x_k - (1/2n)(\sum_{i=1}^{n} x^3 + k), \quad k = 1 \text{ to } n;$

(2) $f_k(x) = .01(\sum_{i=k-1}^{k+1} x_i + 1)^3, \quad k = 1 \text{ to } n;$

(3) $f_k(x) = \exp(\cos(k \sum_{k=1}^{n} x_i)), \quad k = 1 \text{ to } n;$

(4) $f_1(x) = x_1 - (\prod_{i=1}^{n} x_i - 1), \quad f_j(x) = x_j - (\sum_{i=1}^{n} x_i + x_j - (n+1)), \quad j = 2 \text{ to } n.$

In all cases, the homotopy used was $H(x,t) = (1-t)x + t(x - F(x))$, where $F(x) = (f_1(x), f_2(x), \ldots, f_n(x))$ and the object was to find fixed points of F.

The results for these functions are presented in Tables 1 to 4, respectively. In each case, the dimension, the number of function evaluations, the number of inner iterations, and the number of function evaluations per inner iteration are given in columns 1 through 4. "Equivalent" function evaluations, given in the fifth column, are computed for Watson's test runs by multiplying the number of Jacobi matrix evaluations Watson's algorithm required by the dimension. The ratio of equivalent function evaluations to function evaluations is given in the last column.

Double precision was used in the Fortran program on a Honeywell 68/80 (36 bit word length). In all cases, $\delta_0 = .1$, $c_i = .99$, and $c_d = .95$. The maximum predictor steplength δ_{max} was not limited except in the fourth problem, for $n = 10$, $n = 20$, $n = 25$, and $n = 45$. This was necessary since H' is ill-conditioned in the fourth problem near $t = 1$.

In all results listed, the fixed point of F was found to at least 12 significant digits.

Except for the extremely nonlinear function (Table 3) and except for certain runs with the ill-conditioned function (Table 4), the derivative-free method showed a definite advantage. In many cases, the total number of function evaluations for an outer iteration was less than that required for a single Jacobi matrix evaluation. Furthermore, roundoff and truncation possibly accumulate in Watson's direct high-order scheme, but such errors are corrected in general predictor-corrector methods.

It should be pointed out that it is often possible to compute an n by $n+1$ Jacobi matrix with less than the equivalent of n evaluations of H. For this reason, our method of comparison would be most valid for complicated functions and functions which are difficult to encode.

In this preliminary version, G' was not updated directly; hence, $(n+1)^3$ were required per inner iteration to solve the algebraic system given in (3).

n	eval.	nit	eval. /nit	eq. eval.	eq. eval. / eval.
10	32	4	8.	500	15.6
20	46	5	9.2	800	17.4
30	57	5	11.4	1020	17.9
40	67	5	13.4	1840	27.5
50	76	5	15.2	1800	23.7
60	86	5	17.2	2280	26.5
70	95	5	19.	3500	36.8
80	105	5	21.	4320	41.1
90	114	5	22.8	5040	44.2
100	132	6	21.	3400	25.8

Table 1. $f_k(x) = x_k - (1/2n)(\sum_{i=1}^{n} x_i^3 + k)$, $k = 1$ to n.

n	eval.	nit	eval. /nit	eq. eval.	eq. eval. /eval.
10	52	4	13.	360	6.9
20	57	4	14.3	720	12.6
30	72	4	18.	1080	15.
40	82	4	20.5	1440	17.6
50	89	4	22.3	1800	20.2
60	99	4	24.8	2160	21.8
70	109	4	27.3	2520	23.1
80	119	4	29.8	2880	24.2
90	129	4	32.3	3240	25.1
100	139	4	34.8	3600	25.9

Table 2. $f_k(x) = .01(\sum_{i=k-1}^{k+1} x_i + 1)^3$, $k = 1$ to n .

n	eval.	nit	eval. /nit	eq. eval.	eq. eval. /eval.
2	71	7	10.	88	1.2
3	536	32	16.8	663	1.2
4	840	56	15.	892	1.1
5	1485	93	16.	2565	1.7
6	1787	117	15.3	7272	4.1
7	3039	183	16.6	13860	4.6
8	4849	294	16.5	23792	4.9
9	5950	371	16.	33210	5.6
10	8078	476	17.	46440	5.7

Table 3. $f_k(x) = \exp(\cos(k \sum_{i=1}^{n} x_i))$, $k = 1$ to n .

n	eval.	nit	eval./nit	eq. eval.	eq. eval./eval.
5	115	6	19.2	260	2.3
10*	451	38	11.8	740	1.6
15	424	14	30.3	1455	3.4
20*	1320	62	19.4	1460	1.1
25*	1371	67	20.5	2025	1.5
30	1000	17	58.8	3240	3.2
35	898	17	52.8	4235	4.7
40	851	16	53.2	4840	5.7
45*	3419	370	9.2	5535	1.6
50	897	17	52.8	6450	7.2

Table 4. $f_1(x) = x_1 - (\prod_{i=1}^{n} x_i - 1)$,

$f_j(x) = x_j - (\sum_{i=1}^{n} x_i + x_j - (n+1))$, $j = 2$ to n.

Special parameters were used for the starred dimensions (see text).

4. Bifurcation

Corrector iteration in Algorithm 2.1 fails when H' becomes ill-conditioned or singular, i.e., when the null space of H' effectively has dimension greater than 1. Indeed, at such (and only such) points y^*, $H(y^*) = \theta$, the manifold structure of $H^{-1}(\theta)$ may break down, and two or more arcs of $H^{-1}(\theta)$ may intersect.

H. B. Keller [13] has proposed several procedures for following all such arcs emanating from y^*. However, these involve evaluation of second partial derivatives of H or other drawbacks. Here, we outline a different general technique which will always work in theory and which can be expected to function reasonably well when the effective dimension of the null space of H' is 5 or less.

Suppose that the solution arcs $\{y(s)\} \subset R^{n+1}$ of $H(y) = \theta$ are smooth except at bifurcation points y^*. It can then be shown that all such arcs intersecting at y^* must be tangent to the tangent space Π to y^* generated by the null space of $H'(y^*)$. If $\{z^1, z^2, \ldots, z^k\}$ is a basis for the null space of $H'(y^*)$ and $\delta > 0$, consider $\Pi_\delta = \{y^* + v \mid v = \sum_{i=1}^{k} a_i z^i,\ -\delta \le a_i \le \delta\}$. For δ

sufficiently small, the local minimum points $\{m^1, m^2, \ldots, m^q\}$ of $\|H\| \mid \partial\Pi_\delta$ correspond to points on solution curves bifurcating from y^*. Thus, predictor directions for such curves at y^* are given by:

$$b^i = (m^i - y^*)/\|m^i - y^*\|, \quad i = 1 \text{ to } q. \tag{6}$$

When $k \leq 5$, finding the m^i involves searching $2k$ cubes of dimension 4 or less. That can possibly be done by using numerical computation of the topological degree and a generalized method of bisection to find roots of grad $(\|H\|)$ within the low dimensional parameter space [9], [10], [11]. This approach will find all such m^i since the Kronecker index of any continuous grad $(\phi) \neq 0$ at minima of ϕ, where its sign depends upon the dimension k of the space. Furthermore, grad $(\|H\|)$ need not be approximated with high accuracy for proper functioning of the bisection algorithm.

Algorithm 4.1 uses the above approach. It consists of two phases: (1) computation of an orthonormal basis for the null space of $H'(y^*)$; and (2) computation of the direction vectors b^i in (5) and their storage, along with other information necessary to continue Algorithm 2.1 from y^*.

The first phase may be executed by computing the eigenvalues and eigenvectors of the symmetric matrix $H'^t H'$. First, the eigenvectors $\{u^i\}_{i=1}^{n+1}$ and corresponding eigenvalues $\{\lambda_i\}_{i=1}^{n+1}$ are computed; the basis $\{v^i\}_{i=1}^k$ for the tangent hyperplane is then obtained by selecting those u^i for which $|\lambda_i| < \varepsilon_e$.

Three interrelated parameters from Algorithm 2.1 govern the selection of $\{v^i\}_{i=1}^k$: δ_{max}, ε_{mat}, and ε_e. If all bifurcation points are to be found, ε_{mat} must be sufficiently large, since the iterates y^i cannot be expected to be closer than $\delta_{max}/2$ to bifurcation points; likewise, ε_e must be sufficiently large to detect the approximate null space. On the other hand, if ε_{mat} and ε_e are too large, false directions may be given or Algorithm 4.1 will be invoked repeatedly in the vicinity of a bifurcation point, reducing efficiency or causing redundant tracing of a single arc. In such cases, all three of δ_{max}, ε_{mat}, and ε_e may need to be reduced.

The second phase can be executed by computation of the topological degree and a generalized method of bisection. (In the case $k = 1$, H' is not really singular; $b \leftarrow \pm v^1$ without further computation.) A good bisection method is documented in [10]; details will not be given here. Search of the $(k-1)$-dimensional parallelopiped:

(7) $$\Pi_{\delta,i} = \{y^* \pm \delta v^i + \sum_{\substack{j=1 \\ j \neq i}}^{k} a_j v^j, \quad -\delta \leq a_j \leq \delta\}$$

is begun by computing the Brouwer degree of grad ($\|H\|$) at θ relative to a canonically shaped simplex S containing $\Pi_{\delta,i}$; if this degree is 0, no relative minima exist on $\Pi_{\delta,i}$. Otherwise, "Whitney" bisection (ibid.) is applied to S, each iteration of which yields simplexes with the same shape as S but with diameters equal to precisely half that of S. A sufficient number of such iterations is done to reduce the diameter, nominally, to $.1\delta$, while all simplexes having non-zero Brouwer degree are considered. The m^i are set equal to the resulting barycenters.

Computing the Brouwer degree requires either choice of a heuristic parameter or coding of a bound on the modulus of continuity of the components of grad ($\|H\|$) whenever $k > 2$.

When an m^i and b^i are computed, y^*, b^i, nit, and δ' are stored in a stack. Here, nit, y, and δ are from Algorithm 2.1, while $\delta' = 20\delta$. We need δ' so large that the next iteration of Algorithm 2.1 will not encounter an effectively singular H', but small enough so that the arc can be followed. (The factor of 20 was found to be satisfactory in test cases.)

Algorithm 4.1

0. Input n, H', y, nit, δ, b, and ε_e.
1. Compute the $n + 1$ by $n + 1$ matrix $H'^t H'$.
2. Compute the eigenvalues and orthonormal set of eigenvectors of $H'^t H'$.
3. Store those eigenvectors $\{v^i\}_{i=1}^k$ whose corresponding eigenvalues λ_i satisfy $|\lambda_i| < \varepsilon_e$.
4. If $k = 1$, do the following:
   (a) $\delta \leftarrow 20\delta$.
   (b) If $v^1 \cdot b > -.7$, $b \leftarrow v^1$; otherwise, $b \leftarrow -v^1$.
   (We make this check to avoid going in the direction opposite to that Algorithm 2.1 traversed to get to y^*.)
   (c) Store δ, nit, y, and b in stack $\mathcal{S}$.
   (d) Return to Algorithm 2.1.
5. If $k > 1$, do Steps 6-9 for $i = 1$ to k and sgn = -1 and sgn = 1:
6. $p \leftarrow \text{sgn} \cdot v^i$.

7. Set $w^j \leftarrow v^j$, $j = 1$ to $i-1$ and set $w^{j-1} \leftarrow v^j$, $j = i+1$ to k.
8. Using $X = (x_1, x_2, \ldots, x_{k-1})$, find roots of $\phi(X) = \text{grad}\,(\| H(z_X) \|)$, where $z_X = y + \delta p + \sum_{j=1}^{k-1} x_j w^j$.
9. Suppose $(x_1^1, x_2^1, \ldots, x_{k-1}^1), \ldots, (x_1^q, x_2^q, \ldots, x_{k-1}^q)$ are the roots found in Step 8. Define:
$b^\ell = (\delta p + \sum_{j=1}^{k-1} x_j^\ell w^j) / \| \delta p + \sum_{j=1}^{k-1} x_j^\ell w^j \|$, $\ell = 1$ to q. If $b \bullet b > -.7$, store 20δ, nit, y, and b^l in stack $\mathscr{S}$, for $l = 1$ to q.
10. Return to Algorithm 2.1.

The stack $\mathscr{S}$ is expanded whenever bifurcation points are encountered in Algorithm 2.1. Algorithm 2.1 follows an arc until it is determined to diverge, terminate at $t = 1$, or terminate at $t = 0$. Upon such termination, the algorithm continues with values of δ, nit, y, and b retrieved from $\mathscr{S}$ via the following:

Algorithm 4.2 (Retrieval)

0. Let $\mathscr{S}$ be the stack of δ, nit, y, and b.
1. If $\mathscr{S}$ is empty, stop.
2. Retrieve δ, nit, y, and b from $\mathscr{S}$.
3. $z \leftarrow y + b$.
4. Initialize the Powell indices: $\omega_i \leftarrow 1$, $d^i \leftarrow e^i$, $i = 1$ to n.
5. $H' \leftarrow H'(z)$, where H' is computed using finite differences.
6. $H \leftarrow H(y)$.
7. Return to Algorithm 2.1 .

Actual detection of the bifurcation points may be effected by monitoring any of several determinants. For example, the determinants of $G(z)$ (formula 3) could be averaged over sequences of $2n + 3$ iterations to obtain approximate values. Alternatively, the reciprocals of the condition numbers of H' (i.e., the ratios of the smallest to largest singular values of H') could be computed directly. The condition number of H' may even be approximated in $o(n^2)$ operations [1].

Detailed algorithms for approximating and refining even-order bifurcation points, in addition to specifics on finding the $m^{(i)}$ in the general tangent hyperplane setting, appear in [12]. There results of several numerical experiments involving bifurcation are given. These results look promising with regard to the handling of arbitrary types of bifurcation points, even in cases where the arcs intersect tangentially; it is feasible to apply the techniques to various classes of problems. However, the algorithms in [12] are subject to improvement; also, additional tests on realistic problems and comparisons with other methods are needed to determine the practical value of each of the individual techniques.

References

1. A.K. Cline, C.B. Moler, G.W. Stewart, and J.H. Wilkinson, "An estimate for the condition number of a matrix," SIAM J. Numer. Anal. 16, #2, 1979.

2. E. Allgower and K. Georg, "Simplicial and continuation methods for approximating fixed points and solutions to systems of equations," SIAM Review 22, no. 1, January, 1980.

3. J.E. Dennis, Jr. and R.B. Schnabel, "Least change secant updates for quasi-Newton methods," SIAM Review 21 no. 4, 1979 (pp. 443-459).

4. C.B. Garcia and T.Y. Li, "On a path-following method for systems of equations," MRC Technical Summary Report #1983, Mathematics Research Center, University of Wisconsin - Madison, July, 1979.

5. C.B. Garcia and W.I. Zangwill, "Finding all solutions to polynomial systems and other systems of equations," Math. Prog. 16, 1979 (pp. 159-176).

6. Kurt Georg, "Numerical integration of the Davidenko equation," these proceedings.

7. Kurt Georg, "On tracing an implicitly defined curve by quasi-Newton steps and calculating bifurcation by local perturbations," to appear in the SIAM Journal on Scientific and Statistical Computing.

8. Harmut Jürgens, Heinz-Otto Peitgen, and Dietmar Saupe, "Topological perturbations in the numerical nonlinear eigenvalue and bifurcation problems," Proc. Conf. Analysis and Computation of Fixed Points, Academic Press, New York-London, 1980, 139-181.

9. R.B. Kearfott, "An efficient degree-computation method for a generalized method of bisection," Numer. Math. 32, 1979 (pp. 109-127).

10. R.B. Kearfott, "An improved program for generalized bisection," to appear.

11. R.B. Kearfott, "A summary of recent experiments to compute the topological degree," Applied Nonlinear Analysis, ed. V. Lakshmikantham, A.P., 1979.

12. R.B. Kearfott, "Some general derivative-free bifurcation techniques," submitted to the SIAM Journal on Scientific and Statistical Computing.

13. H.B. Keller, "Numerical solution of bifurcation and nonlinear eigenvalue problems." Applications of Bifurcation Theory, ed. P.H. Rabinowitz, Academic Press, New York, 1979, (pp. 359-384).

14. T.Y. Li and J.A. Yorke, "A simple reliable numerical algorithm for following homotopy paths," Technical Summary Report #1984, Mathematics Research Center, University of Wisconsin - Madison, Wisconsin 53706.

15. M. J. D. Powell, "A Fortran subroutine for solving systems of nonlinear algebraic equations," Numerical Methods for Nonlinear Algebraic Equations, ed. P. Rabinowitz, Gordon and Breach, 1970.

16. L. T. Watson, "A globally convergent algorithm for computing fixed points of C^2 maps," Appl. Math. Comput., to appear.

AN INTRODUCTION TO VARIABLE DIMENSION ALGORITHMS
FOR
SOLVING
SYSTEMS OF EQUATIONS

BY

M. KOJIMA

Department of Information Sciences
Tokyo Institute of Technology
Meguro, Tokyo 512
Japan

AN INTRODUCTION TO VARIABLE DIMENSION ALGORITHMS FOR SOLVING SYSTEMS OF EQUATIONS

Masakazu Kojima

1. INTRODUCTION

This paper deals with variable dimension algorithms (abbreviated by vd algorithms), a class of computational methods which was originally proposed by van der Laan and Talman [14] for approximating Brouwer's fixed points and later extended by van der Laan and Talman [15] and Todd [25] to solutions of systems of equations. Several vd algorithms have been developed so far, and various interpretations have been given (Kojima [10], van der Laan [13], van der Laan and Talman [16, 17, 18], Reiser [21], Talman [23], Todd [26], Todd and Wright [27], Wright [28], etc.). Recently Freund [6] and Kojima and Yamamoto [11, 12] have proposed unified theories on which we can design various vd algorithms. In the papers listed above, vd algorithms were explained in terms of simplicial approximation methods. Roughly speaking, the vd algorithms given in those papers start from a single point, a zero dimensional simplex and generate a sequence of simplices of varying dimensions to attain a simplex which contains an approximation of a fixed point or a solution of a system of equations. To learn them, we are required to become familiar with terminologies of simplicial approximation methods such as triangulation (simplicial subdivision), piecewise linear approximation and complementary pivoting. Understanding such technical terminologies is of course necessary to develop a practical computational procedure of a specified vd algorithm. However, it often causes difficulty in finding out a basic idea of the vd algorithms. The author feels that this has been an obstacle to vd algorithms being popular though they have been attracting much attention.

This paper has two purposes closely related with each other. One is to give a simple explanation of a basic idea of vd algorithms. For this purpose, we try to describe vd algorithms in the framework of homotopy continuation methods (see, for example, Allgower and Georg [3]) without referring to triangulation and piecewise linear approximation. The other purpose is to establish a foundation for "piecewise smooth vd algorithms". The possibility of such vd algorithms was indicated by Kojima [10]. The second purpose seems to fit the first one. To carry out these two purposes, we propose a class of piecewise smooth vd algorithms for solving systems of equations, which we will outline below.

Consider the system of equations:

$$f(x) = 0 ,$$

where f is a continuous map from the n-dimensional Euclidean space R^n into itself. Except the last section, we shall assume that $f : R^n \to R^n$ is C^1 (continuously differentiable). In the construction of our vd algorithm, a primal-dual pair of subdivided manifolds (abbreviated by PDM), whose notion was first introduced by [11], will play an essential role. A PDM induces an n-dimensional manifold L which consists of a finite or countable number of n-dimensional polyhedral sets Z_i $(i \geq 0)$ in R^{2n} and has a certain special structure. Specifically, $(0,0) \in Z_0$.

Let $w^0 \in R^n$. When we know a rough approximation of a solution of the system $f(x) = 0$, we choose it as w^0. Let R_+ denote the set of all nonnegative numbers. Now we consider the system of equations:

$$(1.1) \qquad y + t\, f(w^0+x) = 0 , \quad (x,y,t) \in L \times R_+ .$$

It follows from $(0,0) \in L$ that $(x^0,y^0,t^0) = (0,0,0)$ is a trivial solution. On the one hand, the left side of the system (1.1) is defined on the (n+1)-dimensional manifold $L \times R_+$ and takes values of the n-dimensional Euclidean space R^n. Hence, under an appropriate regularity condition, the solution set turns out to be a disjoint union of curves (1-dimensional manifolds). Since the (n+1)-dimensional manifold $L \times R_+$ is the union of the (n+1)-dimensional polyhedral sets $Z_i \times R_+ \subset R^{2n+1}$ $(i \geqq 0)$ on each of which the left side of the system (1.1) is C^1, each solution curve is piecewise smooth (see Fig.1). Let S denote the solution curve which contains (x^0,y^0,t^0). Then we see

$$y/t + f(w^0+x) = 0 \quad \text{for every} \quad (x,y,t) \in S \quad \text{with } t > 0.$$

Hence, if the curve S "converges" to some $(\bar{x},\bar{y},+\infty)$ then we have $f(w^0+\bar{x}) = 0$. Thus, in this case, tracing the curve S from the trivial solution (x^0,y^0,t^0) of the system (1.1), we get an approximation of the solution $w^0+\bar{x}$ of the system $f(x) = 0$.

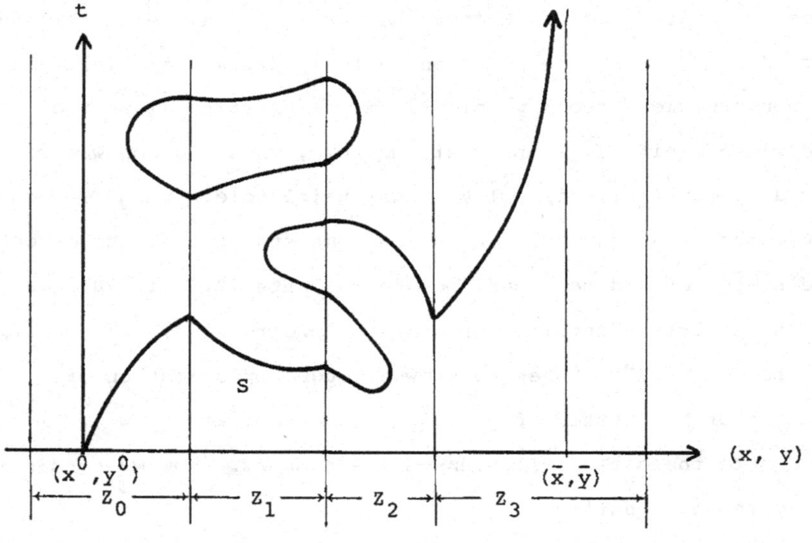

Fig. 1

The readers who are familiar with the homotopy continuation methods given by Allexander and Yorke [2], Chow, Mallet-Paret and Yorke [4], Kellogg, Li and Yorke [9], etc. (see also Allgower and Georg [3]) may understand that the vd algorithm outlined above is similar to them. An important feature of the vd algorithm to be pointed out is that the piecewise smooth curve S has a certain variable dimension property which is induced by the special structure of the n-dimensional manifold L. This property can be effectively utilized in making an efficient computational procedure for tracing the curve S.

In Section 2, after presenting the terminology and notation, we state a theorem on piecewise smooth continuation methods which will provide us with a fundamental tool in our succeeding discussion. In Section 3, we present the definition of a PDM and its properties in accordance with the paper [11], and then illustrate three examples of PDM's. The class of piecewise smooth vd algorithms is described in detail in Section 4, and a sufficient condition for the success of the vd algorithms is given in Section 5.

In Section 6, we focus our attention to a special case of the proposed vd algorithms in which one of the PDM's given in the examples of Section 3 is employed. We show that the predictor-corrector procedure (see, for example, [3]) developed for smooth continuation methods is applicable to the computation of the piecewise smooth solution curve S of the system (1.1), and how we utilize the variable dimension property to increase the computational efficiency.

In Section 7, we briefly present a transformation of the piecewise smooth vd algorithms into simplicial vd algorithms. This will be done by replacing the C^1-map $f: R^n \to R^n$ by a piecewise linear approximation $F : R^n \to R^n$ of the map f on a triangulation of R^n in the system (1.1). The resultant simplicial vd algorithms cover some of the existing ones, the n+1 algorithms given by van der Laan and Talman [15] and Todd [25], the 2n algorithm by van der Laan and Talman [18] and the checkerboard algorithm by Kojima and Yamamoto [12]. The last one has a close relation with well-known Merrill's method (see Section 4 of [12]).

2. PRELIMINARIES

By a cell, we mean a convex polyhedral set (the intersection of a finite number of closed halfspaces) in some Euclidean space. For each cell C, we define $\dim C$ (the dimension of C) to be the dimension of $\operatorname{aff} C$, the affine subspace spanned by C. A cell C with a dimension m is said to be an m-cell. We use the notation $\operatorname{rel\,int} C$ and ∂C for the interior and the boundary of C relative to $\operatorname{aff} C$, respectively. If a cell B is a face of a cell C, we write $B < C$.

Let M be a finite or countable collection of m-cells in some Euclidean space R^k. Let $\bar{M}$ denote the collection of all faces of $C \in M$, i.e., $\bar{M} = \{ B : B < C , C \in M \}$, and $|M|$ the union of all m-cells of M. M is said to be a subdivided m-manifold if it satisfies the following three conditions:

(2.1) For each pair of $B, C \in M$, either $B \cap C = \phi$ or $B \cap C$ is a common face of both B and C.

(2.2) Each (m-1)-cell of $\bar{\mathcal{M}}$ lies in at most two m-cells of $\mathcal{M}$.

(2.3) $\mathcal{M}$ is locally finite; each point $x \in |\mathcal{M}|$ has a neighborhood which intersects with only a finite number of m-cells of $\mathcal{M}$.

See Fig.2.

Let M be a subset of R^k, and $\mathcal{M}$ a subdivided m-manifold in R^k. If $M = |\mathcal{M}|$, we say that M is an m-manifold and that $\mathcal{M}$ is a subdivision of M. If $m \geqq 1$, M has a lot of subdivisions. The boundary of $\mathcal{M}$, denoted by $\partial\mathcal{M}$, is defined to be the collection of all (m-1)-cells of $\bar{\mathcal{M}}$ which lie in exactly one m-cell of $\mathcal{M}$, and ∂M (the boundary of M) the union of all (m-1)-cells of $\partial\mathcal{M}$. ∂M is independent of the choice of a subdivision $\mathcal{M}$ of M (see, for example, Eaves [5]).

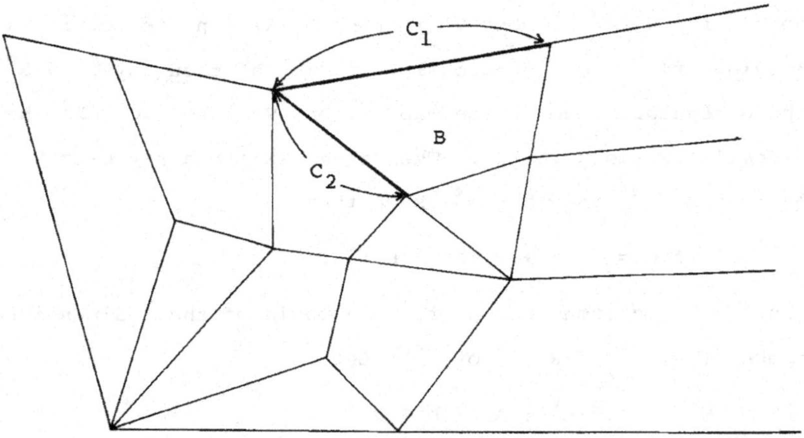

Fig. 2. $C_1 \in \partial\mathcal{M}$, $C_2 \notin \partial\mathcal{M}$.

By a unit interval, we mean $[0,1]$, $[0,1)$ or $(0,1) \subset R$, and by the unit circle, the subset $\{(x_1,x_2) : x_1^2 + x_2^2 = 1\}$ of R^2 . It is well-known that a connected 1-manifold S is homemorphic to either a unit interval or the unit circle. S is said to be a path in the former case, and a loop in the latter case. If S is a path and g is a homeomorphism from a unit interval onto S then the image of a boundary point of the interval under the map g is said to be an endpoint of the path S .

Let M be a subdivided m-manifold in R^k . A map h from $|M|$ into R^n is said to be PC^1 (piecewise continuously differentiable) on M if it is continuous on $|M|$ and the restriction of h to each m-cell C can be extended to a C^1-map defined on an open subset of R^k which contains C .

Let f be a C^1-map from an open subset U of R^k into R^n and V a subset of U . We denote the Jacobian matrix of f at each $x \in U$ by $Df(x)$. A point $c \in R^n$ is said to be a regular value of $f : U \to R^n$ on V if rank $Df(x) = n$ for every $x \in V$ satisfying $f(x) = c$. Obviously, if $k < n$ then any $c \in R^n$ can not be a regular value of the map f on V . Let $C \subset R^k$ be an ℓ-cell contained in U . Then there exists a $k \times \ell$ matrix A with rank $A = \ell$ and $b \in R^k$ such that

$$\text{aff } C = \{ Az + b : z \in R^\ell \} .$$

The set of the columns of A forms a basis of the ℓ-dimensional subspace $\{ Az : z \in R^\ell \}$ of R^k. Let

$$W = \{ z \in R^\ell : Az + b \in U \} ,$$

$$D = \{ z \in R^\ell : Az + b \in C \} ,$$

Then we see that W is an open subset of R^ℓ and that $D \subset R^\ell$ is an ℓ-cell. Define the C^1-map $g : W \to R^n$ by

$$g(z) = f(Az + b) \quad \text{for every } z \in W .$$

We say that a point $c \in R^n$ is a regular value of $f \mid C$, the restriction of the map f to the ℓ-cell C, if c is a regular value of the map $g : W \to R^n$ on D, i.e.,

$$(2.4) \qquad \text{rank } Df(x) A = n \quad \text{for every } x \in X ,$$

where $X = \{ x \in C : f(x) = c \}$. Since $Df(x) A$ is an $n \times \ell$ matrix, any $c \in R^n$ can not be a regular value of $f \mid C$ if $\ell = \dim C < n$. It is easily verified that this definition does not depend on the choice of a matrix A. Let B be a $k \times (k-\ell)$ matrix whose columns form a basis of $\{ x \in R^k : A^T x = 0 \}$, the orthogonal complement of the ℓ-dimensional subspace $\{ Az : z \in R^\ell \}$ of R^k. Then the condition (2.4) can be also written as

$$(2.5) \qquad \text{rank} \begin{bmatrix} Df(x) \\ \\ B^T \end{bmatrix} = n + (k-\ell) \quad \text{for every } x \in X .$$

Let M be a subdivided m-manifold in R^k, and $h : |M| \to R^n$ a PC^1-map on M. A point $c \in R^n$ is said to be a regular value of the PC^1-map $h : |M| \to R^n$ if $c \in R^n$ is a regular value of $h|C$ for every cell C of $\bar{M}$. We shall be only concerned with a special case where $m = n + 1$. In this case, if $c \in R^n$ is a regular value of h then $h^{-1}(c) = \{ z \in |M| : h(z) = c \}$ does not intersect with any cell of $\bar{M}$ with dimension less than n. Futhermore, using the implicit funtion theorem, one can prove the following result:

<u>Theorem 2.1.</u> Let M be a subdivided (n+1)-manifold in R^k, and $h : |M| \to R^n$ a PC^1-map on M. Suppose that $c \in R^n$ is a

regular value of the PC^1-map h. Then $h^{-1}(c)$ is a disjoint union of paths and loops such that

(2.6) every loop $h^{-1}(c)$ does not intersect with $\partial|M|$,

(2.7) $z \in h^{-1}(c)$ is an endpoint of a path iff $z \in |M|$.

If, in addition, $|M|$ is closed, every path in $h^{-1}(c)$ which is homeomorphic to either $[0, 1)$ or $(0, 1)$ is unbounded (see Fig. 3).

The proof is omitted here. See Allexander [1] for more general discussion.

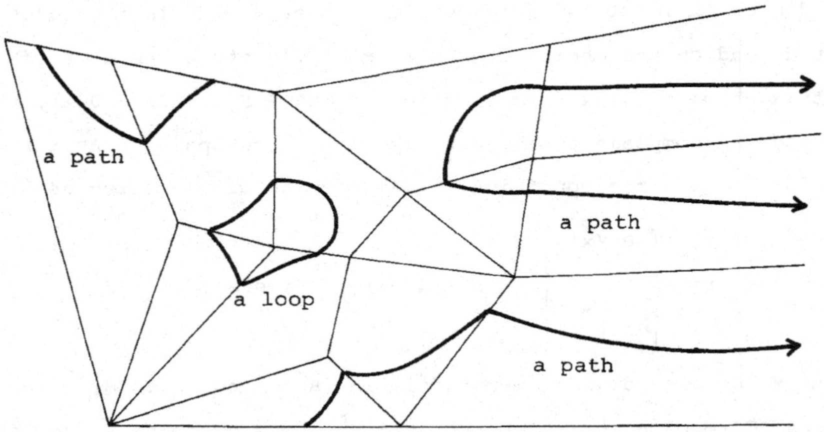

Fig. 3

3. PRIMAL-DUAL PAIRS OF SUBDIVIDED MANIFOLDS

The following definition of a primal-dual pair of subdivided manifolds (abbreviated by PDM) is due to [11]. Let n be a positive integer. A triplet $(\mathcal{P}, \mathcal{D}; d)$ is said to be a PDM with degree n if the following conditions (3.1) - (3.5) are satisfied:

(3.1) $\mathcal{P}$ and $\mathcal{D}$ are subdivided manifolds.

(3.2) For every $X \in \bar{\mathcal{P}}$, either $X^d \in \bar{\mathcal{D}}$ or $X^d = \phi$.

(3.2)' For every $Y \in \bar{\mathcal{D}}$, either $Y^d \in \bar{\mathcal{P}}$ or $Y^d = \phi$.

(3.3) If $Z \in \bar{\mathcal{P}} \cup \bar{\mathcal{D}}$ and $Z^d \neq \phi$ then $(Z^d)^d = Z$.

(3.4) If X_1, $X_2 \in \bar{\mathcal{P}}$, $X_1 < X_2$, $X_1^d \neq \phi$ and $X_2^d \neq \phi$ then $X_2^d < X_1^d$

(3.4)' If Y_1, $Y_2 \in \bar{\mathcal{D}}$, $Y_1 < Y_2$, $Y_1^d \neq \phi$ and $Y_2^d \neq \phi$ then $Y_2^d < Y_1^d$

(3.5) If $Z \in \bar{\mathcal{P}} \cup \bar{\mathcal{D}}$ and $Z^d \neq \phi$ then $\dim Z + \dim Z^d = n$.

$\mathcal{P}$ (or $\mathcal{D}$) is said to be the primal (or dual) subdivided manifold, and $|\mathcal{P}|$ (or $|\mathcal{D}|$) the primal (or dual) manifold. We call d a dual operator, and Z^d the dual of Z for every $Z \in \bar{\mathcal{P}} \cup \bar{\mathcal{D}}$

Fig. 4 illustrates an example of a PDM with degree 3. Here each X_i (or Y_i) for $i = 1,2,\ldots,6$ is a cone with the apex at X_0 (or Y_0) and

$$\mathcal{P} = \{ X_2 , X_4 , X_6 \} , \quad \bar{\mathcal{P}} = \{ X_i : 0 \leq i \leq 6 \} ,$$

$$\mathcal{D} = \{ Y_1 , Y_3 , Y_5 \} , \quad \bar{\mathcal{D}} = \{ Y_i : 0 \leq i \leq 6 \} ,$$

$$X_i^d = Y_i , \quad Y_i^d = X_i \quad (1 \leq i \leq 6) , \quad X_0^d = Y_0^d = \phi .$$

Note that $\mathcal{P}$ and $\mathcal{D}$ are subdivisions of R^2 .

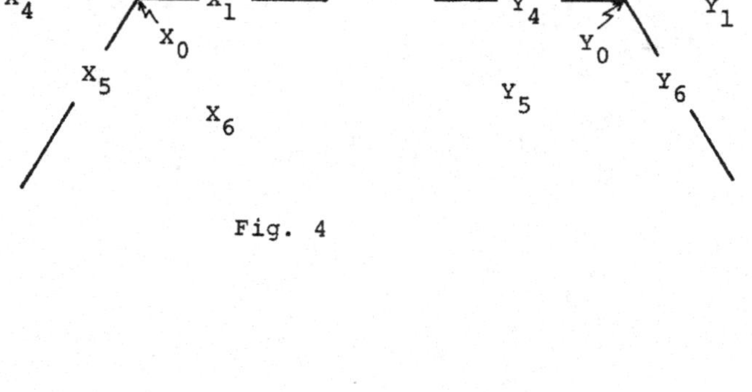

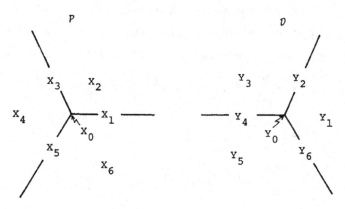

Fig. 4

For a PDM $(\mathcal{P}, \mathcal{D}; d)$ with degree n, we define

(3.6) $$< \mathcal{P}, \mathcal{D}; d > = \{ X \times X^d : X \in \bar{\mathcal{P}} \text{ and } X^d \neq \phi \}.$$

Using (3.3), we immediately see that

(3.6)' $$< \mathcal{P}, \mathcal{D}; d > = \{ Y^d \times Y : Y \in \bar{\mathcal{D}} \text{ and } Y^d \neq \phi \}.$$

One can establish the following properties of a PDM (see Section 3 of [11] for their proof).

<u>Theorem 3.1.</u> Let $(\mathcal{P}, \mathcal{D}; d)$ be a PDM with degree n and $L = < \mathcal{P}, \mathcal{D}; d >$. Suppose that $B = X \times Y$ is an (n-1)-cell of $\bar{L}$, where $X \in \bar{\mathcal{P}}$ and $Y \in \bar{\mathcal{D}}$. If an n-cell C of L has B as its face, i.e., $B < C$ then we have either $C = X \times X^d$ or $C = Y^d \times Y$.

<u>Theorem 3.2.</u> Let $(\mathcal{P}, \mathcal{D}; d)$ be a PDM with degree n. Then $L = < \mathcal{P}, \mathcal{D}; d >$ is a subdivided n-manifold.

<u>Theorem 3.3.</u> Let $(\mathcal{P}, \mathcal{D}; d)$ be a PDM with degree n and $L = < \mathcal{P}, \mathcal{D}; d >$. Suppose that $B = X \times Y$ is an (n-1)-cell of $\bar{L}$, where $X \in \bar{\mathcal{P}}$ and $Y \in \bar{\mathcal{D}}$. Then $B \in \partial L$ if and only if either $X^d = \phi$ or $Y^d = \phi$.

We now give three examples of PDM's with degree n which will be utilized in our succeeding sections.

<u>Example 3.1.</u> Let e^i denote the i-th unit vector of R^n. Define the $2(n+1)$ vectors p^i, d^i $(0 \leq i \leq n)$ as follows:

$$p^0 = -\sum_{j=1}^{n} e^j, \quad p^i = e^i \quad (1 \leq i \leq n),$$

$$d^0 = \sum_{j=1}^{n} e^j,$$

$$d^i = \sum_{j=1}^{n} e^j - (n+1)\, e^i \quad (1 \leq i \leq n).$$

Let $N^* = \{0,\ldots,n\}$. We define, for every nonempty subset I of N^*

$$X(I) = \{ \sum_{i \in I} \lambda_i p^i : \lambda_i \geqq 0 \quad (i \in I) \},$$

$$Y(I) = \{ \sum_{i \in I} \lambda_i d^i : \lambda_i \geqq 0 \quad (i \in I), \sum_{i \in I} \lambda_i = 1 \},$$

and specifically $X(\phi) = \{ 0 \}$. Let

$$\mathcal{P} = \{ X(I) : I \subsetneqq N^* \text{ and } \#I = n \},$$

$$\mathcal{D} = \{ Y(N^*) \},$$

where $\#I$ denotes the number of elements in I. Then we see

$$\bar{\mathcal{P}} = \{ X(I) : I \subsetneqq N^* \text{ or } I = \phi \},$$

$$\bar{\mathcal{D}} = \{ Y(I) : I \subset N^* \}.$$

Defining the dual operator d by

$$X(I)^d = Y(N^* \sim I) \quad \text{for every} \quad I \subsetneqq N^* \quad \text{and} \quad I = \phi,$$

$$Y(I)^d = X(N^* \sim I) \quad \text{for every} \quad I \subset N^*,$$

we obtain a PDM $(\mathcal{P}, \mathcal{D}; d)$ with degree n, where $N^* \sim I = \{ i \in N^* : i \notin I \}$. Fig. 5 illustrates the two dimensional case.

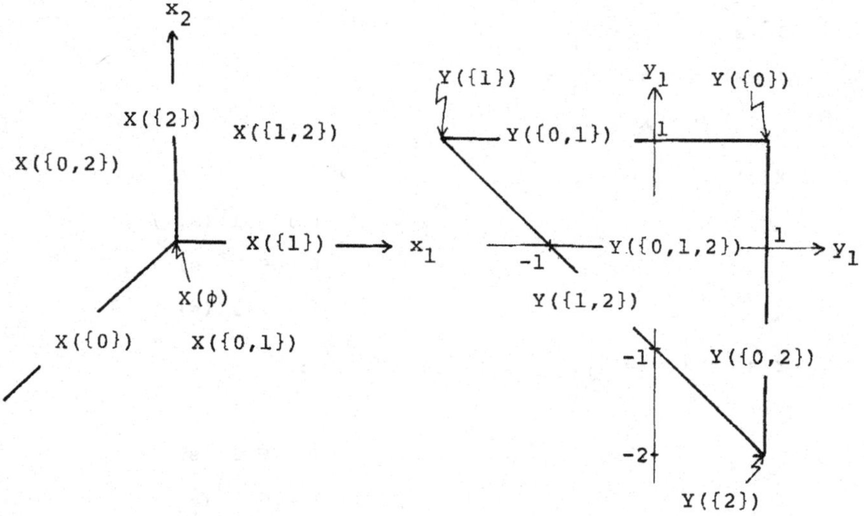

Fig. 5

It can be readily verified that the PDM $(\mathcal{P}, \mathcal{D}; d)$ constructed above satisfies the following properties:

(3.7) $|\mathcal{P}| = R^n$.

(3.8) $|\mathcal{D}| \subset R^n$.

(3.9) there exist some $X_0 \in \bar{\mathcal{P}}$ and $Y_0 \in \bar{\mathcal{D}}$ such that $0 \in \text{int}\, Y_0$, $X_0 = \{ 0 \}$ and $Y_0 = X_0^d$

(3.10) $|L|$ is a closed subset of R^{2n} , where $L = \langle \mathcal{P}, \mathcal{D}; d \rangle$.

(3.11) X^d is nonempty for every $X \in \bar{\mathcal{P}}$.

(3.12) Y^d is nonempty for every $Y \in \bar{\mathcal{D}}$.

(3.13) $\partial |L| = \phi$ (this follows from (3.11) and (3.12)).

This PDM was implicitly used in the vd algorithm proposed by Todd [25], and also has a close relation with the PDM given in Section 5.2 of Kojima and Yamamoto [11].

Example 3.2. Let $N = \{ 1,2,\ldots,n \}$, and let Σ be the set of all n-dimensional vectors $s = (s_1, s_2, \ldots, s_n)$ such that $s_i \in \{ -1, 0, 1 \}$ for every $i \in N$. For each $s \in \Sigma$, define

$$I^-(s) = \{ i : s_i = -1 \} ,$$
$$I^+(s) = \{ j : s_j = 1 \} ,$$
$$I^0(s) = \{ k : s_k = 0 \} ,$$

$$X(s) = \{ x = (x_1, x_2, \ldots, x_n) \in R^n : \quad x_i \leq 0 \quad (i \in I^-(s)) , \quad x_j \geq 0 \quad (j \in I^+(s)) , \quad x_k = 0 \quad (k \in I^0(s)) \} ,$$

$$Y(s) = \{ y = (y_1, y_2, \ldots, y_n) \in R^n : \quad y_i = -1 \quad (i \in I^-(s)) , \quad y_j = 1 \quad (j \in I^+(s)) , \quad -1 \leq y_j \leq 1 \quad (k \in I^0(s)) \} .$$

It follows that $\dim X(s) = n - \#I^0(s)$ and $\dim Y(s) = \#I^0(s)$

for every $s \in \Sigma$; hence $\dim X(s) + \dim Y(s) = n$. Let

$$\mathcal{P} = \{ X(s) : s \in \Sigma \text{ and } I^0(s) = \phi \}$$

$$\mathcal{D} = \{ Y(0) \} .$$

Then we see $\mathcal{P}$ is a subdivision of R^n consisting of all the orthants, and that

$$\bar{\mathcal{P}} = \{ X(s) : s \in \Sigma \} ,$$

$$\bar{\mathcal{D}} = \{ Y(s) : s \in \Sigma \} ,$$

Defining the dual operator d by

$$X(s)^d = Y(s) \text{ for every } s \in \Sigma ,$$

$$Y(s)^d = X(s) \text{ for every } s \in \Sigma ,$$

we obtain a PDM $(\mathcal{P}, \mathcal{D}; d)$ with degree n . It is easily verified that all the properties (3.7) - (3.13) are satisfied. Fig. 6 illustrates the two dimensional case. This PDM was implicitly used in van der Laan and Talman [18].

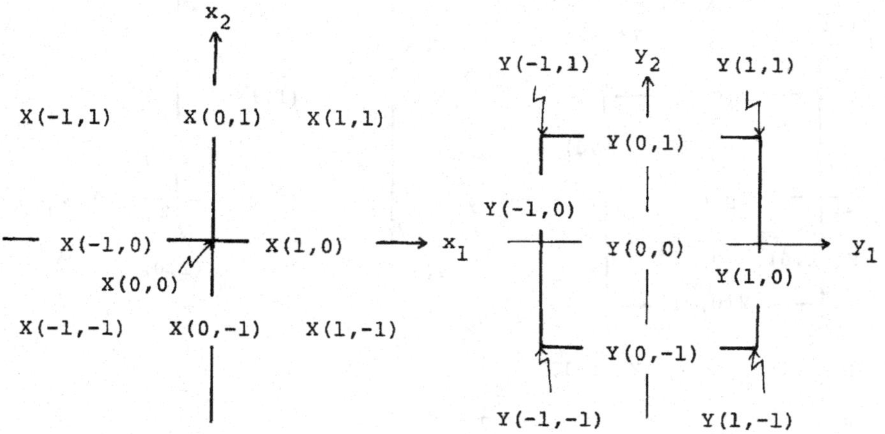

Fig. 6

Example 3.3. Let Q denote the set of all n-dimensional vectors $q = (q_1, q_2, \ldots, q_n)$ whose components are integer. For every $q \in Q$, let

$$I_e(q) = \{ i : q_i \text{ is even } \},$$

$$I_o(q) = \{ j : q_j \text{ is odd } \},$$

$$X(q) = \{ x = (x_1, x_2, \ldots, x_n) \in R^n : \quad x_i = q_i \quad (i \in I_e(q)), \quad q_j - 1 \leqq x_j \leqq q_j + 1 \quad (j \in I_o(q)) \},$$

$$Y(q) = \{ y = (y_1, y_2, \ldots, y_n) \in R^n : \quad q_i - 1 \leqq y_i \leqq q_i + 1 \quad (i \in I_e(q)), \quad y_j = q_j \quad (j \in I_o(q)) \}.$$

Fig. 7 illustrates a two dimensional case. We obviously see that

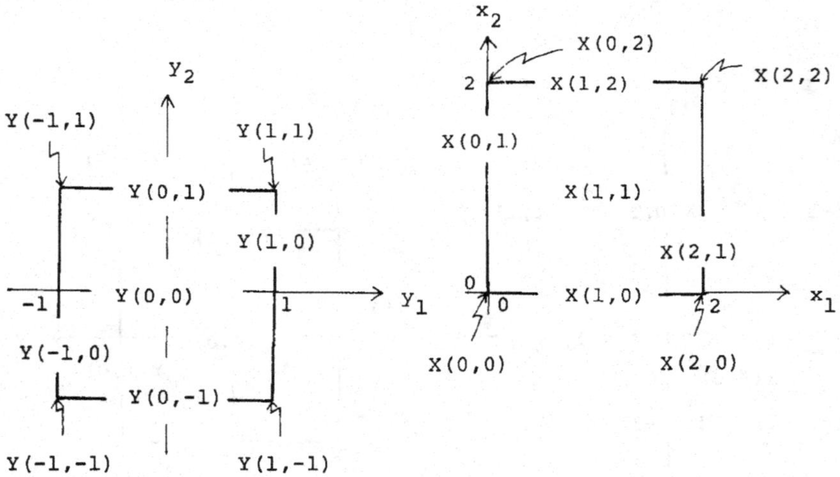

Fig. 7

$\dim X(q) = \# I_o(q)$ and $\dim Y(q) = \# I_e(q)$ for every $q \in Q$. Hence, $\dim X(q) + \dim Y(q) = n$ for every $q \in Q$. Specifically, if all the components of $q \in Q$ are odd (or even), $X(q)$ (or $Y(q)$) turns out to be an n-dimensional hypercube with the center q, i.e., $X(q)$ (or $Y(q)$) $= \{ x \in R^n : q_i - 1 \leq x_i \leq q_i + 1 \ (1 \leq i \leq n) \}$. Thus, the collection of such $X(q)$'s , i.e.,

(3.14) $\quad \mathcal{P} = \{ X(q) : q \in Q \text{ and } I_e(q) = \phi \}$

and the collection of such $Y(q)$'s , i.e.,

(3.15) $\quad \mathcal{D} = \{ Y(q) : q \in Q \text{ and } I_o(q) = \phi \}$

are both hypercubical subdivisions of R^n with grid size 2 . When we take $n = 2$, $\mathcal{P}$ and $\mathcal{D}$ are just like checkerboards. So we call them checkerboard subdivisions.

It follows directly from the definitions above that

$$\bar{\mathcal{P}} = \{ X(q) : q \in Q \},$$

$$\bar{\mathcal{D}} = \{ Y(q) : q \in Q \}.$$

To complete a PDM $(\mathcal{P}, \mathcal{D}; d)$ with degree n , we finally define the dual operator d by

$$X(q)^d = Y(q) \quad \text{for every } q \in Q,$$

$$Y(q)^d = X(q) \quad \text{for every } q \in Q.$$

This PDM also satisfies the properties (3.7) - (3.13).

The PDM given in this example is due to Kojima and Yamamoto [12] where they derived it from the Union Jack triangulation of $R^n \times [0, 1]$ (Todd [24]) which has been often used in Merrill's method [19]. The definition here is slightly different from the original one; $\mathcal{P}$ defined by (3.14) corresponds to the dual subdivided manifold given in Section 4.1 of [12], and $\mathcal{D}$ by (3.15) to the primal one.

4. PIECEWISE SMOOTH VD ALGORITHMS.

We are now ready to describe the class of piecswise smooth vd algorithms outlined in Introduction. Consider the system of equations:

$$(4.1) \qquad f(x) = 0 , \quad x \in R^n .$$

We shall assume that $f : R^n \to R^n$ is a C^1-map. Let $(\mathcal{P}, \mathcal{D}; d)$ be a PDM with degree n which satisfies the conditions (3.7) - (3.13). As we have seen, we can employ all the PDM's given in Examples 3.1 - 3.3. Let $\mathcal{L} = < \mathcal{P}, \mathcal{D}; d>$, and

$$(4.2) \qquad \mathcal{M} = \{ Z \times R_+ : Z \in \mathcal{L} \} .$$

Then we see that $\mathcal{M}$ is a subdivided (n+1)-manifold in R^{2n+1} . Since $\partial|\mathcal{L}| = \phi$,

$$(4.3) \qquad \partial|\mathcal{M}| = |\mathcal{L}| \times \{ 0 \}$$

holds. Let $w^0 \in R^n$. When a rough approximation of a solution of the system (4.1) is known, we take it as w^0 . We now define the PC^1-map $h : |\mathcal{M}| \to R^n$ on $\mathcal{M}$ by

$$(4.4) \qquad h(x,y,t) = y + t f(w^0+x) \quad \text{for every} \quad (x,y,t) \in |\mathcal{M}| ,$$

and consider the system of equations

$$(4.5) \qquad h(x,y,t) = 0 , \quad (x,y,t) \in |\mathcal{M}| .$$

Let $(x^0,y^0,t^0) = (0,0,0) \in R^{2n+1}$. By (3.9), we have $(x^0,y^0) \in X_0 \times Y_0 \in \mathcal{L}$. Hence, (x^0,y^0,t^0) is a solution of the system (4.5). This solution will serve as a starting point for our vd algorithm. In the remainder of the paper, we shall assume that 0 is a regular value of the PC^1-map $h : |\mathcal{M}| \to R^n$.

<u>Remark.</u> If $f : R^n \to R^n$ is C^2 then, by using Sard's theorem

(see, for example, Ortega and Rheinboldt [20]), we can prove that almost every $c \in R^n$ is a regular value of the PC^1-map h. Hence, if the assumption was not satisfied, we could choose a $c \in R^n$ which lies in int Y_0 (see (3.9)) and is a regular value of h. In such a case, the initial point $(0,0,0)$ would be replaced by $(0,c,0)$.

By Theorem 2.1, we see that the solution set $h^{-1}(0)$ of the system (4.5) is a disjoint union of paths and loops. Let S denote the connected component of $h^{-1}(0)$ which contains the point (x^0,y^0,t^0). Since $(x^0,y^0,t^0) = (0,0,0)$ lies on the boundary of $|M|$, S must be a path which is homeomorphic to either $[0, 1]$ or $[0, 1)$.

We shall show that S is homeomorphic to $[0, 1)$. Assume on the contrary that it is homeomorphic to $[0, 1]$. Then the path S has an endpoint $(\bar{x},\bar{y},\bar{t})$ which is distinct from (x^0,y^0,t^0). By (2.1) of Theorem 2.1, we have $(\bar{x},\bar{y},\bar{t}) \in \partial|M|$, and by (4.3) $\bar{t} = 0$. On the other hand, it follows from $(\bar{x},\bar{y},\bar{t}) \in S \subset h^{-1}(0)$ that $h(\bar{x},\bar{y},\bar{t}) = \bar{y} + \bar{t}\, f(W^0+\bar{x}) = 0$; $\bar{y} = 0$. Since $\bar{y} = 0$ lies in the interior of Y_0, Y_0 is the unique cell of $\mathcal{D}$ which contains $\bar{y} = 0$. Hence $(\bar{x},\bar{y},\bar{t}) \in |M|$ or $(\bar{x},\bar{y}) \in |L|$ implies $(\bar{x},\bar{y}) \in Y_0^d \times Y_0 = X_0 \times Y_0 = \{ 0 \} \times Y_0$; $\bar{x} = 0$. Thus we have shown $(\bar{x},\bar{y},\bar{t}) = (0,0,0)$, a contradiction to $(\bar{x},\bar{y},\bar{t}) \neq (x^0,y^0,t^0)$. Therefore S is homeomorphic to $[0, 1)$.

By (3.10), we further observe that $|M|$ is closed. Hence, by the last assertion of Theorem 2:1, we see that the path S is unbounded. Suppose that there exists a bounded open subset U of R^{2n} such that

$$(x^0,y^0) = (0,0) \in U \quad \text{and} \quad S \subset U \times R_+ \tag{4.6}$$

Since the path S is unbounded, for every $t > 0$, there is an $(x,y) \in R^{2n}$ such that

$$(x,y,t) \in S \subset h^{-1}(0) \quad \text{or} \quad f(w^0+x) = -y/t .$$

Hence, if $t > 0$ is sufficiently large then w^0+x is an approximate solution of system (4.1). Thus, in this case, tracing the path S from the initial point (x^0,y^0,t^0), we can compute approximate solutions of (4.1).

In the succeeding sections, we shall give a sufficient condition on the map $f : R^n \to R^n$ for the existence of a bounded open subset U of R^{2n} satisfying (4.6). The remainder of this section is devoted to showing a certain variable dimension property of the path S. For this purpose, we first decompose the system (4.5) into the family of systems of equations (see (4.2)):

(4.7-X) $\quad y + t\, f(w^0+x) = 0 \; , \; (x,y,t) \in X \times X^d \times R_+$

$(X \in \bar{P})$. This decomposition naturally induces a decomposition of the path S into a finite or countable number of smooth pieces S_p $(p \geq 0)$. Namely,

$$S = \cup S_p ,$$

S_p is a connected component of $S \cap \{ X_p \times X_p^d \times R_+ \}$ for some $X_p \in \bar{P}$,

(4.8) $\quad (x^0,y^0,t^0) \in S_0$,

$S_{p-1} \cap S_p = (x^p,y^p,t^p)$ for some $(x^p,y^p,t^p) \in R^{2n+1}$ and every $p \geq 1$,

$X_{p-1} \neq X_p$.

Here the union of the first equality is taken from 0 to a finite number or to the infinity. If $p \geq q + 2$, it can happen that $X_p = X_q$.

(a)

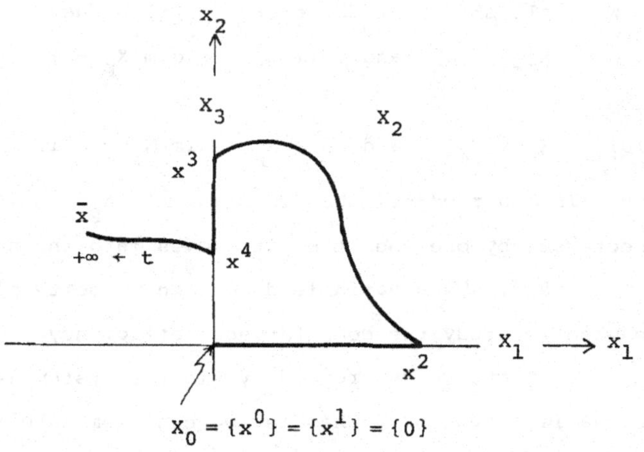

(b)

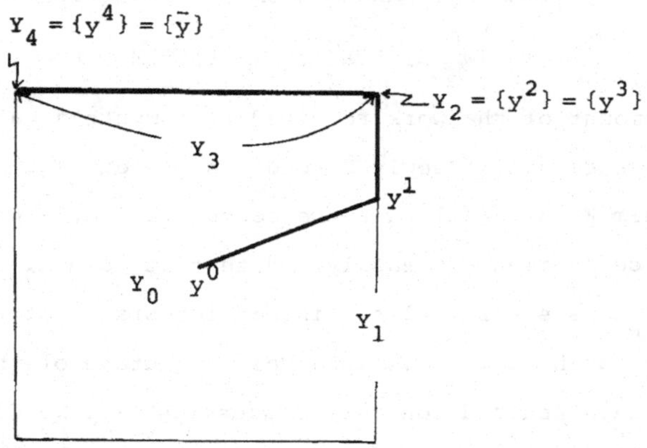

Fig. 8

Fig. 8 shows a case where we have chosen the PDM with degree 2 given in Example 3.2 as the PDM under consideration. The thick lines of (a) and (b) illustrates the projection of the path S on the X_p's and Y_p's, respectively.

We shall show that for every $p \geqq 1$ either

(4.9) $\quad X_{p-1} < X_p$ and $\dim X_{p-1} = \dim X_p - 1$

or

(4.10) $\quad X_p < X_{p-1}$ and $\dim X_p = \dim X_{p-1} - 1$

holds. In other words, the dimension of X_{p-1} either increases or decreases by one when the path moves into the new cell $X_p \times X_p^d \times R_+$ of M . This variable dimension property plays an essential role when we study the computational efficiency of the vd algorithm. In fact, we can regard y in the system (4.7-X) as a slack variable vector. That is, we can eliminate the variable vector y from the system to obtain an equivalent one

$$- t\, f(w^0+x) \in X^d , \quad (x,t) \in X \times R_+ .$$

The amount of the work required to compute a path of solutions of the system above deeply depends on the dim X . One can say that if dim X is small then the computation of a path of solutions will be carried out easily. Note that $\dim X_0 = 0$. Hence, the $\dim X_p$ is small at least in earlier stages of the computation of the path S . This is a main advantage of the vd algorithms. More detailed and concrete discussion will be done in Section 6.

For simplicity of notation, let

$$z^p = (x^p, y^p, t^p) \quad \text{and} \quad Z_p = X_p \times X_p^d \times R_+ \quad \text{for every} \quad p \geqq 0 .$$

It follows from (4.8) that $z^p \in S$ lies in a common proper face of $Z_p \cap Z_{p-1}$ of the (n+1)-cells Z_{p-1} and Z_p of the subdivided (n+1)-manifold M . On the one hand, we have been assuming that 0 is a regular value of the PC^1-map $h : |M| \to R^n$, so that $S \subset h^{-1}(0)$ does not meet any cell of $\bar{M}$ with dimension less than n. Hence we obtain

$$\dim Z_{p-1} \cap Z_p = n \quad \text{for every} \quad p \geqq 1.$$

Let $p \geqq 1$ be fixed, and $Z = Z_{p-1} \cap Z_p$. Then Z can be written as

$$Z = X \times Y \times R_+$$

where $X = X_{p-1} \cap X_p$ and $Y = X^d_{p-1} \cap X^d_p$. Furthermore we see that

$$\begin{aligned} &X < X_{p-1}\ , \quad X < X_p\ , \quad Y < X^d_{p-1}\ , \quad Y < X^d_p\ , \\ (4.11) \qquad &\dim X + \dim Y = n - 1\ , \\ &\dim X_{p-1} + \dim X^d_{p-1} = \dim X_p + \dim X^d_p = n\ . \end{aligned}$$

Hence $X \times Y \in \bar{L}$ is an (n-1)-cell and a common face of the two cells $X_{p-1} \times X^d_{p-1}$ and $X_p \times X^d_p$ of L . By Theorem 3.1, we obtain either

$$(4.12) \qquad X_{p-1} = X \quad \text{and} \quad X_p = Y^d$$

or

$$(4.13) \qquad X_{p-1} = Y^d \quad \text{or} \quad X_p = X\ .$$

Assume that (4.12) holds. Taking (4.11) and $X_{p-1} \neq X_p$ into account, we see that (4.9) holds. Similarly, if (4.13) occurs then we have (4.10). Thus we have shown that for every $p \geqq 1$ either (4.9) or (4.10) holds.

5. GLOBAL CONVERGENCE

In this section, we give a condition which ensures the existence of a bounded open subset U of R^{2n} such that the path $S \subset h^{-1}(0)$ starting from $(x^0,y^0,t^0) = (0,0,0)$ is contained in $U \times R_+$. As we have shown in the previous section, under such a condition, we can get an approximate solution of the system $f(x) = 0$ by tracing the path S . First we need to introduce two assumptions on the PDM's which we employ in defining the subdivided (n+1)-manifold M and the PC^1-map $h : |M| \to R^n$ (see (4.2) and (4.4)). One is that

(5.1) there exist α , $\beta > 0$ such that $x \cdot y \geq \alpha \| x \| \cdot \| y \|$ for every $(x,y) \in |L|$ satisfying $\| x \| \geq \beta$, where $L = < \mathcal{P}, \mathcal{D}; d >$.

The other one is that

(5.2) if $W \subset R^n$ is bounded then so is the set $\{ (x,y) \in |L| : x \in W \}$.

It is easily verified that these two assumptions are satisfied by the PDM's given in Examples 3.1 - 3.3.

Theorem 5.1. In addition to (3.7) - (3.13), suppose that (5.1) and (5.2) hold. Suppose that for some $\mu > 0$ and every $x \in R^n$ with $\| x \| \geq \mu$ there exists an $\hat{x} \in R^n$ such that

$$\| \hat{x} \| \leq \mu \quad \text{and} \quad (x - \hat{x}) \cdot f(x) > 0$$

(a weaker version of Merrill's condition, see [19]). Then there exists a bounded open subset U of R^{2n} such that $U \times R_+$ contains the path S .

Proof. Let

(5.3) $$\gamma = \max \{ \| w^0 \| + \mu , \ (\| w^0 \| + \mu) / \alpha , \ \beta \},$$
$$V = \{ x \in R^n : \| x \| \leq \gamma \}.$$

By (5.2), we can find a bounded open set $U \subset R^{2n}$ such that

(5.4) $$\{ (x,y) \in |L| : x \in V \} \subset U .$$

Assume on the contrary that the path S does not lie in the set $U \times R_+$. Then we can find a point $(x,y,t) \in S$ such that $(x,y) \notin U$ and $t > 0$. It follows from $(x,y,t) \in S \subset h^{-1}(0)$ that $(x,y) \in |L|$. Hence, by (5.4), we see $x \notin V$ or $\| x \| > \gamma \geq \beta$, and that

$$\| w^0 + x \| \geq \| x \| - \| w^0 \| \geq \gamma - \| w^0 \| \geq \mu .$$

The last inequality follows from (5.3). By the assumption of the theorem, corresponding to the point $w^0 + x$ with $\| w^0 + x \| \geqq \mu$, there exists an $\hat{x} \in R^n$ such that

$$\| \hat{x} \| \leqq \mu \quad \text{and} \quad (w^0 + x - \hat{x}) \cdot f(w^0+x) > 0 .$$

It also follows from $(x,y,t) \in S$ that

$$y = - t\, f(w^0+x) .$$

Thus we see

$$\begin{aligned} 0 &< (w^0 + x - \hat{x}) \cdot t\, f(w^0+x) && (\text{since } t > 0) \\ &= (w^0 + x - \hat{x}) \cdot (-y) \\ &= x \cdot (-y) + (w^0 - \hat{x}) \cdot (-y) \\ &\leqq - \alpha \| x \| \cdot \| y \| + \| w^0 - \hat{x} \| \| y \| && (\text{by } (5.1)) \\ &\leqq \{ - \alpha \gamma + \| w^0 \| + \mu \} \| y \| \\ & \qquad\qquad (\text{since } \| \hat{x} \| \leqq \mu \text{ and } \| x \| \geqq \gamma) \\ &\leqq 0 && (\text{by } (5.3)). \end{aligned}$$

This is a contradiction. Q.E.D.

6. COMPUTATION OF THE PIECEWISE SMOOTH PATH S

To make our argument simple and concrete, we shall confine ourselves to a special case where we employ the PDM $(\mathcal{P}, \mathcal{D}; d)$ given in Example 3.2 and the subdivided n-manifold induced from it in the definitions of the subdivided (n+1)-manifold M and the PC^1-map $h : |M| \to R^n$ (see (4.2) and (4.4)). Furthermore, we may assume without loss of generality that $w^0 = 0$, because if $w^0 \neq 0$, we only have to replace the map $f : R^n \to R^n$ by the map $f' : R^n \to R^n$ defined by $f'(x) = f(w^0+x)$ for every $x \in R^n$. Thus we have

$$L = \{ X(s) \times Y(s) : s \in \Sigma \} ,$$

$$h(x,y,t) = y + t\, f(x) \quad \text{for every } (x,y,t) \in |L| \times R_+ ,$$

and the family of systems (4.7-X) ($X \in P$) , which is a decomposition of the system $h(x,y,t) = 0$, turns out to be

$$(6.1\text{-}s) \qquad y + t\, f(x) = 0 , \quad (x,y,t) \in X(s) \times Y(s) \times R_+$$

($s \in \Sigma$) . See Example 3.2 for the definitions of Σ , $X(s)$ and $Y(s)$. As we have shown in Section 4, the path $S \subset h^{-1}(0)$ starting from $z^0 = (x^0,y^0,t^0) = (0,0,0)$ consists of a finite or countable number of smooth pieces S_p ($p \geqq 0$) , each of which is a path consisting of solutions to (6.1-s) for some $s \in \Sigma$ (see (4.8)). Therefore, for the computation of the path S , it suffices to show how to approximate each piece S_p .

Let p be fixed, and s an element of Σ such that S_p is a smooth path consisting of solutions to (6.1-s). The path S_p has an endpoint $z^p = (x^p,y^p,t^p)$ in the relative interior of some n-dimensional face Z of $X(s) \times Y(s) \times R_+$. When $p = 0$, we see $z^0 = (0,0,0)$ and $Z = X(0) \times Y(0) \times \{0\}$. For simplicity, we assume that

$$I^-(s) = \{ i : s_i = -1 \} = \{ 1,2,\dots,\ell \} ,$$

$$I^+(s) = \{ j : s_j = 1 \} = \{ \ell+1,\dots,m \}$$

$$I^0(0) = \{ k : s_k = 0 \} = \{ m+1,\dots,n \} .$$

Then the set $X(s) \times Y(s)$ is written as the set of solutions of the linear inequalities

$$- x_i \geqq 0 , \qquad y_j = -1 \quad (1 \leqq i \leqq \ell) ,$$

$$x_j \geqq 0 , \qquad y_j = 1 \quad (\ell+1 \leqq j \leqq m) ,$$

$$x_k = 0 , \quad - 1 \leqq y_k \leqq 1 \quad (m+1 \leqq k \leqq n) .$$

Eliminating the variables which are fixed to -1 , 0 or 1 and the slack variables y_k ($m+1 \leqq k \leqq n$) , we can transform the system (6.1-s) into the system consisting of equalities and inequalities:

$$(6.2)\quad \begin{cases} \text{(a)} & \begin{cases} -1 + tf_i(x_1,\dots,x_m,0,\dots,0) = 0 & (1 \le i \le \ell), \\ 1 + tf_j(x_1,\dots,x_m,0,\dots,0) = 0 & (\ell+1 \le j \le m), \end{cases} \\ \text{(b)} & \begin{cases} -x_i \ge 0 & (1 \le i \le \ell), \\ x_j \ge 0 & (\ell+1 \le j \le m), \\ 1 - tf_k(x_1,\dots,x_m,0,\dots,0) \ge 0 & (m+1 \le k \le n), \\ 1 + tf_k(x_1,\dots,x_m,0,\dots,0) \ge 0 & (m+1 \le k \le n), \\ t \ge 0 . \end{cases} \end{cases}$$

There is a natural one-to-one corespondence between solutions of (6.1-s) and solutions of (6.2). That is, a solution (x,y,t) of (6.1-s) corresponds to the solution $u = (x_1,\dots,x_m,t)$ of (6.2), and conversely, a solution $u' = (x_1',\dots,x_m',t')$ of (6.2) to the solution (x,y,t) of (6.1-s) determined by

$$\begin{aligned} &x_i = x_i', \quad y_i = -1 && (1 \le i \le \ell), \\ &x_j = x_j', \quad y_j = 1 && (\ell+1 \le j \le m), \\ &x_k = 0, \quad y_k = -t'f(x_1',\dots,x_m',0,\dots,0) && (m+1 \le k \le n), \\ &t = t' . \end{aligned}$$

Hence the path S_p corresponds to a path

$$T_p = \{ u = (x_1,\dots,x_m,t) \in R^{m+1} : (x,y,t) \in S_p \},$$

which consists of solutions to (6.2). Specifically, the endpoint $z^p = (x^p,y^p,t^p)$ of S_p corresponds to the endpoint $u^p = (x_1^p,\dots,x_m^p,t^p)$ of T_p. In what follows, we shall assume that u^p (or an approximation of u^p) is known and show how to approximate the path T_p starting from it.

We denote the left side equations (a) of (6.2) by $g : R^{m+1} \to R^m$, i.e.,

$$g_i(u) = -1 + u_{m+1} f_i(u_1,\dots,u_m,0,\dots,0) \quad (1 \le i \le \ell),$$

$$g_j(u) = 1 + u_{m+1} f_j(u_1,\dots,u_m,0,\dots,0) \quad (\ell+1 \le j \le m)$$

for every $u = (u_1,\dots,u_m,u_{m+1})$. Then (a) can be rewriten as

(a)' $\quad g(u) = 0$.

Let V denote the solution set of the inequalities (b) of (5.2). Note that $g : R^{m+1} \to R^m$ is C^1. By using the assumption that 0 is a regular value of the PC^1-map $h : |M| \to R^m$, we can easily prove that 0 is a regular value of the map $g : R^{m+1} \to R^m$ on some open neighborhood V' of V . Hence, by applying Theorem 2.1, we see $g^{-1}(0) \cap V'$ is a disjoint union of paths and loops. Apparently, the path T_p is a part of a path (or loop), denoted by $\bar{T}_p$, of $g^{-1}(0) \cap V'$. See Fig. 9.

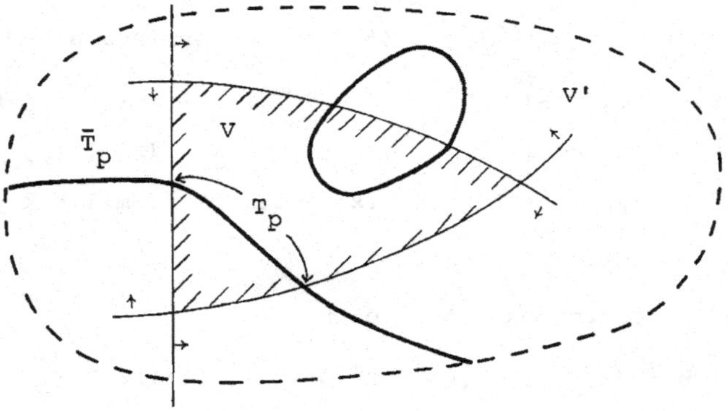

Fig. 9

Since g is C^1, we can apply the well-known the predictor-corrector procedure (see, for example, [3]) to the system (a)' for tracing the path $T_p \subset \bar{T}_p$. The initial point will be $v^0 = u^p \in T_p$. We need to calculate the initial predictor direction vector $d^0 \in R^{m+1}$, which must satisfy the two requirements. One is that d^0 is a tangent vector of the path $\bar{T}_p$ at the point v^0

$\in \bar{T}_p$. Hence, it satisfies

$$\nabla g_i(v^0) \cdot d^0 = 0 \quad (1 \leqq i \leqq m) ,$$

$$\| d \| = 1 .$$

The other requirement is that the vector d^0 point into the interior of the set V determined by the inequalities (b) of (5.2). Recall that $z^p = (x^p, y^p, t^p)$ lies in the relative interior of the n-dimensional face Z of $X(s) \times Y(s) \times R_+$. This implies that exactly one of the inequalities (b) of (6.2) holds with equality at $v^0 = u^p = (x_1^p, \ldots, x_m^p, t^p)$. We denote that inequality by $a(u) \geqq 0$, where $u = (x_1, \ldots, x_m, t) \in R^{m+1}$. Then the second requirement turns to be

$$(6.4) \qquad \nabla a(v^0) \cdot d^0 > 0$$

From the assumption that 0 is a regular value of the PC^1-map $h : |M| \to R^n$, we can derive that the set of the $m + 1$ gradient vectors $\nabla g_i(v^0)$ $(1 \leqq i \leqq m)$ and $\nabla a(v^0)$ is linearly independent. Hence the conditions (6.3) and (6.4) uniquely determine the initial direction vector d^0 . See Fig. 10.

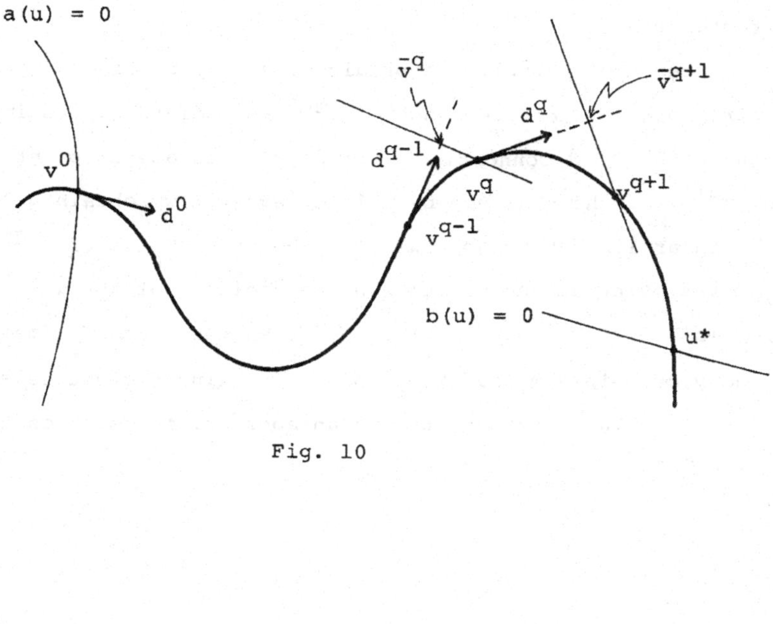

Fig. 10

Generally, the q-th predictor direction d^q ($q \geqq 1$) is determined by

$$\nabla g_i(v^q) \cdot d^q = 0 \quad (1 \leqq i \leqq m) ,$$

$$d^{q-1} \cdot d^q > 0 ,$$

$$\| d^q \| = 1 ,$$

where v^q denotes the q-th point (approximately) on the path T_p generated by the (q-1)-th corrector step which will be described below. Let

$$\bar{v}^{q+1} = v^q + \delta_q d^q ,$$

where $\delta_q > 0$ is a step length.

The point $\bar{v}^{q+1}$ gives us a rough approximation of the (q+1)-th point on the path T_p . Taking $\bar{v}^{q+1}$ as an initial point, we then apply Newton-Raphson method to the system of equations

$$g_i(v) = 0 \qquad (1 \leqq i \leqq m)$$

$$d^q \cdot (v - \bar{v}^{q+1}) = 0$$

to get a point v^{q+1} (approximately) on the path T_p (the corrector step).

We repeatedly apply the predictor step and the corrector step in turn to generate a sequence $\{ v^q \}$ of points (approximately) on the path T_p . Connecting each pair of adjacents points v^q and v^{q+1} of the sequence by a line segment, we obtain a piece-wise linear line which approximates the path T_p . If v^{q+1} is the first point of the sequence which violates of the inequalities (b) of (6.2), say $b(u) \geqq 0$, then the path T_p meet the boundary of V in some neighborhood of the line segment between v^q and v^{q+1} . In this case, taking an appropriate point on the line segment as an initial point, we shall apply Newton-Raphson method to the sytem

$$g_i(u) = 0 \qquad (1 \leq i \leq m) , \qquad b(u) = 0$$

to get an approximate solution u^* . The point u^* is an approximation of the endpoint u^{p+1} of the path T_p which corresponds to the endpoint $(x^{p+1}, y^{p+1}, t^{p+1})$ of the path S_p . Thus the path S will move into a new piece $X(s') \times Y(s') \times R_+$ of M for some $s' \in \Sigma$. To compute the path S_{p+1} , we shall replace (6.2) by the new system of equalities and inequalities associated with the $s' \in \Sigma$ and apply the same procedure to the new system.

It should be noted that the size of the system of equations solved in the predictor step or the corrector step is small when $m = \dim X(s)$ is small. Therefore, if $\dim X(s)$ is small, the above procedure will be carried out very efficiently. This is a main advantage of the vd algorithm. Since we start from $X(0) = \{ 0 \}$, the earlier stages of the vd algorithm always enjoy this advantage.

The computational procedure outlined above for tracing the path S is a prototype. Recently, there have been developed some techniques (Georg [7], Kearfott [8], Saupe [22]) to increase the computational efficiency of smooth continuation methods. In order to develop a more efficient computational procedure, we need to employ those techniques.

7. SIMPLICIAL VD ALGORITHMS

So far we have assumed the continuous differentiability of the map $f : R^n \rightarrow R^n$, and studied a class of piecewise smooth vd algorithms for solving the system $f(x) = 0$. These vd algorithms can be transformed into simplicial vd algorithms as will be shown below.

Let $(\mathcal{P}, \mathcal{D}; d)$ be a PDM with degree n which satisfies the conditions (3.7) - (3.13). Let $\mathcal{Q}$ be a simplicial refinement of the subdivided n-manifold $\mathcal{P}$; $\mathcal{Q}$ is a subdivision of $|\mathcal{P}| = R^n$ such that each $\sigma \in \mathcal{Q}$ is an n-dimensional simplex contained in some $X \in \mathcal{P}$. See Fig. 11. Then we construct a PL (piecewise linear) approximation $F : |\mathcal{Q}| \to R^n$ of the map $f : R^n \to R^n$ on the simplicial subdivision $\mathcal{Q}$ of R^n . That is, for each $x \in \sigma \in \mathcal{Q}$, we define

$$F(x) = \sum_{i=0}^{n} \lambda_i f(v^i) ,$$

where v^i $(0 \leq i \leq n)$ are vertices of the n-simplex σ and

$$x = \sum_{i=0}^{n} \lambda_i v^i , \quad \sum_{i=0}^{n} \lambda_i = 1 , \quad \lambda_i \geq 0 \quad (0 \leq i \leq n)$$

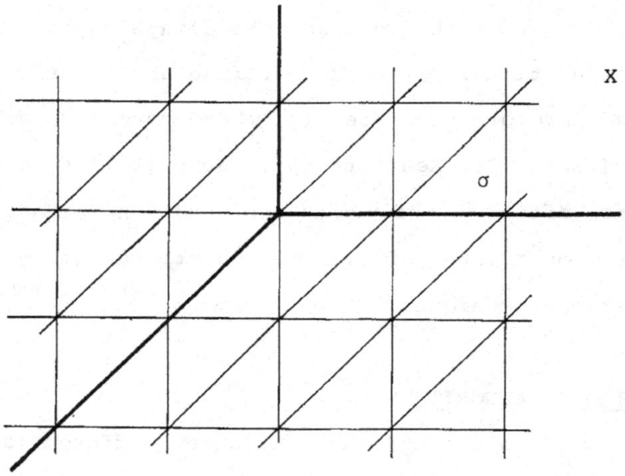

Fig. 11. A refinement of $\mathcal{P}$

On the other hand, the simplicial refinement $\mathcal{Q}$ of P naturally induces a simplicial subdivision of each $X \in \bar{P}$, which we will denote by $\mathcal{Q} \mid X$. Specifically, $\mathcal{Q} \mid X_0 = \{ 0 \}$. Define a refinement L' of $L = < P , \mathcal{D}; d > = \{ X \times X^d : X \in P \}$ by

$$L' = \{ \tau \times X^d : \tau \in \mathcal{Q} \mid X , \ X \in P \} ,$$

and a refinement M' of $M = \{ Z \times R_+ : Z \in L \}$ by

$$M' = \{ \sigma \times R_+ : \sigma \in L' \} .$$

We assume $w^0 = 0$ as in the previous section. Now we replace the map $f : R^n \to R^n$ appeared in the definition (4.4) of PC^1-map $h : |M| \to R^n$ by $F : |\mathcal{Q}| \to R^n$ to obtain

$$H(x,y,t) = y + t\,F(x) \quad \text{for every} \quad (x,y,t) \in |M'| \times R_+ ,$$

and consider the system of equations

$$(7.1) \qquad H(x,y,t) = 0 , \quad (x,y,t) \in |M'| \times R_+ .$$

The mapping $H : |M'| \to R^n$ defined above is PC^1 (but not PL generally) on the subdivided (n+1)-manifold M'. Suppose that 0 is a regular value of the PC^1-map $H : |M'| \to R^n$. Then, by Theorem 2.1, the solution set $H^{-1}(0)$ of the system (7.1) is a disjoint union of paths and loops. Since $(0, 0) \in (\bar{\mathcal{Q}} \mid X_0) \times Y_0 \in L'$, $(x^0,y^0,t^0) = (0,0,0)$ is a trivial solution of (7.1). The path which will be traced by the simplicial vd algorithm is the one starting from (x^0,y^0,t^0). We will denote it by S'. The path S' can be regarded as an approximation of the path S consisting of solutions to (4.5) when $f : R^n \to R^n$ is C^1. In order to ensure the convergence of the path to $(x',y',+\infty)$, where x' is a solution to $F(x) = 0$ (hence, an approximate solution of $f(x) = 0$), we need to assume slightly stronger conditions than the ones in Theorem 5.1.

The system (7.1) itself is not PL (but PC^1 generally); hence so is the path S' . However, by using a similar argument as given in Section 4.2 of [12], we can convert this system into a system of PL equations, and represent the resultant PL system as a family of systems of linear equations, to which we can apply the complementary pivoting for generating a path of solutions of the PL system which corresponds to the path S' . Here we shall derive the family of systems of linear equations directly from the system (7.1).

By the construction, we can decompose the system into the family of systems of equations

$$(7.2\text{-}\tau) \qquad y + t\, f(x) = 0\ , \quad (x,y,t) \in \tau \times X^d \times R_+$$

($\tau \in Q \mid X$, $X \in \bar{P}$) . Let $\tau \in Q \mid X$, $X \in P$ be fixed. We represent each $x \in \tau$ as a convex combination of the vertices v^i ($0 \leqq i \leqq m$) of τ , where $\dim \tau = m$, i.e.,

$$x = \sum_{i=0}^{m} \lambda_i\, f(v^i)\ , \quad \sum_{i=0}^{m} \lambda_i = 1\ , \quad \lambda_i \geqq 0 \quad (0 \leqq i \leqq m) .$$

Then the system (7.2-τ) can be written as

$$y + t \sum_{i=0}^{m} \lambda_i\, f(v^i) = 0\ , \qquad \sum_{i=0}^{m} \lambda_i = 1$$

$$\lambda_i \geqq 0 \quad (0 \leqq i \leqq m)\ , \qquad y \in X^d\ , \quad t \geqq 0$$

Furthermore, letting $\mu_i = t\,\lambda_i$ ($0 \leqq i \leqq m$) , we obtain the system of linear equations

$$(7.3\text{-}\tau) \qquad y + \sum_{i=0}^{m} \mu_i\, f(v^i) = 0\ , \quad \mu_i \geqq 0 \quad (0 \leqq i \leqq m)\ , \quad y \in X^d .$$

Therefore a path of solutions of the system (7.2-τ) corresponds to a path of solutions of (7.3-τ) and vice versa. We can apply the complementary pivoting to the family of systems (7.3-τ)

$(\tau \in \mathcal{Q} \mid X , X \in \bar{P})$, where v^i $(0 \leq i \leq m)$ are the vertices of τ and $m = \dim \tau$ for tracing the path S' . The details are omitted here.

If we employ the PDM's given in Examples 3.1 - 3.3, the simplicial vd algorithm described above is essentially equivalent to the (n+1) algorithm by Todd [25], the 2n algorithm by van der Laan and Talman [18] and the checkerboard algorithm by Kojima and Yamamoto [12], respectively. See Kojima and Yamamoto [11,12] for more detailed discussion on simplicial vd algorithms with the use of PDM's.

ACKNOWLEDGMENT

This research was partially done while the author was visiting University of Bremen in the summer of 1980. He wishes to thank Professor Dr. H. O. Peitgen for his friendly support. He is also indebted to Mrs. A. Tsukada and Mr. T. Akita for their excellent typing.

REFERENCES

[1] J. C. Allexander, "The topological theory of an embedding method", in: H. Wacker, ed., Continuation Methods (Academic Press, New York, 1978) pp.36-67.

[2] J. C. Allexander and J. A. Yorke, "The homotopy continuation methods, Numerically implementable topological procedures", Trans. Amer. Math. Soc. 242 (1978) 271-284.

[3] E. Allgower and K. Georg, "Simplicial and continuation methods for approximating fixed points and solutions to systems of equations", SIAM Review 22 (1980) 28-85.

[4] S. N. Chow, J. Mallet-Paret and J. A. Yorke, "Finding zeros of maps: Homotopy methods that are constructive with probability one", Math. Comp. 32 (1978) 887-899.

[5] B. C. Eaves, "A short course in solving equations with PL homotopies", SIAM-AMS Proc. (1976) 73-143.

[6] R. M. Freund, "Variable-dimension complexes with applications", Tech. Rept. SOL 80-11, Dept. of Operations Research, Stanford University, Stanford, California, June 1980.

[7] K. Georg, "On tracing an implicitly defined curve by quasi-Newton steps and calculating bifurcation by local perturbations", University of Bonn, Bonn, Jan. 1980.

[8] R. B. Kearfott, "A derivative-free arc continuation method and a bifurcation technique", University of South Louisiana, 1980.

[9] R. B. Kellogg, T. Y. Li and J. A. Yorke, "A constructive proof of the Brouwer fixed point theorem and computational results", SIAM J. Numer. Anal. 4 (1976) 473-483.

[10] M. Kojima, "A note on 'A new algorithm for computing fixed points' by van der Laan and Talman", in: Forster, ed.,

Numerical Solution of Highly Nonlinear Problems, Fixed Point Algorithms and Complementarity Problems (North-Holland, New York, 1980) pp.37-42.

[11] M. Kojima and Y. Yamamoto, "Variable dimension algorithms, Part I: Basic theory", Res. Rept. B-77, Dept. of Information Sciences, Tokyo Institute of Technology, Tokyo, Dec. 1979.

[12] M. Kojima and Y. Yamamoto, "Variable dimension algorithms, Part II: Some new algorithms and triangulations with continuous refinement of mesh size", Res. Rept. B-82, Dept. of Information Sciences, Tokyo Institute of Technology, Tokyo, May 1980.

[13] G. van der Laan, "Simplicial fixed point algorithms", Ph. D. Dissertation, Free University Amsterdam, 1980.

[14] G. van der Laan and A. J. J. Talman, "A restart algorithm for computing fixed points without extra dimension", Math. Programming 17 (1979) 74-84.

[15] G. van der Laan and A. J. J. Talman, "A restart algorithm without an artificial level for computing fixed points on unbounded regions", in: H. O. Peitgen and H. O. Walther, ed., Functional Differential Equations and Approximation of Fixed Points, Lecture Notes in Mathematics 730 (Springer, Berlin, 1979) pp.247-256.

[16] G. van der Laan and A. J. J. Talman, "Convergence and properties of recent variable dimension algorithms", in: W. Forster, ed., Numerical Solution of Highly Nonlinear Problems, Fixed Point Algorithms and Complementarity Problems (North-Holland, New York, 1980) pp.3-36.

[17] G. van der Laan and A. J. J. Talman, "On the computation of of fixed points in the product space of the unit simplices and

an application to non cooperative n-person games", Free University, Amsterdam, Oct. 1978.

[18] G. van der Laan and A. J. J. Talman, "A class of simplicial subdivisions for restart fixed point algoritms without an extra dimension", Free University, Amsterdam, Dec. 1980.

[19] O. H. Merrill, "Applications and extentions of an algorithm that computes fixed points of a certain upper semi-continuous point to set mapping", Ph. D. Dissertation, Dept. of Industrial Engineering, University of Michigan, 1972.

[20] J. M. Ortega and W. C. Rheinboldt, Iterative Solutions of Nonlinear Equations in Several Variables (Academic Press, New York, 1970).

[21] R. M. Reiser, "A modified integer labelling for complementarity algorithms", Institut für Operations Research der Universtät Zürich, June 1978.

[22] D. Saupe, "Predictor-corrector methods and simplicial continuation algorithms", presented at the conference on Numerical Solutions of Nonlinear Equations, Simplicial & Classical Methods, University of Bremen, July 1980.

[23] A. J. J. Talman, "Variable dimension fixed point algorithms and triangulations", Ph. D. Dissertation, Free University, Amsterdam, 1980.

[24] M. J. Todd, "Union Jack triangulations", in S. Karamardian, ed., Fixed Points: Algorithms and Applications (Academic Press, New York, 1977).

[25] M. J. Todd, "Fixed-point algorithms that allow restarting without an extra dimension", Tech. Rept. No.379, School of Operations Research and Industrial Engineering, Cornell University, Ithaca, New York, Sept. 1978.

[26] M. J. Todd, "Global and local convergence and monotonicity results for a recent variable dimension simplicial algorithm", in: W. Forster, ed., Numerical Solution of Highly Nonlinear Problems, Fixed Point Algorithms and Complementarity Problems (North-Holland, New York, 1980) pp. 43-69.

[27] M. J. Todd and A. H. Wright, "A variable-dimension simplicial algorithm for antipodal fixed-point theorems", Tech. Rept. No.417, School of Operations Research and Industrial Engineering, Cornell University, Ithaca, New York, April 1979.

[28] A. H. Wright, "The octahedral algorithm, a new simplicial fixed point algorithm", Mathematics rept. No.61, Western Michigan University, Oct. 1979.

Author's Address: Department of Information Sciences
Tokyo Institute of Technology
Meguro, Tokyo 152, Japan

LABELLING RULES AND ORIENTATION: ON SPERNER'S LEMMA AND BROUWER DEGREE

BY

G. V.D. LAAN*

AND

A.J.J. TALMAN**

*) Interfaculteit der Actuariële Wetenschappen en Econometrie, Vrije Universiteit, Amsterdam

**) Yale School of Organization and Management
New Haven, Connecticut 06520
USA

LABELLING RULES AND ORIENTATION : ON SPERNER'S LEMMA AND BROUWER DEGREE*)

G. van der Laan

Interfaculteit der Actuariële
Wetenschappen en Econometrie,
Vrije Universiteit, Amsterdam

A.J.J. Talman

Yale School of Organization
and Management, New Haven

Abstract

In this paper we consider two labelling rules used in simplicial fixed point algorithms. The first one is the standard labelling rule from an n-dimensional set to the set of integers $\{1,\ldots, n+1\}$. The second one is a labelling to the set $\{\pm i \mid i = 1,\ldots,n\}$. The main purpose of the paper is to compare the two rules. We define the orientation of a completely labelled simplex and give some generalizations of the lemma of Sperner and the related lemma of Knaster, Kuratowski and Mazurkiewicz. Also, for both labelling rules it is shown that the Brouwer degree can be obtained from the completely labelled simplices.

1. Introduction

The classical lemma of Sperner [1928] is well-known and has wide applications. Generalizations of the lemma were given by several authors. In Fan [1970] an n-dimensional simplex is triangulated and Sperner's lemma is generalized given a labelling function from the vertices of the triangulation to the set of integers $\{\pm i \mid i=1,\ldots,n+1\}$. The generalization evolves as a special case of a theorem on an n-dimensional pseudo-manifold. In the proof of this theorem a result of Fan [1967] is used. A special case

*) This research has been done while the first author was visiting CORE, Louvain-la-Neuve, Belgium and the second author stayed at IAWE, Vrije Universiteit, Amsterdam.

of the latter result is a generalization of a lemma given by Tucker [1945] (see also Lefschetz [1949] and Fan [1952]). The lemma of Tucker was used to prove antipodal theorems generalizing the classical theorems of Borsuk-Ulam and Lusternik-Schnirelman.

In Wolsey [1977] two cubical versions of Sperner's lemma, due to Kuhn [1960] and Fan [1960] are proved constructively.

In this paper we consider two labelling rules on the vertices of a triangulation of the unit cube C^n. The first one is a labelling from C^n to the set of integers $\{1,\ldots,n+1\}$. The other one is a labelling from C^n to the set $\{\pm i \mid i = 1,\ldots, n\}$. These two labelling rules can be seen as the two extreme cases of a class of labelling rules on C^n introduced by Van der Laan and Talman [1981]. They used such labelling rules in simplicial algorithms to compute a fixed point of a continuous function from R^n into itself (see also Reiser [1978], Van der Laan and Talman [1979,1980], Todd and Wright [1979], Todd [1980], and Talman [1980]). A closely related labelling rule was given by Van der Laan and Talman [1978] on the product space of unit simplices.

The main purpose of the paper is to compare the two labelling rules. It is organised as follows. In Section 2 some preliminaries are given. In particular, a completely labelled n-simplex is defined for both labelling rules. In Section 3 the orientation of completely labelled simplices is defined. A result which is well-known for the standard labelling rule is generalized to the 2n-labelling. Moreover it is proved that there exists a relationship between several types of completely labelled simplices. In Section 4 some generalizations of Sperner's lemma are given by stating conditions which guarantee the existence of a completely labelled simplex. Moreover the related lemma of Knaster, Kuratowski and Mazurkiewicz [1929] is generalized. Finally, in Section 5 the relation between the number of the completely labelled simplices and the Brouwer degree is discussed for both labelling rules. Again, a result known for the standard labelling is generalized for the case of 2n-labelling.

2. Preliminaries

A t-dimensional simplex or t-simplex, denoted by σ, is the convex hull of t+1 affinely independent points $w^1, \ldots, w^{t+1}$ of R^n ($t \le n$). We write $\sigma = \sigma(w^1, \ldots, w^{t+1})$. The points $w^1, \ldots, w^{t+1}$ are called the vertices of σ. A k-simplex $\tau(w^1, \ldots, w^{k+1})$ is a face of a t-simplex $\sigma (k \le t)$ if all the vertices of τ are vertices of σ. A (t-1)-face of a t-simplex σ is called a facet of σ. The facet τ of σ is said to be opposite to the vertex w^i if w^i is the vertex of σ not in τ. Two different simplices are adjacent if they share a common facet. Let C be an m-dimensional convex subset of R^n. A finite collection G of m-simplices is a triangulation of C if

a) C is the union of all simplices in G;

b) the intersection of two simplices in G is either empty or a common face. The mesh of a triangulation G is defined by mesh G = $\sup_{\sigma\in G}\max_{i,j}\|w^i-w^j\|$. The set C^n is the n-dimensional unit cube $\{x\in R^n \mid -1\le x_i\le 1,\ i=1,\ldots,n\}$. Let G be a triangulation of C^n. Then $\bar{G}$ is the collection of all one-dimensional faces of the simplices of G. The set of integers $\{-n,-n+1,\ldots,-1,1,\ldots,n-1,n\}$ is denoted by K_n. The set of integers $\{1,\ldots,n\}$ is denoted by I_n.

Definition 2.1. A function ℓ from R^n into I_{n+1} is called a standard labelling rule. A function ℓ from R^n into K^n is called a 2n-labelling rule.

Definition 2.2. A standard labelling rule from C^n into I_{n+1} is called proper if $\ell(x)\neq i$ if $x_i=1$ and $\ell(x)\neq n+1$, if, for some i, $x_i=-1$. A 2n-labelling from C^n into K_n is called proper if $\ell(x)\neq i$ if $x_i=1$ and $\ell(x)\neq -i$ if $x_i=-1$.

Definition 2.3. Let ℓ be a standard or 2n-labelling. Then an n-simplex $\sigma(w^1,\ldots,w^{n+1})$ is called completely labelled if $\ell(w^i)\neq\ell(w^j)$ for any two vertices $w^i\neq w^j$ of σ.

3. The orientation of simplices.

An orientation or index theory was introduced for bimatrix games by Shapley [1974] and was generalized by Lemke and Grotzinger [1976]. Deeper and more abstract discussions can be found in Eaves and Scarf [1976] and Todd [1976]. We define the orientation of a completely labelled simplex with respect to a standard labelling as follows.

Definition 3.1. Let ℓ be a standard labelling and $\sigma(w^1,\ldots,w^{n+1})$ a completely labelled simplex. Let $(j_1,\ldots,j_{n+1})$ be the permutation of the elements of I_{n+1} such that $\ell(w^{j_i})=i$, $i=1,\ldots,n+1$. Then the orientation of σ, denoted by id σ, is defined by

$$\text{id}\ \sigma = \text{sign det}\begin{bmatrix} 1 & \cdots & 1 \\ w^{j_1} & \cdots & w^{j_{n+1}} \end{bmatrix}.$$

We also define the orientation of an (n-1)-simplex in bd C^n whose vertices carry all the labels of the set $I_{n+1}\setminus\{k\}$, for some $k\in I_{n+1}$.

Definition 3.2. Let ℓ be a standard labelling and $\tau(w^1,\ldots,w^n)$ be an (n-1)-simplex in bd C^n such that for some $k \in I_{n+1}$, $\{\ell(w^i) | i=1,\ldots,n\} = I_{n+1}\setminus\{k\}$. Let σ be the unique simplex of the triangulation of C^n such that τ is a facet of σ. Let $(j_1,\ldots,j_{n+1})$ be a permutation of the elements of I_n such that $\ell(w^{j_i})=i$, $i \in I_{n+1}\setminus\{k\}$ and w^{j_k} is the vertex of σ opposite τ. Then the orientation of τ, denoted by id τ, is defined by

$$\text{id}\ \tau = \text{sign det} \begin{bmatrix} 1 & \cdots & 1 \\ w^{j_1} & \cdots & w^{j_{n+1}} \end{bmatrix}.$$

The following lemma can be found in Prüfer and Siegberg [1979], see also Sperner [1980] and Forster [1980].

Lemma 3.3. *Let ℓ be a standard labelling,* H *the set of completely labelled* n*-simplices in* C^n *and* H_k *the set of* (n-1)*-simplices in* bd C^n *with labelset* $I_{n+1}\setminus\{k\}$. *Then*

$$\sum_{\sigma \in H} \text{id}\ \sigma = \sum_{\tau \in H_k} \text{id}\ \tau \qquad k=1,\ldots,n+1.$$

Now we consider a 2n-labelling rule. Observe that from Definition 2.3 we have that any completely labelled simplex has at least one 1-face $\tau(w^1,w^2)$ such that $\ell(w^1)+\ell(w^2)=0$. However, in contrast with a standard labelling, the labelset of a completely labelled simplex with respect to a 2n-labelling is not unique.

Definition 3.4. Let ℓ be a 2n-labelling and σ a completely labelled simplex. Let $s \in R^n$ be a sign vector, i.e., $s_i \in \{+1, -1\}$ for all i. Then σ is s-complete if

$$\{s_i i | i=1,\ldots,n\} \subset \{\ell(w^i) | i=1,\ldots,n+1\}.$$

Remarks 1. If σ is s-complete and $h \in K_n$ is the label of a vertex of σ not in the set $\{s_i i | i=1,\ldots,n\}$, then σ is also $\hat{s}$-complete with $\hat{s}_i = s_i$, $i \neq |h|$ and $\hat{s}_{|h|} = -s_{|h|}$.

2. A completely labelled simplex σ does not imply that there exists an s such that σ is s-complete.

Now we define the orientation of an s-complete simplex.

Definition 3.5. Let $\sigma(w^1,\ldots,w^{n+1})$ be s-complete and let $(j_1,\ldots,j_{n+1})$ be the permutation of the elements of I_{n+1} such that $\ell(w^{j_i})=s_i i$, $i=1,\ldots,n$ and $\ell(w^{j_{n+1}}) \neq s_i i$, $i=1,\ldots,n$. Then

$$\text{id } \sigma = \prod_{i=1}^{n} s_i \text{ sign det} \begin{bmatrix} 1 & \cdots & 1 \\ w^{j_1} & \cdots & w^{j_{n+1}} \end{bmatrix}.$$

Observe that the orientation of an s-complete simplex is independent with respect to s and $\hat{s}$.

<u>Definition 3.6</u>. Let τ be a facet of an n-simplex $\sigma(w^1,\ldots,w^{n+1})$. For some sign vector $s \in R^n$, τ is s-complete if

$$\{s_i i \mid i=1,\ldots,n\} = \{\ell(w^i) \mid w^i \text{ is a vertex of } \tau\}.$$

We also define the orientation of an s-complete facet τ of a simplex σ.

<u>Definition 3.7</u>. Let τ be an s-complete facet of a simplex $\sigma(w^1,\ldots,w^{n+1})$ and let $(j_1,\ldots,j_{n+1})$ be the permutation of the elements of I_{n+1} such that $w^{j_{n+1}}$ is the vertex of σ opposite τ and $\ell(w^{j_i})=s_i i$, $i=1,\ldots,n$. Then the orientation of τ with respect to σ, denoted by $\text{id}_\sigma \tau$, is

$$\text{id}_\sigma \tau = \prod_{i=1}^{n} s_i \text{ sign det} \begin{bmatrix} 1 & \cdots & 1 \\ w^{j_1} & \cdots & w^{j_{n+1}} \end{bmatrix}.$$

<u>Corollary 3.8</u>. *If τ is an s-complete facet of an s-complete simplex σ, then*

$$\text{id } \sigma = \text{id}_\sigma \tau.$$

<u>Corollary 3.9</u>. *Let σ be an s-complete simplex and τ_1,τ_2 the two facets of σ such that τ_1 is s-complete and τ_2 is $\hat{s}$-complete with for some $h \in I_n$, $\hat{s}_i = s_i$, $i \neq h$ and $\hat{s}_h = -s_h$. Then*

$$\text{id}_\sigma \tau_1 = \text{id}_\sigma \tau_2.$$

<u>Definition 3.10</u>. Let τ be an s-complete (n-1)-simplex in bd C^n and let σ be the unique simplex in C^n having τ as a facet. Then the orientation of τ is defined by

$$\text{id } \tau = \text{id}_\sigma \tau.$$

The next two lemmas can easily be proved (see e.g. Allgower and Georg [1980] and Van der Laan [1980], where they are proved for a standard labelling.)

Lemma 3.11. *Let* τ_1 *and* τ_2 *be two* s-*complete facets of a simplex* σ. *Then*

$$\mathrm{id}_\sigma\, \tau_1 = -\, \mathrm{id}_\sigma\, \tau_2.$$

Lemma 3.12. *Let* τ *be the common* s-*complete facet of two adjacent simplices* σ_1 *and* σ_2. *Then*

$$\mathrm{id}_{\sigma_1}\, \tau = -\, \mathrm{id}_{\sigma_2}\, \tau.$$

Now we are ready to prove the analogon of Lemma 3.3.

Theorem 3.13. *For some* s, *let* H(s) *be the set of* s-*complete simplices and let* $\overline{H}(s)$ *be the set of* s-*complete* (n-1)-*simplices in* bd C^n. *Then*

$$\sum_{\sigma\in H(s)} \mathrm{id}\, \sigma = \sum_{\sigma\in \overline{H}(s)} \mathrm{id}\, \tau.$$

Proof. Using the "door in-door out" pivoting scheme, there are, for some $s \in R^n$, exactly three types of simplicial chains, such that the common facet of two adjacent simplices is s-complete.

a) Starting with a simplex having an s-complete facet τ on bd C^n as a facet, a path of simplices is generated until an s-complete simplex σ of G is found. From Definition 3.10, Lemmas 3.11 and 3.12 and Corollary 3.8 it follows that id σ = id τ.

b) Starting with a simplex having an s-complete τ_1 on bd C^n as a facet, a simplex is generated having an s-complete τ_2 on bd C^n as a facet. By Definition 3.10, Lemmas 3.11 and 3.12 we have that id τ_2 = - id τ_1.

c) Starting with an s-complete simplex σ_1, a path of simplices is generated until an s-complete simplex σ_2 is found. From Corollary 3.8 and the Lemmas 3.11 and 3.12 we obtain id σ_2 = - id σ_1.

The theorem then follows by combining these three cases. □

For the next theorem we need an assumption on the boundary of C^n.

Definition 3.14. A 2n-labelling has the boundary property is for any one-simplex $\tau(w^1,w^2)$ in bd C^n we have

$$\ell(w^1) + \ell(w^2) \neq 0.$$

Theorem 3.15. *If a 2n-labelling ℓ has the boundary property, then*

$$\sum_{\tau\in\overline{H}(s^1)} \text{id}\,\tau = \sum_{\tau\in\overline{H}(s^2)} \text{id}\,\tau$$

for every two sign vectors s^1 *and* s^2.

Observe that bd C^n is a finite (n-1)-dimensional pseudomanifold with an empty boundary. Therefore, if the boundary property is satisfied, the conditions stated in Theorem 1 of Fan [1970] are fulfilled on bd C^n. Hence, Theorem 3.15 follows as an application of Theorem 1 of Fan by observing that the (n-1)-dimensional pseudomanifold bd C^n has no (n-2)-dimensional boundary facets and that our orientation is defined slightly differently. The theorem can also be found in Krasnosel'skii [1964]. In our case, Krasnosel'skii's proof is based on counting the sum of the orientations of all (n-2)-simplices $\sigma(w^1,\dots,w^{n-1})$ in bd C^n, such that, for some sign vector s and some index $j \in I_n$, $\{\ell(w^i) \mid i=1,\dots,n-1\} = \{s_i i \mid i \neq j\}$. Below, we will give a simple and constructive proof of the theorem, based on generating paths of simplices. Therefore we need the following lemma.

Lemma 3.16. *Let* G *and* G' *be two triangulations of* C^n *which yield the same* (n-1)-*simplices on* bd C^n. *Let* τ *be an s-complete facet in* bd C^n *and let* σ (σ') *be the unique simplex of* G (G') *having* τ *as a facet. Then*

$$\text{id}_\sigma\,\tau = \text{id}_{\sigma'}\,\tau$$

Proof. Let w (w') be the vertex of σ (σ') opposite τ. Then the lemma follows immediately from the fact that w and w' are on the same side of the hyperplane spanned by the vertices of τ. □

Proof of Theorem 3.15. It is sufficient to prove the theorem for two sign vectors s^1 and s^2 such that for some h, $s^1_h = -s^2_h$ and $s^2_i = s^1_i$ for all $i \neq h$. First we define a triangulation G' of C^n which is induced by G. Let V be the set of all vertices on the boundary of C^n. Then G' is defined as the triangulation of C^n obtained from connecting all elements of V with the zero vector (see Figure 1 for n = 2). Clearly each (n-1)-simplex τ of G in bd C^n is an (n-1)-simplex of G' in bd C^n and conversely. By Lemma 3.16 and Definition 3.10 it follows that the orientation of an s-complete boundary facet is the same for both G and G'. Now we label the zero vector which is the only vertex of G' in int C^n, artificially with $\ell(\underline{0}) = h$.

Without loss of generality we assume that $s_h^1 = 1$. Now we generate sequences of simplices of G' such that the common facets are s^1-complete. The following types of paths can occur.

a. Starting with a simplex of G' having an element τ_1 of $\overline{H}(s^1)$ as a facet, a path of simplices can be generated with common s^1-complete facets. Since $\ell(w^1) + \ell(w^2) \neq 0$ for any boundary face $\tau(w^1,w^2)$, we must have that the sequence terminates with a simplex of G' such that the facet τ_2 on bd C^n (the facet opposite $\underline{0}$) is either s^1-complete or s^2-complete. By Definition 3.10, Lemma 3.11 and 3.12 and Corollary 3.9 we have

(i) If τ_2 is s^1-complete then id τ_2 = − id τ_1

(ii) If τ_2 is s^2-complete then id τ_2 = id τ_1.

b. Starting with a simplex of G' having an element τ_1 of $\overline{H}(s^2)$ as a facet, the path terminates with a simplex of G' having again an s^1- or s^2-complete boundary facet τ_2. Then we have

(i) If τ_2 is s^1-complete then id τ_2 = id τ_1

(ii) If τ_2 is s^2-complete then id τ_2 = − id τ_1

Hence it follows that

$$\sum_{\tau\in\overline{H}(s^1)} \text{id } \tau = \sum_{\tau\in\overline{H}(s^2)} \text{id } \tau.$$

The proof of Theorem 3.15 is illustrated in Figure 1 for n = 2, $s^1 = (1,1)$ and $s^2 = (-1,1)$.

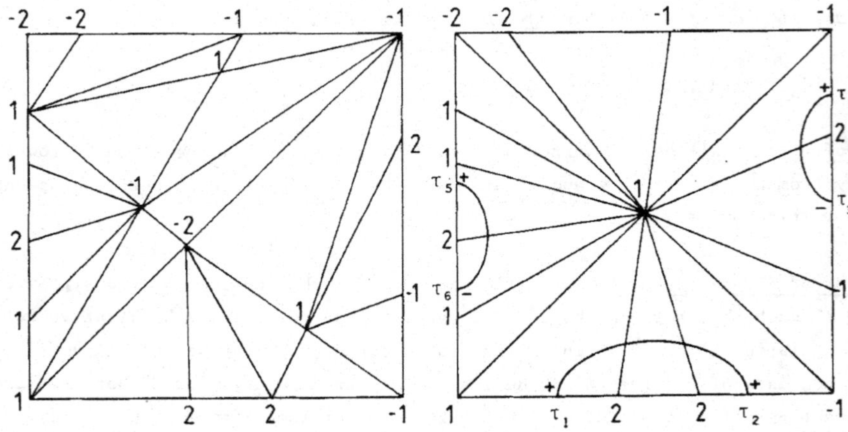

G triangulation G' triangulation

Figure 1. n = 2, $s^1 = (1,1)$, $s^2 = (-1,1)$, id τ_1 = id τ_2 = 1,
id τ_3 = − id τ_4 = −1, id τ_5 = − id τ_6 = 1.

Corollary 3.17. *If a 2n-labelling rule has the boundary property then*

$$\sum_{\sigma\in H(s^1)} \text{id}\,\sigma = \sum_{\sigma\in H(s^2)} \text{id}\,\sigma$$

for every two sign vectors s^1 *and* s^2.

4. Combinatorial lemmas.

In this section we generalize the lemmas of Sperner and Knaster, Kuratowski and Mazurkiewicz in the sense that both for the standard and 2n-labelling related lemmas on C^n instead of a simplex are given.

Theorem 4.1. *Let* G *be a triangulation of* C^n *and* ℓ *a proper standard labelling. Then*

$$\sum_{\sigma\in H} \text{id}\,\sigma = 1.$$

Proof. We only give a sketch of the proof. First we define for each $I \subset I_n$ the set

$$A(I) = \{x \in C^n \mid x_i = -1,\ i \notin I\}.$$

In particular we have $A(\emptyset) = \{(-1,\ldots,-1)^T\}$ and $A(I_n) = C^n$. Following Van der Laan [1980] we can define the orientation of a t-simplex in $A(I)$ with labelset $I \cup \{k\}$ for some $k \notin I$, where $t = |I|$. Then it can be proved that

$$\sum_{|I|=h} \sum_{\sigma\in H(I)} \text{id}\,\sigma = 1 \quad \text{for } h = 1,\ldots, n,^{*)}$$

where $H(I)$ is the set of all t-simplices in $A(I)$ with labelset $I \cup \{k\}$ for some $k \notin I$. In particular for $h = n$ we have

$$\sum_{\sigma\in H(I_n)} \text{id}\,\sigma = \sum_{\sigma\in H} \text{id}\,\sigma = 1.$$

□

---

*) For the case of a standard labelling on the unit simplex an analogous formula is proved by Van der Laan [1980, page 129].

Corollary 4.2. *Let ℓ be a proper standard labelling rule on C^n. Then there exists an odd number of completely labelled simplices.*

In Wolsey [1977] and Kuhn [1960] a weaker result was obtained. Moreover in both cases the labelling rules were proper in another sense. Observe that our definition of a proper labelling evolved from fixed point algorithms. Now we give the analog of the lemma of Knaster, Kuratowski and Mazurkiewicz.

Lemma 4.3. *Let $\{D_i | i = 1,\ldots,n+1\}$ be n+1 closed subsets of C^n such that*

(i) $$C^n = \bigcup_{i=1}^{n+1} D_i$$

(ii) $$\bigcap_{i\in I} C_i \subset \bigcup_{i\notin I} D_i$$ *for all proper subsets $I \subset I_{n+1}$ where*

$$C_i = \{x \in C^n | \; x_i = 1\} \quad i = 1,\ldots,n \text{ and}$$

$$C_{n+1} = \{x \in C^n | \; x_i = -1 \text{ for some } i \in I_n\}.$$

Then

$$\bigcap_{i=1}^{n+1} D_i \neq \emptyset.$$

The proof of the lemma is straightforward and is analogous to the proof of the clàssical lemma of K-K-M on the simplex (see e.g. Todd [1976]).

We consider now the 2n-labelling.

Theorem 4.4. *Let G be a triangulation of C^n and let ℓ be a proper 2n-labelling, which satisfies the boundary property. Then, for any sign vector $s \in R^n$,*

$$\sum_{\sigma\in H(s)} \text{id } \sigma = 1.$$

Proof. From Corollary 3.17 it follows that it is sufficient to consider only one sign vector. So, let s be the vector with $s_i = 1$ for all $i \in I_n$. Again we define the sets $A(I)$, $I \subset I_n$, as in the proof of Theorem 4.1. Following the lines of the proof of Theorem 7.3.10 of Van der Laan [1980] we consider for all $I \subset I_n$, the t-simplices in $A(I)$ with labelset $I \cup \{k\}$ for some $k \notin I$. If $|I| = n$ such a simplex is s-complete. If $|I| < n$ we have that for all $i \notin I$, $\ell(x) \neq -i$ for all $x \in A(I)$, since the labelling rule is proper. Moreover, if a simplex $\sigma(w^1,\ldots,w^{t+1})$ in $A(I)$ has the labelset $I \cup \{k\}$ for some $k \notin I$, then for all $i \in I$, there exists an index j_i such that $\ell(w^{j_i}) = i$. Hence, because of the boundary property, $k \neq -i$ for all $i \in I$. Consequently, if $\sigma(w^1,\ldots,w^{t+1})$ in $A(I)$ has labelset $I \cup \{k\}$ then k is positive. Now, in the same way as for Theorem 4.1, it can be proved that

$$\sum_{|I|=h} \sum_{\sigma \in H(I)} \text{id}\,\sigma = 1 \qquad \text{for } h=1,\ldots,n,$$

where $H(I)$ is the set of all simplices in $A(I)$ with labelset $I \cup \{k\}$ for some (positive) $k \notin I$. Hence, for $h = n$ we obtain

$$\sum_{|I|=n} \sum_{\sigma \in H(I)} \text{id}\,\sigma = \sum_{\sigma \in H(I_n)} \text{id}\,\sigma = \sum_{\sigma \in H(s)} \text{id}\,\sigma = 1.$$

□

Corollary 4.5. *Let ℓ be a proper 2n-labelling with boundary property. Then there exists an odd number of s-complete simplices for any sign vector s.*

Corollary 4.6. *Let ℓ be a proper 2n-labelling. Then there exists a one face $\tau(w^1,w^2)$ such that $\ell(w^1) + \ell(w^2) = 0$.*

In Van der Laan and Talman [1981] an algorithm is presented to generate such a one-face. Corollary 4.6 is related to a theorem of Tucker [1945] (see also Fan [1952]). However, the labelling rule considered by Tucker and Fan is not proper, but satisfies an antipodal property. On C^n, Fan [1952] states the following.

Theorem 4.7. (Fan). *Let G be a centrally symmetric subdivision of C^n and let $\ell: C^n \to \{\pm i \mid i=1,\ldots,n+1\}$ be a labelling rule, such that*

(i) $\ell(v) = -\ell(-v)$ *for any vertex v of G in* bd C^n.
(ii) $\ell(w^1) + \ell(w^2) \neq 0$ *for each one face* $\tau(w^1,w^2)$ *of* $\bar{G}$.

Then $\alpha(1, -2, 3,\ldots,(-1)^n (n+1)) + \alpha(-1,2,\ldots,(-1)^{n+1}(n+1))$ *is odd where* $\alpha(j_1,\ldots,j_{n+1})$ *is the number of n-simplices with labelset* $\{j_1,\ldots,j_{n+1}\}$.

From this theorem the next corollary follows immediately, since for a 2n-labelling $\alpha(1,\ldots,(-1)^n(n+1)) + \alpha(-1,\ldots,(-1)^{n+1}(n+1)) = 0$.

Corollary 4.8. (Tucker). *Let* G *be a centrally symmetric subdivision of* C^n *and let* ℓ *be a* 2n-*labelling such that* $\ell(v) = -\ell(-v)$ *for any vertex* v *of* G *in* bd C^n. *Then there exists a one-face* $\tau(w^1,w^2)$ *such that*

$$\ell(w^1) + \ell(w^2) = 0.$$

In fact this corollary is stronger than the result of Tucker, which was stated in terms of a cubical decomposition of C^n. In Freund and Todd [1979] the variable dimension fixed point algorithm of Reiser [1978], see also Van der Laan and Talman [1981], is modified to generate a facet $\tau(w^1,w^2)$ with $\ell(w^1) + \ell(w^2) = 0$ for a 2n-labelling on C^n with the antipodal property. So, Freund and Todd give a constructive proof of Corollary 4.8.

Now we give some covering lemmas related to the corollaries 4.5. and 4.6.

Lemma 4.9. *Let* $\{D_i \mid i\in K_n\}$ *be a collection of* 2n *closed subsets of* C^n *such that*

(i) $C^n = \bigcup_{i\in K_n} D_i$

(ii) *For any subset* $I\subset K_n$: $\bigcap_{i\in I} C_i \subset \bigcup_{i\notin I} D_i$

where $C_i = \{x\in C^n \mid x_{|i|} = \text{sgn } i\} \quad i\in K_n$

(iii) $\{D_i \cap D_{-i}\} \cap \text{bd } C^n = \emptyset \quad i=1,\dots,n.$

Then, for all sign vectors $s\in K^n$

$$\{\bigcap_{j\in I(s)} D_j\} \cap \{\bigcup_{j\notin I(s)} D_j\} \neq \emptyset,$$

where $I(s) = \{s_i i \mid i=1,\dots,n\}$.

Proof. Define a 2n-labelling function ℓ such that

$$\ell(x) = j \text{ for some } j \in I_x,$$

where $I_x = \{h\in K_n \mid x\in D_h\}$. Clearly, by condition (ii) ℓ can be defined in such a way that ℓ is proper. Let G be a triangulation of C^n. Clearly, by condition (iii), ℓ has the boundary property if mesh G is small enough. Let $\{G_k,\ k=1,2,\dots\}$ be a sequence of triangulations of C^n such that ε_k = mesh $G_k \to 0$ if $k\to\infty$ and assume ℓ has the boundary property for all G_k. Furthermore, assume ℓ is proper. Then, by Corollary 4.5. for any s, there exists an s-complete simplex $\sigma_k(s)$ in G_k. Let

$w_k^i(s)$ be the vertex of $\sigma_k(s)$ such that $\ell(w_k^i(s)) = s_i i$, $i \in I_n$ and $\ell(w_k^{n+1}(s)) \neq s_i i$, $i \in I_n$. Then, for some subsequence $\{k_j, j=1,2,\ldots\}$ of integers with $k_j \to \infty$ if $j \to \infty$, we have that the sequence

$$\{w_{k_j}^1(s), \; j=1,2,\ldots\}$$

converges to a point $x^*(s)$ in C^n. Since $\varepsilon_k \to 0$ if $k \to \infty$ it follows that $\lim_{j\to\infty} w_{k_j}^i(s) = x^*(s)$ for all $i=1,\ldots,n+1$. Hence

$$x^*(s) \in D_j \quad \text{for all } j \in I(s)$$

and

$$x^*(s) \in \bigcup_{j \notin I(s)} D_j$$

which proves the lemma. □

Lemma 4.10. *Let* $\{D_i | i \in K_n\}$ *be a collection of* 2n *closed subsets of* C^n *such that* (i) *and* (ii) *of Lemma* 4.9 *hold. Then there exists an index* i *such that*

$$D_i \cap D_{-i} \neq \emptyset .$$

The proof of this lemma follows by using Corollary 4.6 in a sequence of triangulations with mesh going to zero.

5. Labelling and Brouwer degree.

In this section we compare standard labelling and 2n-labelling, if both rules are induced by some continuous function f from C^n into R^n.

Definition 5.1. (Standard labelling). Let f be a continuous function from C^n into R^n and let $h \in I_n$ be the smallest index such that $f_h(x) \geq f_i(x)$, $i=1,\ldots,n$. Then the standard labelling rule $\hat{\ell}$ induced by f is defined by

$$\hat{\ell}(x) = h \qquad \text{if } f_h(x) \geq 0$$
$$\qquad = n+1 \qquad \text{otherwise.}$$

Definition 5.2. (2n-labelling). Let f be a continuous function from C^n into R^n. Let $J_x^+ \subseteq I_n$ be the set of indices such that $f_h(x)=\max_{i\in I_n} |f_i(x)|$ if $h\in J_x^+$ and let $J_x^- \subseteq \{-i \mid i\in I_n\}$ be the set of indices such that $-f_h(x)=\max_{i\in I_n} |f_i(x)|$ if $-h\in J_x^-$. Then the 2n-labelling rule $\tilde{\ell}$ induced by f is defined by

$\tilde{\ell}(x)$ = h where h is the smallest index of the set $J_x^- \cup J_x^+$.

Usually, the Brouwer degree is defined at first for continuously differentiable functions as the algebraic sum of preimages of a regular value of the function. Using approximation steps the degree can be defined to continuous functions. In the following we assume that f is a continuous function from C^n into R^n such that $f(x)\neq \underline{0}$ for all $x \in$ bd C^n. Then, assuming that $\underline{0}$ is a regular value of f, the Brouwer degree of f on int C^n with respect to $\underline{0}$, denoted by deg(f, int C^n, $\underline{0}$) can be stated in terms of completely labelled simplices if the mesh of the triangulation of C^n is small enough. Therefore we need the following "smallness conditions" on bd C^n.

Smallness condition I (SC I). A triangulation G of C^n satisfies SC I if for any (n-1)-simplex τ, $\tau \in$ bd C^n there exists an index $i\in I_{n+1}$ such that

$$\hat{\ell}(x) \neq i$$

for all $x\in\tau$.

Smallness condition II (SC II). A triangulation G of C^n satisfies SC II if for any (n-1)-simplex τ, $\tau \in$ bd C^n,

$$\tilde{\ell}(x^1) + \tilde{\ell}(x^2) \neq 0$$

for all $x^1, x^2 \in \tau$.

Using the standard labelling rule $\hat{\ell}$ we have the following theorem. Recall that H is the set of completely labelled simplices.

Theorem 5.3. *If* f *is a continuous frunction from* C^n *into* R^n *with* $f(x)\neq 0$ *for all* $x \in$ bd C^n *and if* G *satisfies* SC I *then*

$$\deg(f, \text{int } C^n, 0) = \sum_{\sigma\in H} \text{id } \sigma.$$

Observe that G satisfies SC I if mesh G is small enough since $f(x)\neq 0$ for all $x \in \mathrm{bd}\, C^n$. The proof of the theorem is analogous to the proof of Theorem 2.5 of Prüfer and Siegberg [1979] (see also Siegberg [1980]). However, observe that they use a slightly different standard labelling. From Theorem 5.3 and Lemma 3.3 the next corollary follows immediately.

Corollary 5.4.

$$\mathrm{def}\,(f, \mathrm{int}\, C^n, 0) = \sum_{\tau\in H_k} \mathrm{id}\,\tau \text{ for all } k=1,\dots,n+1.$$

We consider now the 2n-labelling $\tilde{\ell}$ induced by f. First we define the 2n regions

$$E_h = \mathrm{cl}\,\{x\in C^n \mid \tilde{\ell}(x) = h\} \qquad h\in K_n.$$

Then we have the following lemma

Lemma 5.5. *If for some sign vector* $s\in R^n$

$$x \in \{\bigcap_{j\in I(s)} E_j\} \cap \{\bigcup_{j\notin I(s)} E_j\}$$

then $f(x) = 0$.

Proof. Clearly, there exists an index $i\in I_n$ such that $x\in E_i$ and $x\in E_{-i}$. Since f is continuous, $x\in E_i$ implies

$$f_i(x)\geq |f_h(x)| \qquad h\in I_n$$

and $x\in E_{-i}$ implies

$$-f_i(x)\geq |f_h(x)| \qquad h\in I_n.$$

Hence $f_h(x)=0$ for all $h\in I_n$. □

Remark. For any sign vector $s\in R^n$, it follows from Lemma 5.5 that an s-complete n-simplex yields an approximation of a point x with $f(x)=0$.

Now we define a standard labelling rule $\overline{\ell}$, induced by the 2n-labelling $\tilde{\ell}$ as follows:

$$\bar{\ell}(x) = \tilde{\ell}(x) \qquad \text{if } \tilde{\ell}(x) = i \qquad i=1,\ldots,n$$

$$\bar{\ell}(x) = n + 1 \qquad \text{if } \tilde{\ell}(x) < 0$$

Defining $E_{n+1} = \bigcup_{i=1}^{n+1} E_{-i}$ it follows from Lemma 5.5 that $x \in \bigcap_{i=1}^{n+1} E_i$ implies that $f(x) = 0$, whereas, if for $s = (1,\ldots,1)^T$, σ is an s-complete n-simplex then σ is completely labelled with respect to the standard labelling $\bar{\ell}$ and conversely. Hence a completely labelled simplex yields an approximation of a point x with $f(x)=0$. Moreover, it can easily be seen that $\bar{\ell}$ satisfies SC I with $\bar{\ell}$ instead of $\hat{\ell}$, if $\tilde{\ell}$ satisfies SC II. Now we have the following lemma, which is analogous to Theorem 5.3.

<u>Lemma 5.6</u>. *If* f *is a continuous function from* C^n *into* R^n *with* $f(x)\neq 0$ *for all* $x \in \text{bd } C^n$ *and if* G *satisfies* SC II *then*

$$\deg (f, \text{int } C^n, 0) = \sum_{\sigma \in H'} \text{id } \sigma$$

where H' *is the set of completely labelled simplices with respect to the standard labelling* $\bar{\ell}$.

The proof of this lemma is again analoguous to the proof of Theorem 2.5 of Prüfer and Siegberg [1979]. In fact the result holds for any standard labelling such that $x \in \bigcap_{i=1}^{n+1} E_i$ implies $f(x)=0$ with $E_i = \text{cl}\{x \in C^n \mid \ell(x)=i\}$, $i=1,\ldots,n+1$.

From lemma 5.6 the next theorem follows immediately:

<u>Theorem 5.7</u>. *Let* f *be a continuous function with* $f(x)\neq 0$ *for all* $x \in \text{bd } C^n$ *and let* G *be a triangulation of* C^n *which satisfies* SC II. *Then*

$$\deg (f, \text{int } C^n, 0) = \sum_{\sigma \in H(s)} \text{id } \sigma$$

for any sign vector $s \in R^n$.

<u>Proof</u>. For $s_i=1$, $i \in I_n$, we have that $H'=H(s)$. Hence the theorem follows from Lemma 5.6 and Corollary 3.17, since $\tilde{\ell}$ has the boundary property if G satisfies SC II.

□

With theorem 3.13 the next corollary follows immediately:

Corollary 5.8. *If* G *satisfies* SC II *then*

$$\deg(f, \text{int } C^n, 0) = \sum_{\tau \in \bar{H}(s)} \text{id } \tau$$

for any sign vector s.

From Corollary 5.4 and Corollary 5.8 it follows that the deg $(f, \text{int } C^n, 0)$ can be calculated by considering the labels of the vertices of G on bd C^n. Krasnosel'skii [1964] proves a similar result for some 2n-labelling, defined on the boundary. Observe, however, that the labelling used by Krasnosel'skii does not allow for the result of Theorem 5.7. Therefore, it can not be used for approximating fixed points. In fact, Krasnosel'skii has to know a priori the fixed points of the function.
The Corollaries 5.4 and 5.8 state that deg $(f, \text{int } C^n, 0) = 0$ if one of the labels does not occur on the boundary. So, using 2n-labelling we obtain the following corollary

Corollary 5.9. *Let* f *be a continuous function from* C^n *into* R^n *with* $f(x) \neq 0$ *for all* $x \in \text{bd } C^n$. *If there exists an index* $i \in I_n$ *such that*

$$f_i(x) < \max_{h \in I_n} |f_h(x)| \text{ or } -f_i(x) < \max_{h \in I_n} |f_h(x)|$$

for all $x \in \text{bd } C^n$ *then* deg $(f, \text{int } C^n, 0) = 0$.

Finally we give the relation between the standard labelling and the 2n-labelling.

Theorem 5.10. *Let* f *be a continuous function with* $f(x) \neq 0$ *for all* $x \in \text{bd } C^n$. *Let* G *be a triangulation of* C^n *such that* SC I *and* SC II *are satisfied. Let* $\hat{\ell}$ *and* $\tilde{\ell}$ *be the standard and* 2n-*labelling induced by* f. *Then, for any sign vector* $s \in R^n$

$$\sum_{\sigma \in H} \text{id } \sigma = \sum_{\sigma \in H(s)} \text{id } \sigma.$$

Proof. The proof follows from Theorems 5.3 and 5.7.

References

Allgower, E.L. and Georg, K. (1980), Simplicial and continuation methods for approximating fixed points and solutions to systems of equations, *SIAM Rev. 22*, 28-85.

Eaves, B.C. and Scarf, H.E. (1976), The solution of systems of piecewise linear equations, *Math. Oper. Res. 1*, 1-27.

Fan, K. (1952), A generalization of Tucker's combinatorial lemma with topological applications, *Ann. of Math. 56*, 431-437.

Fan, K. (1960), Combinatorial properties of certain simplicial and cubical vertex maps, *Arch. Math. 11*, 368-377.

Fan, K. (1967), Simplicial maps from an orientable n-pseudomanifold into S^m with the octohedral triangulation, *J. Combinatorial Theory 2*, 588-602.

Fan, K. (1970), A combinatorial property of pseudomanifolds and covering properties of simplexes, *J. Math. Anal. Appl. 31*, 68-80.

Forster, W. (1980), An application of the generalized Sperner lemma to the computation of fixed points in arbitrary complexes, in W. Forster (ed.), *Numerical Solution of Highly Nonlinear Problems*. North-Holland, 219-232.

Freund, R.M. and Todd, M.J. (1979), A constructive proof of Tucker's combinatorial lemma, Cornell University, Ithaca.

Knaster, B., Kuratowski, C. and Mazurkiewicz, S. (1929), Ein Beweis der Fixpunktsatzes für n-dimensionale Simplexe, *Fund. Math. 14*, 132-137.

Krasnosel'skii, M.A. (1964), *Topological Methods in the Theory of Nonlinear Integral Equations*, Pergamon Press.

Kuhn, H.W. (1960), Some combinatorial lemmas in topology. *I.B.M. J. Res. Develop. 4*, 518-524.

Laan, G. van der (1980), Simplicial fixed point algorithms, Dissertation. Free University, Amsterdam.

Laan, G. van der and Talman, A.J.J. (1978), On the computation of fixed points in the product space of unit simplices and an application to non cooperative N-person games, Free University, Amsterdam.

Laan, G. van der and Talman, A.J.J. (1979), Interpretation of the variable dimension fixed point algorithm with an artificial level, Free University, Amsterdam.

Laan, G. van der and Talman, A.J.J. (1980), Convergence and properties of recent variable dimension algorithms, in: W. Forster (ed.), *Numerical Solution of Highly Nonlinear Problems*. North-Holland, 3-36.

Laan, G. van der and Talman, A.J.J. (1981), A class of simplicial subdivisions for restart fixed point algorithms without an extra dimension, *Mathematical Programming 20*, 33-48.

Lefschetz, S. (1949), *Introduction to Topology*, Princeton University Press.

Lemke, C.E. and Grotzinger, S.J. (1976), On generalizing Shapley's index theory to labelled pseudo manifolds, *Math. Prog. 10*, 245-262.

Prüfer, M. and Siegberg, H.W. (1979), On computational aspects of topological degree in R^n, in: H.O. Peitgen and H.O. Walther (eds.), *Functional Differential Equations and Approximation of Fixed Points*. Springer-Verlag, 410-433.

Reiser, P.M. (1978), Ein hybrides Verfahren zur Lösung von nichtlinearen Komplimentaritätsproblemen und seine Konvergenzeigenschaften, Dissertation, Eidgenössischen Technischen Hochschule, Zürich.

Shapley, L.S. (1974), A note on the Lemke-Howson algorithm, *Math. Prog. Stud. 1*, 175-189.

Siegberg, H.W. (1980), Brouwer degree: history and numerical computation, in: W. Forster (ed.), *Numerical Solution of Highly Nonlinear Problems*. North-Holland, 389-416.

Sperner, E. (1928), Neuer Beweis für die Invarianz der Dimensionszahl und des Gebietes, *Abh. Math. Sem. Univ. Hamburg 6*, 265-272.

Sperner, E. (1980), Fifty years of further development of a combinatorial lemma, in: W. Forster (ed.), *Numerical Solution of Highly Nonlinear Problems*, North-Holland, 183-218.

Talman, A.J.J. (1980), Variable dimension fixed point algorithms and triangulations, Dissertation, Free University, Amsterdam.

Todd, M.J. (1976a), *The Computation of Fixed Points and Applications*, Springer-Verlag.

Todd, M.J. (1976b), Orientation in complementary pivot algorithms, *Math. Oper. Res. 1*, 54-66.

Todd, M.J. (1980), Global and local convergence and monotonicity results for a recent variable-dimension simplicial algorithm, in: W. Forster (ed.), *Numerical Solution of Highly Nonlinear Problems*. North-Holland, 43-70.

Todd, M.J. and Wright, A.H. (1979), A variable-dimension simplicial algorithm for antipodal fixed point theorems, Cornell University, Ithaca.

Tucker, A.W. (1945), Some topological properties of disk and sphere, *Proc. First Canadian Math. Congress* (Montreal), 285-309.

Wolsey, L.A. (1977), Cubical Sperner lemmas as application of generalized complementary pivoting, *J. Comb. Theory A 23*, 78-87.

ON THE NUMERICAL SOLUTION OF CONTACT PROBLEMS

BY

H.D. MITTELMANN

Abteilung Mathematik
Universität Dortmund
D-4600 Dortmund 50

ON THE NUMERICAL SOLUTION OF CONTACT PROBLEMS

H.D. Mittelmann

Abteilung Mathematik der Universität, Postfach 500500, D-4600 Dortmund 50, Bundesrepublik Deutschland

Summary

We consider finite element discretizations of variational problems which correspond to quasilinear elliptic boundary value problems with linear constraints. A modified block-relaxation method and a preconditioned conjugate gradient algorithm are presented which generalize known methods for bound-constraints to more general restrictions. Global convergence proofs are given and an application to the contact problem for two membranes.

1. Introduction

Recently the efficient solution of the nonlinear algebraic systems which arise from the discretization of quasilinear elliptic boundary value problems has found some interest and also considerable progress was made in that field. For the optimization problems which occur in connection with nonlinear variational inequalities respectively free boundary problems this is not true to the same extent. For fine discretizations in lower dimensional problems and even for coarser grids in higher dimensions, however, it is necessary to keep the computational work as low as possible and already a reduction of it by a factor of 1/2 may justify the use of a somewhat more sophisticated algorithm.

For the case that the constraints are merely bounds for the solution respectively the variables in the discrete problem a rather simple method is projected relaxation. This algorithm was considered frequently in the literature and a convergence proof for it may be found e. g. in [4]. In computations this method has the well-known advantages of being extremely easy to implement and that of approximating the proper subspace of active variables relatively fast. Then however the linear and often very slow convergence of this method shows up. In [6] it was therefore suggested to use an algorithm of this type only as phase I in a two-phase method. In the second phase a suitable preconditioned cg-algorithm was used there in order to obtain faster convergence. A completely different idea to increase the efficiency namely that of relaxing the constraints by an $\varepsilon > 0$ and using a rather great ε in the beginning of the computations was proposed in [9].

If the constraints are linear but not of bound type it is well-known that point relaxation cannot be used and block relaxation was suggested for such problems. An efficient algorithm, however, can only be expected if the matrix defining the constraints is sparse and an additional simplification would be caused by a certain separation of the variables. In the next section we define for a certain class of such problems which occur e. g. in connection with two-body contact problems a suitable block relaxation and prove its convergence.

A generalization of the cg-SMOR-N algorithm of [6] to these problems is given section 3 and a global convergence proof in the subsequent section. Finally some numerical results are presented in the last section.

2. A block-relaxation method

In the following we shall consider the problem

$$(2.1)\qquad f(x) = \underset{S}{\mathrm{Min}}\ ,\ f : \mathbb{R}^M \to \mathbb{R}$$
$$S = \{x \in \mathbb{R}^M,\ G^T x + g_o \geq 0\},\ G \in \mathbb{R}^{M,p},\ g_o \in \mathbb{R}^p$$

which may be obtained by the discretization of a variational inequality. Inequalities between vectors are to be understood componentwise. In order to simplify the notation we assume that G in (2.1) has the form

$$(2.2)\qquad G = \begin{bmatrix} g_1 & & & & 0 \\ & g_2 & & & \\ & & \cdot & & \\ & & & \cdot & \\ 0 & & & & g_p \end{bmatrix}$$

where $g_i \in \mathbb{R}^{m_i}$, $\sum_{i=1}^{p} m_i = M$. Generalizations are possible but will not be discussed in detail here. It is further assumed as in [5,6] that the gradient of f may be written as

$$(2.3)\qquad \nabla f(x) = A(x)x - b(x).$$

For the problems under consideration there is often a canonical decomposition of this form (cf.[5]).

In the following we shall decompose any vector $x \in \mathbb{R}^M$ according to(2.2) and write $x^T = (x_1^T, \ldots, x_p^T)$, $x_i \in \mathbb{R}^{m_i}$ and analogously $A = (A_{ij})$, $A_{ij} \in \mathbb{R}^{m_i, m_j}$ for any $A \in \mathbb{R}^{M,M}$. For any $x \in \mathbb{R}^M$ let $I(x) := \{i \in \{1, \ldots, p\}, g_i^T x_i + g_{oi} = 0\}$ and denote by G_I the matrix obtained from G by omitting the columns corresponding to $i \notin I$. Let further $Q_I = E_M - G_I P_I$, where E_M is the M×M identity matrix and $P_I = (G_I^T G_I)^{-1} G_I^T$. For $x = x^{(k)}$ denote $I_k = I(x^{(k)})$, $G_k = G_{I_k}$ and P_k, Q_k analogously.

We can state now the algorithm.

<u>The projected block-MOR-N method</u>

Let $x^{(o)} \in S$ be arbitrary. Iterate according to

$$x_i^{(k+1)} = x_i^{(k)} - \gamma_{i,k} t_{i,k},$$

$$(2.4) \quad \gamma_{i,k} = \begin{cases} \omega_{i,k} & \text{if } i \in I(x_{i,k}) \text{ and } g_i^T \, (\nabla f(x_{i,k}))_i < 0, \\ 0 & \text{if } i \in I(x_{i,k}) \text{ and } g_i^T \, (\nabla f(x_{i,k}))_i \geq 0, \\ \min(\omega_{i,k}, \bar{\gamma}_{i,k}) & \text{otherwise,} \end{cases}$$

$$t_{i,k} = \begin{cases} \lambda_{i,k}^{-1} (\nabla f(x_{i,k}))_i, & \text{if } i \in I(x_{i,k}), \\ (A_{ii}(x_{i,k}))^{-1} (\nabla f(x_{i,k}))_i & \text{otherwise,} \end{cases}$$

$$x_{i,k} = (x_1^{(k+1)T}, \ldots, x_{i-1}^{(k+1)T}, x_i^{(k)T}, \ldots, x_p^{(k)T})^T,$$

$$0 < \varepsilon \leq \omega_{i,k} \leq \bar{\omega}_{i,k} - \varepsilon,$$

$$\bar{\gamma}_{i,k} = \begin{cases} (g_i^T x_i^{(k)} + g_{oi}) / g_i^T t_{i,k} & \text{if } g_i^T t_{i,k} > 0, \\ + \infty & \text{otherwise.} \end{cases}$$

Here $\gamma_{i,k}$ is the steplength, $\bar{\gamma}_{i,k}$ the maximal admissible steplength, $t_{i,k}$ the direction vector, $\omega_{i,k}$ the relaxation parameter with bound $\bar{\omega}_{i,k}$ and $\lambda_{i,k}$ has to be chosen as an upper bound for the eigenvalues of $A_{ii}(x_{i,k})$.

The following theorem may be proved for this algorithm.

<u>Theorem 1</u>

Let $x^{(o)} \in S$ be arbitrary and define $S_o = \{x \in S,\ f(x) \leq f(x^{(o)})\}$. Assume that f is twice continuously differentiable on S_o and that the Hessian matrix F(x) of f as well as A(x) are uniformly positive definite on S_o. The sequence $\{x^{(k)}\}$, $k = 0,1,\ldots$ generated by the algorithm (2.4) converges to the unique solution $x^* \in S_o$ of (2.1) for suitable choices of $\bar{\omega}_{i,k}$.

Proof We shall only sketch the proof. As in [4,6] we can utilize a general result of [8] and it remains to show that:

$f(x_{i+1,k}) = f(x_{i,k})$ iff $x_{i,k}$ minimizes f
with respect to all values satisfying
$g_i^T x_i + g_{oi} \geq 0$ and
$f(x_{i+1,k}) < f(x_{i,k})$ iff this is not the case.

Let $x_{i,k}$ be minimizing. Then the Kuhn-Tucker conditions, which are also sufficient here under the assumptions of the theorem, hold

$$(\nabla f(x_{i,k}))_i + \mu g_i = 0 , \quad \mu \in \mathbb{R}$$
$$\mu (g_i^T x_i^{(k)} + g_{oi}) = 0 , \quad \mu \leq 0$$

hence $g_i^T (\nabla f(x_{i,k}))_i \geq 0$. If the i-th constraint is active then $\gamma_{i,k} = 0$ and $(\nabla f(x_{i,k}))_i = 0$ otherwise. In both cases we have $x_{i+1,k} = x_{i,k}$. If $x_{i,k}$ is not minimizing then either $\mu > 0$ if the i-th constraint is active or $(\nabla f(x_{i,k}))_i \neq 0$ otherwise. We have in both cases $(\nabla f(x_{i,k}))_i \neq 0$ and a simple Taylor series expansion along the lines of those in [4,5] yields $f(x_{i+1,k}) < f(x_{i,k})$ for sufficiently small $\bar{\omega}_{i,k}$.

Remark Algorithm (2.4) was not designed for an efficient computation of the approximate solution but essentially for the use as first phase as described above.

3. A preconditioned conjugate gradient method

In this section we shall describe a modification of the projected cg-SMOR-N method of [6] which can be applied to problem (2.1), (2.2).

First we rewrite the gradient in (2.3)

$$\nabla f(x) = \tilde{A}(x)x - \tilde{b}(x) \tag{3.1}$$

where $\tilde{A}_{ii}(x^{(k)}) = \mathrm{diag}\{\lambda_{ik},\ldots,\lambda_{ik}\}$ if $i\in I(x^{(k)})$ and $\tilde{A}_{ij}(x^{(k)})=A_{ij}(x^{(k)})$ otherwise, where λ_{ik} is here an upper bound for the eigenvalues of $A_{ii}(x^{(k)})$, and $\tilde{b}(x) = b(x) + (A(x) - \tilde{A}(x))x$. This modification is similar to that in (2.4) for the case of active constraints. Obviously $\tilde{A}$ is positive definite if A has this property. We consider the auxiliary quadratic minimization problem

$$\tilde{f}_k(x) = \frac{1}{2}\, x^T\tilde{A}(x^{(k)})x - \tilde{b}^T(x^{(k)})x = \underset{S}{\mathrm{Min}} \tag{3.2}$$

and the following symmetric projected relaxation method

$$\begin{aligned} \tilde{x}_i^{(k)} &= x_i^{(k)} - \omega_k\,(Q_k)_{ii}\tilde{A}_{ii}(x^{(k)})^{-1}\nabla\tilde{f}_k(\tilde{x}_{i,k}), \quad i = 1,\ldots,p, \\ \approx\!\!x_i^{(k)} &= \tilde{x}_i^{(k)} - \omega_k\,(Q_k)_{ii}\tilde{A}_{ii}(x^{(k)})^{-1}\nabla\tilde{f}_k(\approx\!\!x_{i,k}), \quad i = p,\ldots,1, \end{aligned} \tag{3.3}$$

where $\tilde{x}_{i,k} = (\tilde{x}_1^{(k)T},\ldots,\tilde{x}_{i-1}^{(k)T},x_i^{(k)T},\ldots,x_p^{(k)T})^T$ and

$\approx\!\!x_{i,k} = (\tilde{x}_1^{(k)T},\ldots,\tilde{x}_i^{(k)T}\approx\!\!x_{i+1}^{(k)T},\ldots,\approx\!\!x_p^{(k)T})^T.$

Now we can give an algorithmic description of

The projected cg-SMOR-N algorithm (3.4)

Let $x^{(0)}\in S$ and $p_{-1}\in\mathbb{R}^M$ be arbitrary. Set $j = 0$, $k = 0$ and from the indicators λ_k, μ_k set $\mu_k = 0$.

<u>Step 1</u> Set $\lambda_k = 0$. Compute $r_k = -\nabla f(x^{(k)})$ and I_k. Terminate the iteration if $\|Q_k r_k\| = 0$, $P_k r_k \le 0$.

Compute $t_{k\ell} = (P_k r_k)_\ell = \max\{(P_k r_k)_i > 0\}$

<u>Step 2</u> If ($\{\|Q_k r_k\| < t_{k\ell}$ and $\mu_k = 0\}$ or $\|Q_k r_k\| = 0$) then set $\tilde{I}_k = I_k - \{\ell\}$, $\lambda_k = 1$, $z_k = \tilde{Q}_k r_k$ and go to <u>Step 3</u> otherwise let $\tilde{I}_k = I_k$.

Compute $z_k = \approx{x}^{(k)} - x^{(k)}$, $\approx{x}^{(k)}$ as obtained from (3.3).

<u>Step 3</u> Compute β_k, p_k according to $p_k = z_k + \beta_k p_{k-1}$,

$$\beta_k = \begin{cases} 0, \text{ if } j = 0 \text{ or } \lambda_k = 1 \text{ or } \mu_k = 1, \\ z_k^T r_k / z_{k-1}^T r_{k-1} \quad \text{otherwise.} \end{cases}$$

<u>Step 4</u> Determine $\bar{\alpha}_k$ as the maximal admissible steplength in direction p_k. Compute $\tilde{\alpha}_k = z_k^T r_k / p_k^T A(x^{(k)}) p_k$ and set $\mu_{k+1} = 0$. If $\tilde{\alpha}_k \ge \bar{\alpha}_k$ then set $\alpha_k = \bar{\alpha}_k$ and $\mu_{k+1} = 1$ while $\alpha_k = \tilde{\alpha}_k$ otherwise.

<u>Step 5</u> Set $x^{(k+1)} = x^{(k)} + \sigma_k \alpha_k p_k$,

$$j = \begin{cases} 0 & \text{if } j = m,\ m > 0, \\ 1 & \text{if } \lambda_k = 1 \\ j+1 & \text{otherwise} \end{cases}$$

$k = k + 1$ and go to <u>Step 1</u>.

This algorithm is a generalization of the algorithm in [6]. It does not, however, include that algorithm as a special case because here it is necessary to insert pure gradient-projection steps after inactivation in <u>Step</u> 2 because z_k may then not be feasible. The use of the auxiliary problem (3.2) with $\tilde{A}$, $\tilde{b}$ instead of A, b as in [6] leads to a symmetric and positive definite matrix connecting z_k and r_k thus assuring the descent property and the possibility of acceleration by conjugate gradients (cf. Lemma 1 in Section 4).

Instead of a steplength formula in connection with a damping parameter also e. g. the Goldstein-Armijo algorithm could be used. In the following theorem we shall restrict ourselves to the algorithm as given above.

Theorem 2 Let f in (2.1) be twice continuously differentiable on $S_o = \{x \in S,\ f(x) \le f(x^{(o)})\}$ and let there be positive constants such that

(3.5) $\quad \eta_o y^T y \le y^T A(x) y \le \eta_1 y^T y$

(3.6) $\quad \bar{\eta}_o y^T y \le y^T F(x) y \le \bar{\eta}_1 y^T y$

for $x \in S_o$, $y \in \mathbb{R}^M$. Assume that the unique minimizer $x^* \in S_o$ also satisfies $P_{I^*} r^* < 0$ where $r^* = -\nabla f(x^*)$. The sequence $\{x^{(k)}\}$, $k = 0,1,\ldots$ generated by the algorithm (3.4) converges to x^* if

(3.7) $\quad o < \varepsilon \le \omega_k \le 2-\varepsilon$

(3.8) $\quad o < \bar{\varepsilon} \le \sigma_k \le 1/\gamma_k$

where γ_k is defined in (4.3).

4. Convergence Proof

Because of the close relationship to the algorithm in [6] we concentrate on those parts of the proof where essential modifications are necessary.

Lemma 1 The sequence $\{z_k\}$, $k = 0,1,\ldots$ in the algorithm (3.4) is uniformly related to the projected negative gradient of f, i. e. there is a constant $\beta > 0$ such that

$$(4.1) \qquad z_k^T r_k = z_k^T \tilde{Q}_k r_k \geq \beta \; \|z_k\| \; \|\tilde{Q}_k r_k\| , \quad k = 0,1,\ldots$$

with $\|\cdot\|$ the Euclidean vector norm.

Proof It remains to compute the matrix connecting z_k and $\tilde{Q}_k r_k$ and we may exclude the case $z_k = \tilde{Q}_k r_k$ (cf. Step 2). An easy computation yields

$$(4.2) \qquad \begin{aligned} z_k &= S_k \tilde{Q}_k r_k \\ S_k &= \omega_k(2-\omega_k)(\tilde{D}_k+\omega_k\tilde{Q}_k\tilde{L}_k^T\tilde{Q}_k)^{-1}\tilde{D}_k^{-1}(\tilde{D}_k+\omega_k\tilde{Q}_k\tilde{L}_k\tilde{Q}_k)^{-1} \end{aligned}$$

where $\tilde{A}(x^{(k)}) = \tilde{D}_k + \tilde{L}_k + \tilde{L}_k^T$ denotes the usual (block-)decomposition into diagonal, lower and upper triangular part. The crucial point allowing this representation of S_k is that $\tilde{D}_k$ and $\tilde{Q}_k$ commute. Using (3.7) and the fact that $\tilde{A}(x^{(k)})$ and hence $\tilde{D}_k$ are uniformly bounded from above and below and that $\tilde{Q}_k$ is a projection matrix uniform lower and upper bounds λ_0, λ_1 for the eigenvalues of S_k may be found. (4.1) is valid with $\beta = \lambda_0/\lambda_1$.

The proofs of the corresponding lemmas in [5,6] may be used without major changes to derive the following results.

Lemma 2 If the parameter σ_k in (3.4) is chosen according to $\sigma_k \leq 1/\gamma_k$

$$(4.3) \qquad \gamma_k = \sup \left\{ \frac{p_k^T F(x^{(k)}+\alpha p_k) p_k}{p_k^T A(x^{(k)}) p_k} , \alpha > 0, f(x^{(k)} + \beta p_k) < f(x^{(k)}), \; x^{(k)} + \beta_k \in S, \; 0 < \beta \leq \alpha \right\}$$

then the following inequalities are valid

$$(4.4) \qquad p_k^T r_k \geq z_k^T r_k \geq 0$$

Lemma 3 For the sequences $\{p_k\}$, $\{z_k\}$, $k = 0,1,\ldots$ in (3.4) there is a constant ρ, $0 < \rho < \infty$, independent of k such that

$$(4.5) \qquad \|p_k\| \leq \rho \; \|z_k\| , \; k = 0,1,\ldots$$

We are now able to state the descent property in

Lemma 4 Let $x^{(k)} \in S_o$ be generated by the algorithm (3.4), $k \geq 0$. Then $x^{(k+1)} \in S_o$ and

$$(4.6) \qquad f(x^{(k)}) - f(x^{(k+1)}) \geq \tilde{c}_k \begin{cases} \|\tilde{Q}_k r_k\|^2, \text{ if } \mu_k = 1, \\ \max\{\|\tilde{Q}_k r_k\|^2, t_{k\ell}\}^2 \text{ otherwise,} \end{cases}$$

where $\tilde{c}_k \equiv c_1 > 0$ for $\mu_{k+1} = 0$ and $\tilde{c}_k = c_2 \bar{\alpha}_k$, $c_2 > 0$, for $\mu_{k+1} = 1$.

Proof Using Lemma 2 and Lemma 14.2.3 in [1o] it may be shown as in [6] that $x^{(k+1)} \in S_o$ and

$$f(x^{(k)}) - f(x^{(k+1)}) \geq \frac{1}{2} \alpha_k \bar{\varepsilon} \gamma_k z_k^T r_k.$$

In case $\alpha_k = \tilde{\alpha}_k$ we estimate further using Lemma 1, Lemma 3, (3.5), (3.6) and (3.8)

$$\geq \frac{1}{2} \frac{\bar{\eta}_o \bar{\varepsilon}^{-2}}{\eta_1^2} \left(\frac{z_k^T r_k}{\|p_k\|}\right)^2$$

$$\geq \frac{1}{2} \frac{\bar{\eta}_o \bar{\varepsilon}^{-2} \beta^2}{\eta_1^2 \rho^2} \|\tilde{Q}_k r_k\|^2$$

while $\alpha_k = \bar{\alpha}_k$ leads to

$$f(x^{(k)}) - f(x^{(k+1)}) \geq \frac{1}{2} \frac{\bar{\eta}_o \bar{\varepsilon}^{-2} \bar{\alpha}_k}{\eta_1} z_k^T r_k$$

$$\geq \frac{1}{2} \frac{\bar{\eta}_o \bar{\varepsilon}^{-2} \lambda_o \bar{\alpha}_k}{\eta_1} \|Q_k r_k\|^2$$

Thus for $\mu_k = 1$ (4.6) is proved. If $\lambda_k = 0$ and $\mu_k = 0$ then $\tilde{Q}_k = Q_k$ and the anti-zig-zagging strategy (cf.[3]) in Step 2 guarantees that $\|\tilde{Q}_k r_k\| \geq t_{k\ell}$. Finally if $\lambda_k = 1$ then we note that

$$t_{k\ell} = \frac{g_\ell^T r_{k\ell}}{g_\ell^T g_\ell}$$

and hence

$$|t_{k\ell}| \leq \frac{\|r_{k\ell}\|}{\|g_\ell\|} \leq \frac{\|\tilde{Q}_k r_k\|}{\min_i \|g_i\|}$$

This completes the proof.

The convergence proof given for the projected cg-SMOR-N algorithm in [6] may now be used replacing the Lagrange multiplier $G_k{}^T r_k$ there by $P_k r_k$.

5. Numerical Results

In the following we shall consider a simple example for problem (2.1), (2.2)

$$\text{(5.1)}\qquad \begin{aligned} & f(y) = f_1(y_1) + f_2(y_2) = \underset{S}{\mathrm{Min}}, \\ & f_i : \mathbb{R}^M \to \mathbb{R},\ i = 1,2, \\ & S = \{y \in \mathbb{R}^{2M},\ y_2 - y_1 + g_0 \geq 0\},\ g_0 \in \mathbb{R}^M . \end{aligned}$$

This problem may e. g. be obtained by discretizing the variational problem for two membranes which are stretched over the same domain $\Omega \subset \mathbb{R}^2$ with initial distance $\varphi(x)$, $x \in \Omega$. Let $u_i(x)$, $i = 1,2$, denote the displacement of the lower and upper membrane and assume that they are pressed by forces $c_1(x) \leq c_2(x)$. For the simplest approximation resulting in a quadratic f cf.[7] for a detailed treatment of contact problems in elasticity cf. [2].

In order to have a nonquadratic test problem we assume that (5.1) is obtained by discretization of

$$\text{(5.2)}\qquad \begin{aligned} & F_1(u_1) + F_2(u_2) = \underset{K}{\mathrm{Min}}, \\ & F_i(u_i) = \int_\Omega (1+|\nabla u_i|^2)^{1/2} dx + \int_\Omega c_i u_i dx, \quad i = 1,2, \\ & K = \{(u_1,u_2) \in W_0^{1,1}(\Omega) \times W_0^{1,1}(\Omega),\ u_1(x) - u_2(x) \leq \varphi(x),\ x \in \Omega\} \end{aligned}$$

and for simplicity we assume $c_i \equiv \text{const}, c_1 < 0 < c_2$. We discretize (5.2) by linear finite elements over a triangulation $\Omega_h = \bigcup_{j=1}^{L} T_j$, T_j closed triangles, of Ω. The integration may be carried out exactly and we get the discrete problem

$$(5.3) \qquad f_h(v_h) = f_{h1}(v_{h1}) + f_{h2}(v_{h2}) = \underset{K_h}{\text{Min}}$$

$$(5.4) \qquad f_{hi}(v_{hi}) = \sum_{j=1}^{L} \gamma_j (1+p_{ij}^2+q_{ij}^2)^{1/2} + c_i \sum_{j=1}^{L} \gamma_j \sum_{k=1}^{3} v_{hi}(P_{jk})$$

$$K_h = \{(v_{h1},v_{h2}) \in C^0(\Omega_h) \times C^0(\Omega_h), v_{hi}|_{T_j} \text{ linear},$$
$$v_{h1}(P_j) - v_{h2}(P_j) \leq \varphi(P_j),\ j=1,\dots,M,$$
$$v_{h1}(P_j) = v_{h2}(P_j) = 0,\ j=M+1,\dots,N\}$$

Here P_j, $j=1,\dots,M\,(M+1,\dots,N)$ denotes the set of interior (boundary) vertices, $p_{ij} = \left.\frac{\partial v_{hi}}{\partial x_1}\right|_{T_j}$, $q_{ij} = \left.\frac{\partial v_{hi}}{\partial x_2}\right|_{T_j}$, $i = 1,2$, $j = 1,\dots,L$,

γ_j = area (T_j) and P_{ji}, $i = 1,3$ are in a different notation the vertices of T_j numbered in mathematical positive sense. Problem (5.3) is obviously of the form (5.1). For an explicit expression of p_{ij}, q_{ij} in terms of the nodal values $v_{hi}(P_j)$, $i = 1,2$, $j = 1,\dots,M$ cf. e. g. [5]. In order to have a problem of the form (2.1), (2.2) we choose

$$(5.5) \qquad y = (v_{h1}(P_1), v_{h2}(P_1),\dots,v_{h1}(P_M),v_{h2}(P_M))^T$$

in (5.1), $g_o = (\varphi(P_1),\dots,\varphi(P_M))^T$ in (2.1) and $g_i = (1,-1)^T$, $i=1,\dots,M$ in (2.2).

We thus have a problem of dimension 2M with 2×2 blocks. In order to be able to apply the algorithms of Section 2 and 3 it is necessary to write ∇f in the form (2.3). The matrix $A_i \in \mathbb{R}^{M,M}$ resulting from the first term in (5.4) was given in [5] and we shall not repeat it here. The second term in (5.4) yields a vector $b_i \in \mathbb{R}^M$. Then the matrix A in (2.3) and the vector b are obtained by rearranging the partitioned matrix and vector

$$\begin{pmatrix} A_1 & 0 \\ 0 & A_2 \end{pmatrix}, \quad \begin{pmatrix} b_1 \\ b_2 \end{pmatrix}$$

according to the order in (5.5).

The results of [5] may then be used to show that the assumptions of Theorem 1 and Theorem 2 are satisfied. For the practical application of the algorithms (2.4) and (3.4) we remark that the matrices A_{ii}, $i = 1,\dots,M$ are diagonal and hence the block-relaxation may be carried out without solving linear subsystems and it is also not difficult to determine the $\lambda_{i,k}$.

In the following we present graphically the results for problem (5.2) with Ω the unit square, $\varphi(x) \equiv 0.2$ and for different constant values of c_1, c_2. The discretization (5.3), (5.4) was used with the standard triangulation obtained by dividing a uniform square mesh. The numbers M resp. N in (5.4) were 324 resp. 4oo. The algorithm (3.4) was used with starting vectors $v_{h1}^{(o)} = v_{h2}^{(o)} \equiv 0$, $\omega_k \equiv \omega = 1.3$ and $m = 8$. In the figures the diagonals of the squares were not plotted and the two surfaces are shown separately. Below each figure the values of c_1, c_2 together with the number M_c of contact points is given.

In Figure 1 the force acting on the lower surface was kept constant and that of the upper was varied, while in Figure 2 the force on the upper surface was fixed on a higher absolute value. Neither theoretically nor practically there would be difficulties to treat more general problems, e.g. varying c_i, φ, other domains or other variational inequalities.

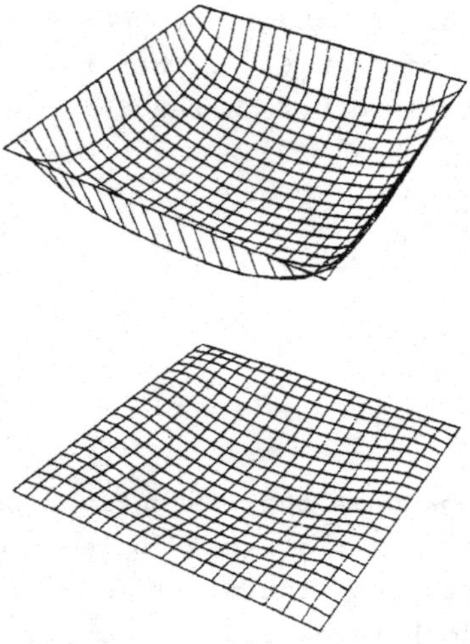

Fig. 1a: $c_1=-2$, $c_2=3.5$, $M_c=280$

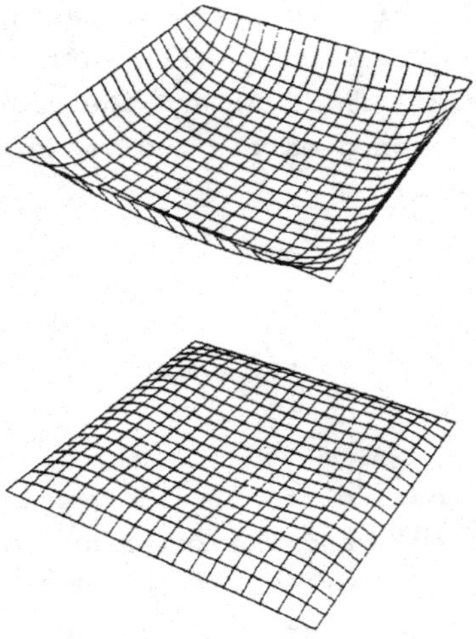

Fig. 1b: $c_1=-2$, $c_2=3$, $M_c=221$

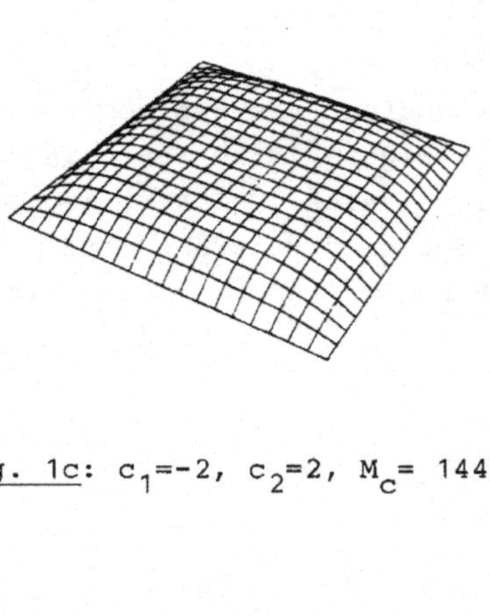

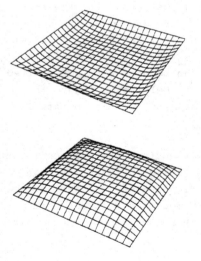

Fig. 1c: $c_1=-2$, $c_2=2$, $M_c=144$

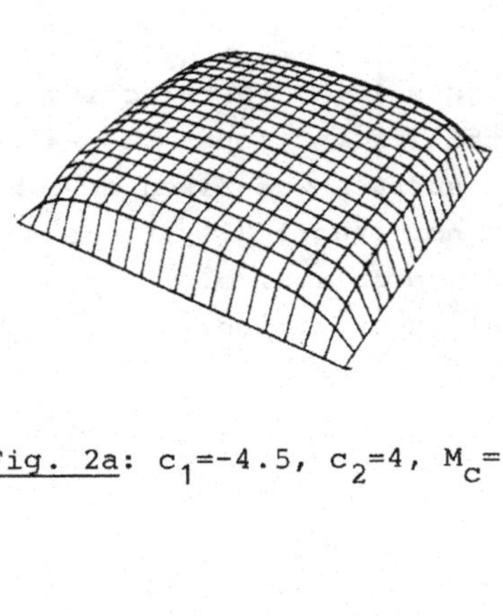

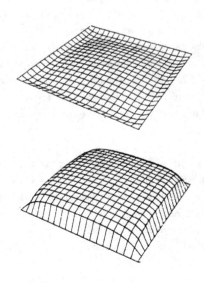

Fig. 2a: $c_1=-4.5$, $c_2=4$, $M_c=282$

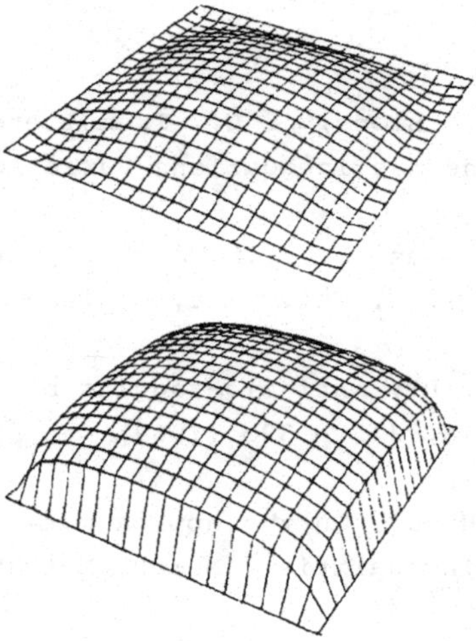

Fig. 2b: $c_1=-5$, $c_2=4$, $M_c= 286$

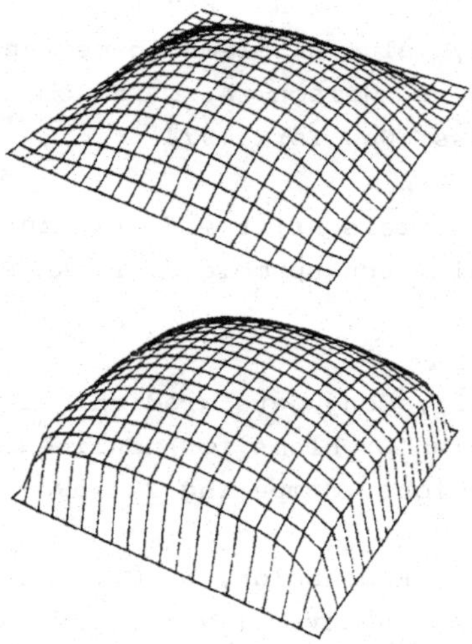

Fig. 2c: $c_1=-6$, $c_2=4$, $M_c= 286$

References

[1] Glowinski, R., Lions, J. L., Tremolieres, R., Approximations des inéquations variationelles. Paris, Dunod 1976

[2] Kikuchi, N., Oden, J. T., Contact porblems in elasticity, TICOM report 79-8, University of Texas at Austin 1979

[3] McCormick, G. P., Anti-zig-zagging by bending, Manag. Sci. 15, 315 - 32o (1969)

[4] Mittelmann, H. D., On the approximate solution of nonlinear variational inequalities, Numer. Math. 29, 451 - 462 (1978)

[5] Mittelmann, H. D., On the efficient solution of nonlinear finite element equations I, to appear in Numer. Math.

[6] Mittelmann, H. D., On the efficient solution of nonlinear finite element equations II, submitted to Numer. Math.

[7] Oden, J. T., Kikuchi, N., Finite element methods for certain free boundary value problems in mechanics, in "Moving boundary problems", D. G. Wilson, A. D. Solomon, P.T. Boggs (eds.), Academic Press, New York 1978

[8] Oettli, W., Einzelschrittverfahren zur Lösung konvexer und dual-konvexer Minimierungsprobleme, Z. Angew. Math. Mech. 54, 334-351 (1974)

[9] O'Leary, D. P., Conjugate gradient algorithms in the solution of optimization problems for nonlinear elliptic partial differential equations. Computing 22, 59 - 77 (1979)

[1o] Ortega, J. M., Rheinboldt, W. C., Iterative solution of nonlinear equations in several variables. Academic Press, New York 197o

POSITIVE AND SPURIOUS SOLUTIONS OF NONLINEAR EIGENVALUE PROBLEMS

BY

H.O. PEITGEN*

AND

K. SCHMITT**

*) Forschungsschwerpunkt "Dynamische Systeme"
Fachbereich Mathematik
Universität Bremen
D-2800 Bremen 33

**) Department of Mathematics
University of Utah
Salt Lake City, Utah 84112
USA

Positive and Spurious Solutions of Nonlinear Eigenvalue Problems

BY

Heinz-Otto Peitgen*

AND

Klaus Schmitt**

1. Introduction

The aim of this paper is to discuss a class of nonlinear elliptic eigenvalue problems and finite dimensional approximations thereof. We shall show how global topological perturbations may be used to prove the existence of solutions of both of these problems by virtually the same set of arguments. In addition we demonstrate that the finite dimensional problems have a solution structure which is much more involved than that of the infinite dimensional problem under consideration, giving rise to solutions which are termed *'numerically irrelevant'*, *'ghost solutions'* or *'spurious'*, i.e. solutions which in no way approximate solutions of the corresponding infinite dimensional problem. In fact we shall be able to distinguish among three different types of spurious solutions of the nonlinear boundary value problem

$$(1.1) \qquad \begin{cases} u'' + \lambda f(u) = 0 \\ u(0) = u(1) = 0 , \end{cases}$$

*Forschungsschwerpunkt "Dynamische Systeme", Fachbereich Mathematik, Universität Bremen, D-2800 Bremen, West Germany. Supported by Stiftung Volkswagenwerk.

**Department of Mathematics, University of Utah, Salt Lake City, Utah 84112, USA. Supported by NSF Research grant MCS 800 1886.

where f is a nonlinearity of the type indicated in Figure 1.

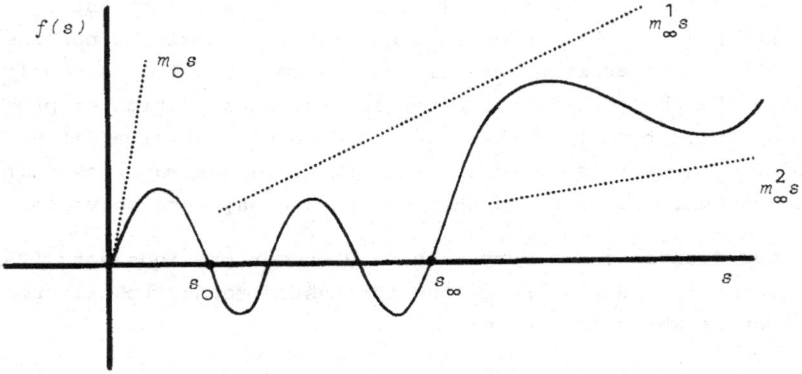

FIGURE 1

Problems of type (1.1) and their elliptic partial differential equations analogues arise in many applications and represent also the steady state part of certain reaction-diffusion processes. Also when studying nonlinear delay-differential equations of the type

$$(1.2) \qquad u'(t) = -\lambda \int_{-1}^{0} f(u(t+a))\,da ,$$

where da is Lebesgue measure and f is an odd nonlinearity special periodic solutions u(t), of period 2, which are such that

$$\begin{cases} u(t+1) = -u(t) \\ u(-t) = -u(t) \end{cases} \qquad 0 \le t \le 1$$

have played an important role. It turns out (see [An]) that such solutions are also solutions of the nonlinear boundary value problem

$$(1.3) \qquad \begin{cases} u'' + 2\lambda f(u) = 0 \\ u(0) = 0 = u'(\tfrac{1}{2}) \end{cases} ,$$

and conversely solutions of (1.3) give rise to such special periodic solutions of (1.2).

Spurious solutions have been discovered in many contexts (see e.g. [A1], [B], [BD], [G], [SS], [K], [SA], [PSS]) and their study is an important part of the analysis and numerical analysis of nonlinear systems. Another interesting example may be obtained from a result of Pohozaev [P] who showed that a certain nonlinear Dirichlet problem has no solutions. If one, on the other hand, employs finite difference methods, such as those discussed in section 4 one may show that the finite dimensional problems thus obtained always are solvable.

In our discussion we very much rely on our previous work [PSS], [PS_1] and [PS_2] and we refer the interested reader to all proofs in these papers which are not given here.

Some of the infinite dimensional results represent extensions of results on nonlinear eigenvalue problems previously considered in [AH]. Our approach to such problems demonstrates the versatility of topological continuation methods both from a mathematical as well as a computational point of view.

This work is organized as follows: We first discuss the infinite dimensional problem, then the corresponding finite dimensional ones, here in particular we show how spurious solutions arise. We then present a series of computational results which were obtained using the constructive piecewise linear continuation methods developed in [PP], [JPS] and [Sa]. H. Jürgens and D. Saupe have implemented these ideas in an interactive code 'SCOUT' (Simplicial COntinuation UTilities) which is designed to trace solution continua of multiparameter problems.

The computations were performed using the PDP 11-60 at the Department of Mathematics of the University of Utah, all computer pictures were obtained by means of the Evans and Sutherland picture system in the computer laboratory of the same department.

We thank Hartmut Jürgens and Dietmar Saupe for their excellent computer work.

2. The Infinite Dimensional Problem and Global Perturbations

Let Ω be a bounded domain in R^n, $n \geq 1$, with boundary $\partial\Omega$ of class $C^{2+\alpha}$, $\alpha \in (0,1)$. Let a_{ij}, b_i, c, $1 \leq i,j \leq n$ be real valued Lipschitz continuous functions on $\bar{\Omega}$ with $c(x) \leq 0$, $x \in \bar{\Omega}$. The matrix $(a_{ij}(x))$ is assumed to be positive definite on $\bar{\Omega}$.

Let L denote the elliptic operator

$$Lu = \sum_{i,j=1}^{n} a_{ij}(x)\frac{\partial^2 u}{\partial x_i \partial x_j} + \sum_{i=1}^{n} b_i(x)\frac{\partial u}{\partial x_i} + c(x)u ,$$

which we shall assume to be self-adjoint.

We consider the nonlinear Dirichlet problem

$$(2.1) \qquad \begin{cases} L u + \lambda f(u) = 0 , & x \in \Omega \\ u = 0 , & x \in \partial\Omega , \end{cases}$$

where $\lambda \geq 0$ is a real parameter and $f\colon R \to R$ is Lipschitz continuous.

We assume that f satisfies in addition the following conditions:

$$(2.2) \qquad \begin{cases} f(s) = m_o s + o(s) , & \text{as} \quad s \to 0+ \\ m_\infty^1 s + o(s) \geq f(s) \geq m_\infty^2 s + o(s) & \text{as} \quad s \to \infty , \end{cases}$$

where, either

$$(2.3) \qquad m_o > 0 , \quad m_\infty^1 \geq m_\infty^2 > 0$$

or

$$(2.4) \qquad m_o < 0 , \quad m_\infty^1 \geq m_\infty^2 > 0 .$$

Throughout we shall be interested in the existence of positive solutions of (2.1), i.e. solutions (λ,u) such that $u(x) \geq 0$, $x \in \bar{\Omega}$.

It will be convenient therefore to extend $f(s)$ for $s < 0$ such that

$$(2.5) \qquad f(s) \geq \alpha s \ , \quad s < 0 \ ,$$

where α is some negative constant. Then, due to the maximum principle, (2.1) will only admit positive solutions.

Because of the smoothness assumptions imposed on the coefficients of L, the nonlinearity f, and the domain Ω, problem (2.1) will, due to regularity theory (see [GT]), be equivalent to an operator equation

$$(2.6) \qquad u = L(\lambda F(u)) \ ,$$

where $L\colon C^0(\overline{\Omega}) \to C^0(\overline{\Omega})$ is a compact positive linear operator and F is the Nemitskii operator defined by f.

The linear problem

$$(2.7) \qquad \begin{cases} Lu + \mu u = 0 \ , & x \in \Omega \\ u = 0 \ , & x \in \partial\Omega \ , \end{cases}$$

has a first characteristic value μ_0 ([GT]) which is positive and geometrically and algebraically simple, the corresponding eigenspace being spanned by an element $u \in C^0_+(\overline{\Omega}) = \{ v : v(x) \geq 0\}$, in fact $u(x) > 0$, $x \in \Omega$.

We let $\lambda_0 = \dfrac{\mu_0}{m_0}$, $\lambda^i_\infty = \dfrac{\mu_0}{m^i_\infty}$, $i = 1,2$,

in case $m_0 > 0$.

The following existence results describe solution sets of problem (2.1).

THEOREM 2.1. *Assume* f *satisfies* (2.2) *and* (2.3). *Then there exist unbounded continua* $\Sigma_0, \Sigma_\infty \subset R \times C^0(\bar{\Omega})$ *of solutions of* (2.1) *with the following properties:*

(i) $(\lambda,u) \in \Sigma_i$, $i = 0,\infty$, $u \neq 0$, *implies* $u(x) > 0$, $x \in \Omega$.

(ii) Σ_0 *bifurcates from* $(\lambda_0, 0)$.

(iii) *For all* $\varepsilon > 0$ *there exists* $\{(\lambda_n, u_n)\} \subseteq \Sigma_\infty$, $\{\lambda_n\} \subseteq [\lambda_\infty^1 - \varepsilon, \lambda_\infty^2 + \varepsilon]$ *such that* $||u_n||_\infty \to \infty$.

(iv) *If* $\{s: f(s) < 0\} \neq \phi$, *then* $\Sigma_0 \cap \Sigma_\infty = \emptyset$.

(v) *If* $s_0 = \min \{s: f(s) < 0\}$, *then* $(\lambda,u) \in \Sigma_0$, *implies* $||u||_\infty < s_0$.

(vi) *If* $s_\infty = \max \{s: f(s) < 0\}$, *then* $(\lambda,u) \in \Sigma_\infty$, *implies* $||u||_\infty > s_\infty$.

(vii) *If* $\{s: f(s) < 0\} \neq \phi$, *then* Σ_0, *respectively* Σ_∞ *contain solutions* (λ,u) *for all* $\lambda \geq \lambda_0$, *respectively* $\lambda > \lambda_\infty^2$.

(viii) *For any* $\varepsilon > 0$ *there exists* $R > 0$ *such that, if* $(\lambda,u) \in \Sigma_\infty$, $\lambda \geq \lambda_\infty^2 + \varepsilon$, *then* $||u||_\infty < R$.

THEOREM 2.2. *Assume that* f *satisfies* (2.2) *and* (2.4). *Then there exists an unbounded continuum* $\Sigma_\infty \subset R \times C^0(\bar{\Omega})$ *of solutions of* (2.1) *such that:*

(i) $(\lambda,u) \in \Sigma_\infty$, *implies* $u(x) > 0$, $x \in \Omega$.

(ii) *For all* $\varepsilon > 0$ *there exists* $\{(\lambda_n, u_n)\} \subseteq \Sigma_\infty$,

$\{\lambda_n\} \subset [\lambda_\infty^1 - \varepsilon, \lambda_\infty^2 + \varepsilon]$ *such that* $||u_n||_\infty \to \infty$.

(iii) $(\lambda, u) \in \Sigma_\infty$, *implies* $||u||_\infty > s_\infty$.

(iv) Σ_∞ *contains solutions* (λ, u) *for all* $\lambda > \lambda_\infty^2$ *and for any* $\varepsilon > 0$ *there exists* R *such that* $||u||_\infty < R$ *for any solution* $(\lambda, u) \in \Sigma_\infty$ *with* $\lambda \geq \lambda_\infty^2 + \varepsilon$.

Before proceeding to the statements of several auxiliary results which are used in verifying these results let us briefly indicate how a lower bound (better than that stated in either theorem 2.1 or theorem 2.2) may be obtained on $||u||_\infty$ for $(\lambda, u) \in \Sigma_\infty$, in the case of problem (1.1), i.e. the ordinary differential equations case.

Thus let us consider problem (1.1) for positive solutions. This is equivalent to finding a solution of

$$\text{(2.8)} \qquad \begin{cases} u' = v \\ v' = -\lambda f(u) \end{cases}$$

which is such that $u(0) = 0 = u(1)$.

Integrating (2.8) with initial conditions $u(t_o) = v_o$ one obtains that along a solution $(u(t), v(t))$ the energy

$$\frac{1}{2}v^2 + \lambda \int_{u_0}^{u} f(s)\,ds = \frac{1}{2}v_0^2$$

is constant.

Because of the structure of the nonlinearity f, there will exist $r_\infty > s_\infty$ such that

$$\int_{s^*}^{r_\infty} f(s)\,ds = 0 \ ,$$

where $s^* = \max \{s\colon f(s) = 0$ and $\int_0^s f(\tau)\,d\tau > 0$ and $s < s_\infty\}$

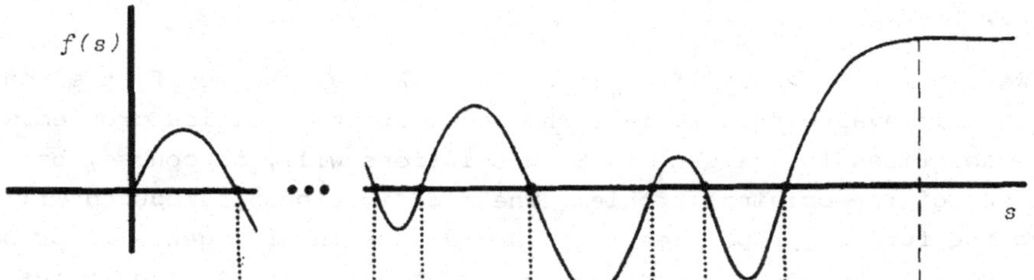

FIGURE 2

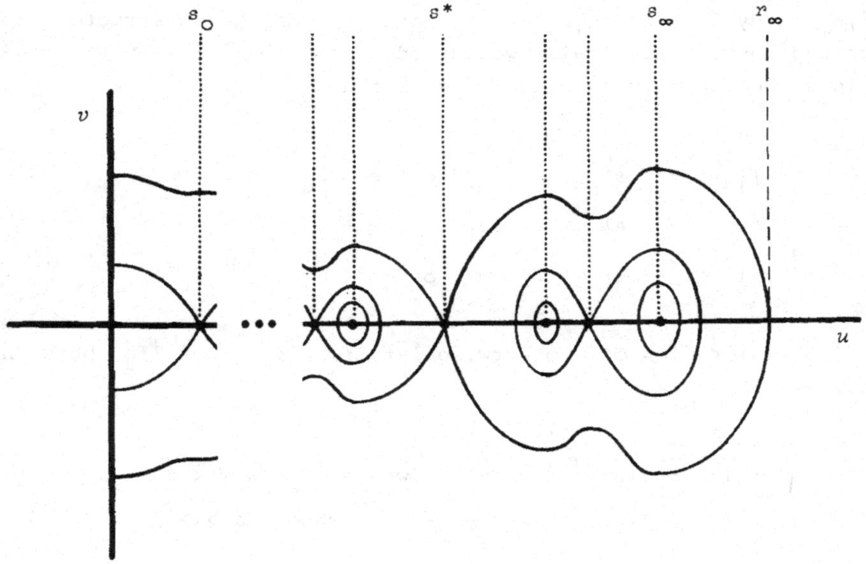

FIGURE 3

The phase portrait of the system will then be as in Figure 3.

It thus follows that if $(\lambda,u) \in \Sigma_\infty$ it must be the case that in fact $||u||_\infty > r_\infty$, where r_∞ is independent of λ.

In case $m_0 < 0$ a similar lower bound may be obtained.

Let us now briefly sketch a proof of theorem 2.1, in proving theorem 2.2 one may proceed similarly. Complete details may be found in [PSS], [PS_1], [PS_2].

We choose $\mu_1 > \max\{\lambda_0,\lambda_\infty^i\}$ and for $\lambda > \mu_1$ we modify the nonlinear eigenvalue problem in such a way that the modified problem will have solutions for $\lambda \leq \mu_1$. Such solutions will, of course, be solutions of the original problem. The modifications introduced will have the further properties that the new nonlinear eigenvalue problem will have an unbounded solution continuum which is bounded in the λ-direction and bifurcates from $(\lambda_0,0)$ and from ∞ in the interval $[\lambda_\infty^1,\lambda_\infty^2]$. By letting $\mu_1 \to \infty$, Σ_0 and Σ_∞ may be constructed from such continua. Specifically we introduce a global perturbation as follows. Define a function f_1 such that

$$\begin{cases} f_1(s) \equiv f(s) , & 0 \leq s \leq s_0 \\ f_1(s) \equiv f(s) , & s \geq s_\infty \\ f_1(s) < 0 , & s_0 < s < s_\infty , \end{cases}$$

if $\{s: f(s) < 0\} \neq \phi$, otherwise let $f_1 \equiv f$. Let f_2 be such that

$$\begin{cases} f_2(s) \equiv f(s) & s \leq \delta , \text{ where } 0 < \delta << 1 \\ f_2(s) \equiv f(s) & s > s_\infty + \Delta , \text{ where } \Delta >> 1 \end{cases}$$

and so that

$$f_2(s) \geq \beta s , \qquad s \geq 0 ,$$

where β is some positive constant.

Choose constants

$$\mu_1 < \mu_2 < \mu_3 < \mu_4 ,$$

$\mu_4 > \frac{\mu_0}{\beta}$, and define the perturbation $\tilde{f}(\lambda,s)$ of $\lambda f(s)$ by

$$\tilde{f}(\lambda,s) = \begin{cases} \lambda f(s) , & 0 \le \lambda \le \mu_1 \\ \lambda f_1(s) , & \mu_2 \le \lambda \le \mu_3 \\ \lambda f_2(s) , & \mu_4 \le \lambda , \end{cases}$$

and for $\lambda \in [\mu_1,\mu_2] \cup [\mu_3,\mu_4]$ $\tilde{f}$ is to be a linear function of λ so that

$$\tilde{f}: [0,\infty) \times R \to R$$

is Lipschitz continuous.

We consider now the perturbed problem

$$(2.10) \qquad u = L\,\tilde{F}(\lambda,u) ,$$

where L is as defined before and $\tilde{F}$ is the Nemitskii operator defined by $\tilde{f}$. It follows that solutions of (2.10) for $\lambda \le \mu_1$ will be solutions of (2.1).

The following set of lemmata may now be established (see [PSS, PS_2]).

LEMMA 2.3. *If* (λ,u) *is a solution of* (2.10), $\lambda \ge 0$, *then* $u \in C^0_+(\bar{\Omega}) = \{u \in C^0(\bar{\Omega}) : u(x) \ge 0\}$ *and if* u *is nontrivial, then* $u(x) > 0$, $x \in \Omega$.

LEMMA 2.4. *If* $\lambda \geq \mu_4$ *, then* (2.10) *has no nontrivial solutions in* $R \times C_+^0(\bar{\Omega})$.

LEMMA 2.5. *If* Λ *is a closed real interval with* $\lambda_0 \notin \Lambda$ *, then there exists* $r > 0$ *such that if* (λ, u) *is a solution of* (2.10), $\lambda \in \Lambda$, $u \not\equiv 0$, *then* $||u||_\infty \geq r$.

COROLLARY 2.6. *The only possible point of bifurcation* $(\lambda, 0)$ *from the trivial solution is for* $\lambda = \lambda_0$.

LEMMA 2.7. *Let* $\varepsilon > 0$ *be given, then there exists* $R = R(\varepsilon)$ *such that* $||u||_\infty < R$ *for all solutions* (λ, u) *of* (2.10) *with* $\lambda \geq \lambda_\infty^2 + \varepsilon$.

LEMMA 2.8. *Let* $\varepsilon > 0$ *be given, then there exists* $R = R(\varepsilon)$ *such that* $||u||_\infty < R$ *for all solutions* (λ, u) *of* (2.10) *with* $\lambda \leq \lambda_\infty^1 - \varepsilon$.

LEMMA 2.9. *If* s *is such that* $\tilde{f}(\lambda, s) < 0$ *, then there does not exist a solution* u *of* (2.10) *such that* $||u||_\infty = s$.

LEMMA 2.10. (i) $\deg_{LS}(id - L\tilde{F}(\lambda, \cdot), B_r, 0) = 1$, $\lambda < \lambda_0$ *and* $r > 0$ *sufficiently small.*

(ii) $\deg_{LS}(id - L\tilde{F}(\lambda, \cdot), B_r, 0) = 0$, $\lambda > \lambda_0$ *and* $r > 0$ *sufficiently small.*

It hence follows from lemma 2.10 that $\lambda = \lambda_0$ must be such that $(\lambda_0, 0)$ is a bifurcation point from trivial solutions for the perturbed problem (2.10).

Moreover the change of degree at $\lambda = \lambda_0$ implies global bifurcation due to the Krasnosel'skii-Rabinowitz bifurcation theorem [R]. Because of the special regions of nonexistence in $R_+ \times C_+^0(\bar{\Omega})$ (indicated by the shaded areas of Figure 4) which have been established by the lemmata above, this yields the existence of an unbounded continuum of positive solutions of (2.10) which bifurcates from infinity in $[\lambda_\infty^1, \lambda_\infty^2]$.

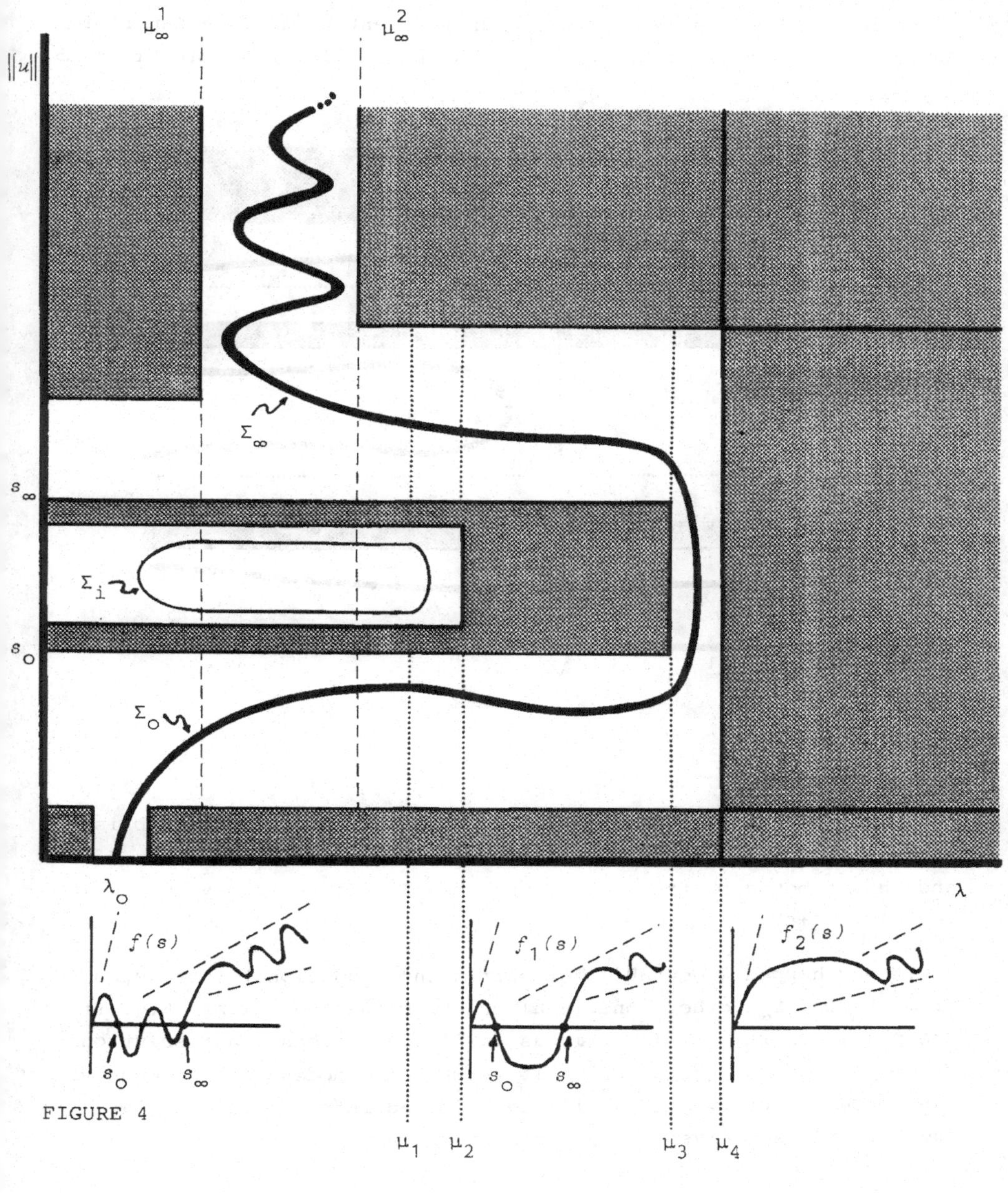

FIGURE 4

In summary: We have deduced the existence of a branch bifurcating from infinity from a branch bifurcating from zero, via the perturbed equation. We emphasize that, since f is not assumed to be asymptotically linear, the bifurcation from infinity could not be obtained by linearization techniques.

We may now let $\mu_1 \to \infty$ and use an argument based on a paper of Stuart [St] to deduce theorem 2.1 which is illustrated in Figure 5.

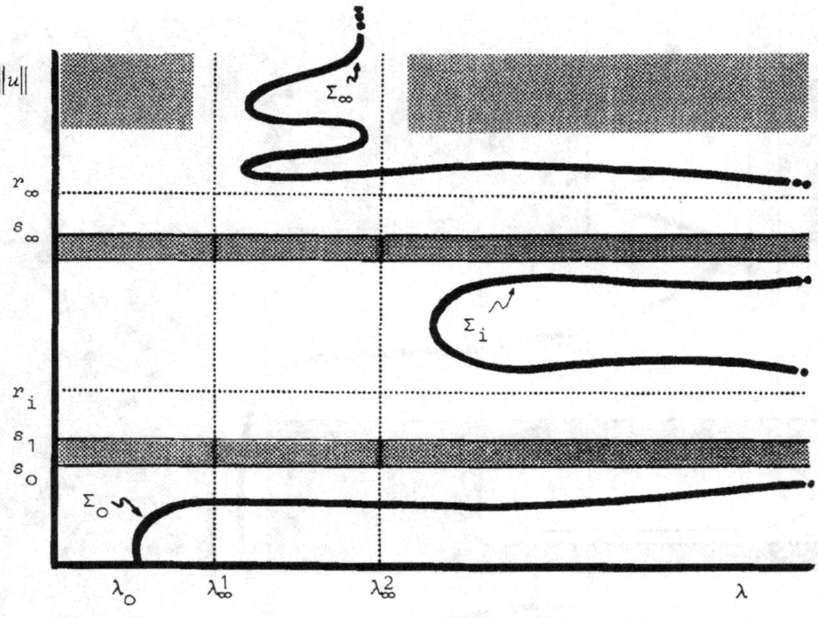

FIGURE 5

In this connection we remark that the a priori bounds established in lemmata 2.7 and 2.8 are in fact independent of μ_i, i=1,...,4 and thus a bound on the u component of Σ_∞ is uniformly available for $\lambda \geq \lambda_\infty^2+\varepsilon$.

It may happen (see [BB] for sufficient conditions) that aside from Σ_0 and Σ_∞ other nontrivial solution branches exist. E.g. if the phase portrait of (1.1) is as in Figure 6, then other solution continua exist (here Σ_i). It seems to be noteworthy to observe that such a continuum Σ_i will be deformed into a 'loop' via the perturbation given by $\tilde{f}$ (see Figure 4).

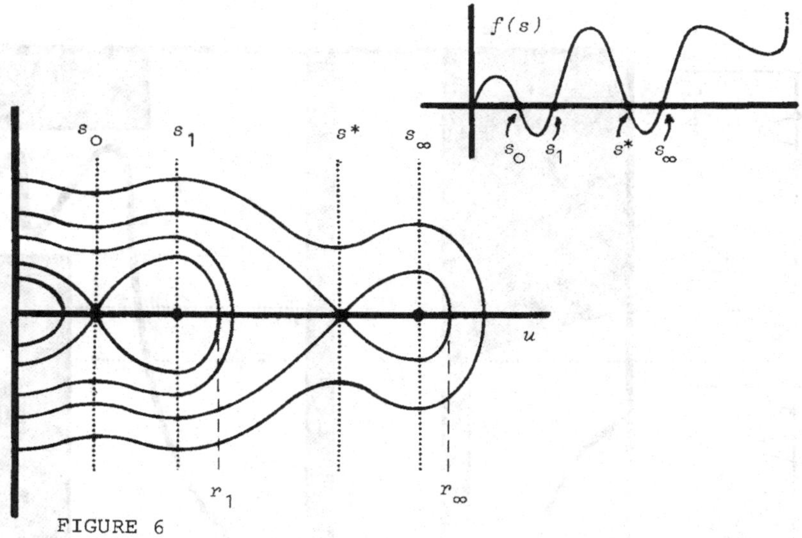

FIGURE 6

These solution continua Σ_i will then, of course, satisfy: $(\lambda,u) \in \Sigma_i$ implies $r_1 < ||u||_\infty < s_*$.

The solution diagram in the case of theorem 2.2 will be as given in Figure 7a. Figure 7b shows a typical nonlinearity with $m_0 < 0$.

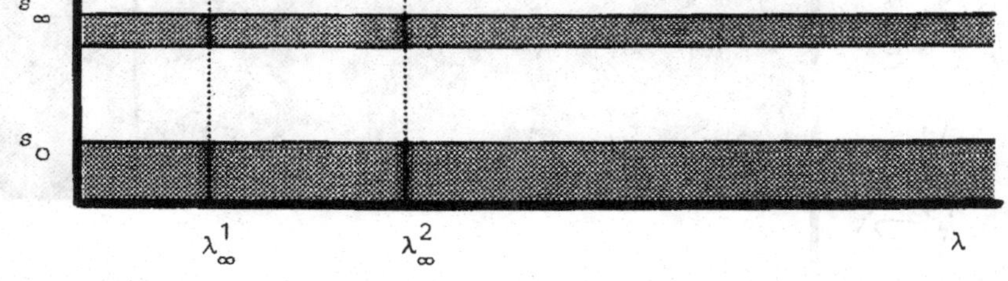

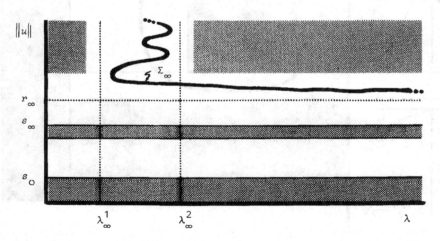

FIGURE 7a

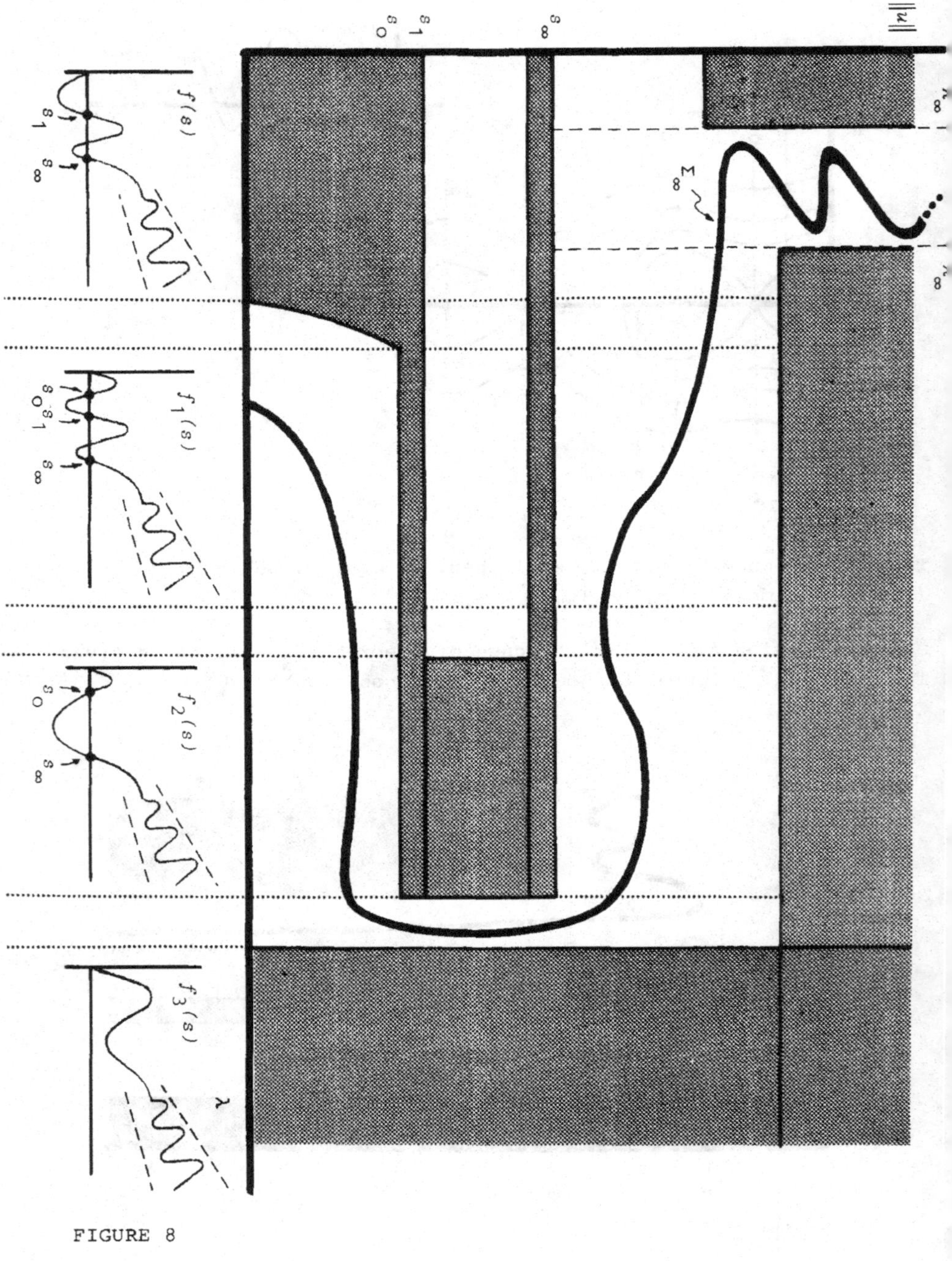

FIGURE 8

An argument based on Figure 8 may be used to deduce theorem 2.2 from theorem 2.1. Namely one perturbes the original problem to a problem of the type discussed before, i.e. a problem which allows for bifurcation from trivial solutions and thus one obtains the branch bifurcating from infinity via an artificially introduced bifurcation from zero.

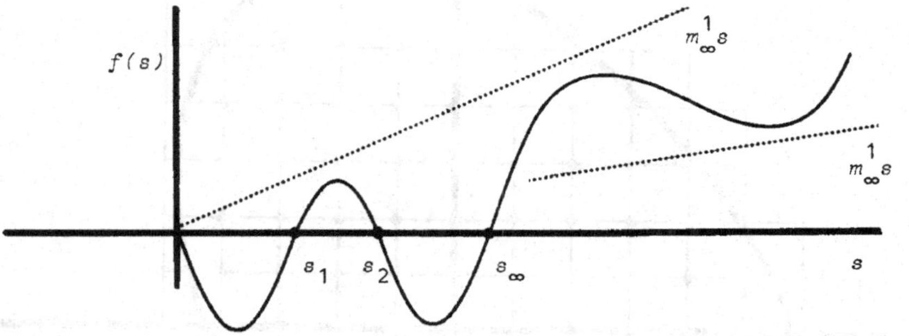

FIGURE 7b

3. The Approximate Problem

We shall consider here finite dimensional approximations of problem (2.1). Many different difference schemes have been treated in the literature (see e.g. [C], [IK]) and are being used to solve elliptic partial differential equations numerically. We restrict ourselves here, for purposes of illustration to a standard 5 point symmetric difference scheme which may be used for numerical treatment of the two-dimensional problem

$$(3.1) \qquad \begin{cases} \Delta u + \lambda f(u) = 0 , & x \in \Omega \\ u = 0 , & x \in \delta\Omega , \end{cases}$$

where $\Omega \subset R^2$ is a bounded domain and Δ is the two-dimensional Laplace operator. One superimposes on R^2 a square grid with edges parallel to the coordinate axes and of width h. One approximates Δu at those grid points which are interior to Ω. We label the grid points using integer double subscripts as indicated in Figure 9.

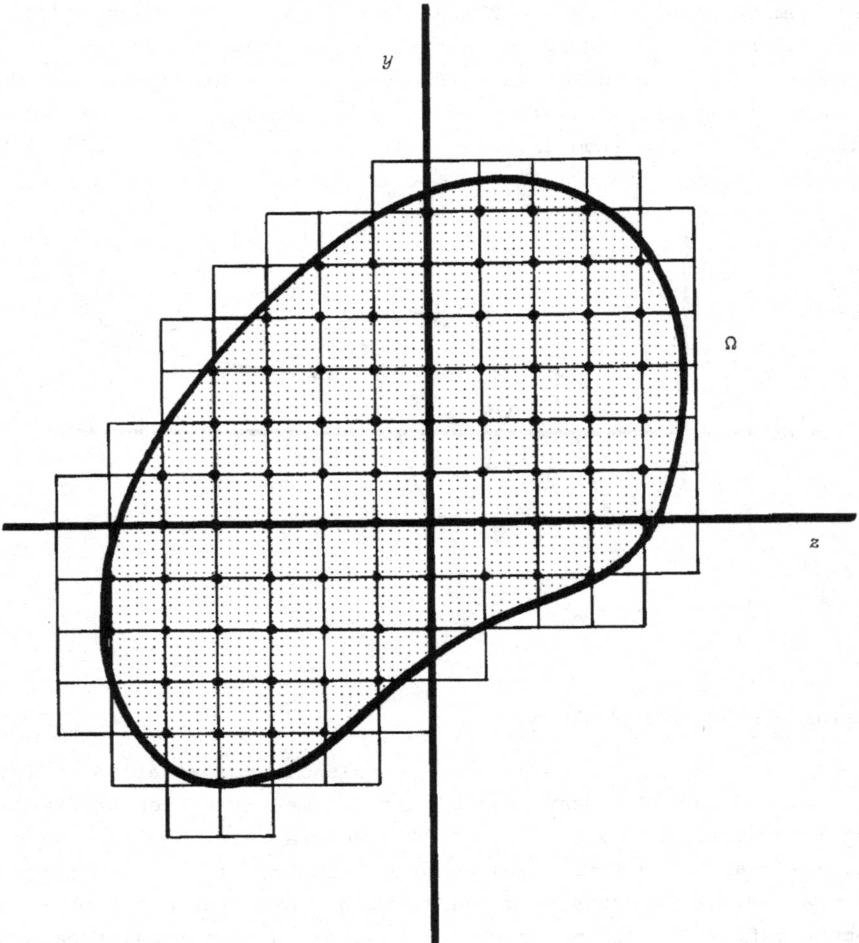

FIGURE 9

Here we put $x = (x_1, x_2)$, $x_1 = z$, $x_2 = y$ $z_i = ih$, $y_i = ih$, $i = 0, \pm 1, \pm 2, \ldots$ and put $u_{i,j} = u(z_i, y_i)$. If $(z_i, y_i) \in \Omega$ we replace Δu by

$$\Delta_{i,j}u := \frac{1}{6h^2}[u_{i-1,j+1} + 4u_{i,j+1} + u_{i+1,j+1} + 4u_{i-1,j}$$

$$- 20u_{i,j} + 4u_{i+1,j} + u_{i-1,j-1} + 4_{i,j-1} + u_{i+1,j-1}],$$

where we put $u_{k,l} = 0$ if $(z_k, y_l) \notin \Omega$.
Finally we replace (3.1) by

$$\Delta_{i,j}u + \lambda f(u_{i,j}) = 0 \tag{3.2}$$

This system may now be rewritten in matrix form by introducing the vector $u = (u_{ij}) = (u_1, \ldots, u_n)$ where n is the number of grid points interior to Ω and the ordering employed is for example that obtained by sweeping out the grid points in a horizontal direction starting at the upper left and ending at the lower right. The system then will become

$$Au = \mu F(u) \ , \tag{3.3}$$

where A is a matrix of the form

$$A = \begin{bmatrix} 20 & a_{12} & \cdots\cdots & a_{1n} \\ a_{21} & 20 & a_{23} \cdots & a_{2n} \\ \vdots & \vdots & \ddots & \vdots \\ a_{n1} & \cdots & a_{n,n-1} & 20 \end{bmatrix}$$

where $a_{ij} \leq 0$ if $i \neq j$ and

$$20 \geq \sum_{\substack{i \neq j \\ j=1}}^{n} |a_{ij}| \ ,$$

with strict inequality holding for at least one i ; further $\mu = 6\lambda h^2$ and $F_i(u) = f(u_i)$.

We have thus obtained a system (3.3) where A is an M-matrix (see [Sc]).

Let now f satisfy (2.2) and (2.3) and let f have $2m$ positive zeros, labelled as

$$0 < s_0 < s_1 < \ldots < s_{2m-1} = s_\infty .$$

In this case the following result is valid.

THEOREM 3.1. *Let* μ_0 *denote the smallest characteristic value of* A *, and let* $\lambda_0 = \frac{\mu_0}{m_0}$, $\lambda_\infty^i = \frac{\mu_0}{m_\infty^i}$. *Then :*

(i) $(\lambda_0,0)$ *is a bifurcation point for* (3.3) *from which an unbounded continuum* Σ_0^o *of nontrivial solutions of* (3.3) *emanates and* $(\mu,u) \in \Sigma_0^o$ *implies* $||u||_\infty < s_0$ *and the projection of* Σ_0^o *onto* R *fills* (λ_0,∞) .

(ii) *There exists an unbounded continuum* Σ_∞^o *of nontrivial solutions of* (3.3) *such that for all* $\varepsilon > 0$ *there exists a sequence* $\{(\mu_n,u_n)\} \subseteq \Sigma_\infty^o$ *with* $\{\mu_n\} \subseteq [\lambda_\infty^1-\varepsilon,\lambda_\infty^2 + \varepsilon]$ *and* $||u_n||_\infty \to \infty$. *Also there exists* $R > 0$, $R = R(\varepsilon)$, *such that* $(\mu,u) \in \Sigma_\infty^o$ *implies* $||u||_\infty < R$. *Further the projection of* Σ_∞^o *onto* R *includes* $(\lambda_\infty^2,\infty)$.

(iii) *If* $(\lambda,u) \in \Sigma_0^0 \cup \Sigma_\infty^0$, *then* $u \in R_+^n$.

(iv) *Equation* (3.3) *has no solutions* (μ,u) *such that*

$$s_{2i} < ||u||_\infty < s_{2i+1} \ , \quad i = 0,\ldots, m-1 \ .$$

This result is verified in much the same way as theorem 2.1 using similar global topological perturbations. In addition a result analogous to theorem 2.2 may be verified for (3.3) in case f satisfies (2.2) and (2.4). We shall not formulate it here.

REMARK: The condition that

$$f(s) \le m_\infty^1 + o(s) \ , \quad s \to \infty \ ,$$

is not needed in the finite dimensional case, i.e. the a priori bounds for $\mu \ge \lambda_\infty^2 + \varepsilon$ may be obtained without this assumption. If this assumption is indeed dropped we conclude that Σ_∞ will bifurcate from ∞ in the interval $[0,\lambda_\infty^2]$.

Figures 10 and 11 below show the numerically computed branches Σ_0^0 and Σ_∞^0 of a nonlinear boundary value problem of the type (1.1) where f is a nonlinearity as in Figure 1. Since solutions of (1.1) will be symmetric about $x = \frac{1}{2}$ we instead of solving (1.1) solve the problem

$$(3.4) \qquad \begin{cases} u'' + \lambda f(u) = 0 \\ u(0) = u'(\frac{1}{2}) = 0 \ , \end{cases}$$

numerically. This is done by subdividing $[0,\frac{1}{2}]$ into n equal intervals of length h and setting $u_i = u(ih)$, proceeding as before one obtains a system of n equations of type (3.3) in the n unknowns $u = (u_1,\ldots,u_n)$

where A is given by

$$\begin{bmatrix} 2 & -1 & 0 & \cdots & \cdots & 0 \\ -1 & 2 & -1 & 0 & \cdots & 0 \\ 0 & \ddots & \ddots & \ddots & & \vdots \\ \vdots & & & & & \vdots \\ 0 & \cdots & 0 & -1 & 2 & -1 \\ 0 & \cdots & \cdots & 0 & -2 & 2 \end{bmatrix};$$

here we have approximated the right boundary condition by setting $u_{n-1} = u_{n+1}$. Using other approximations other possible choices of A can be obtained.

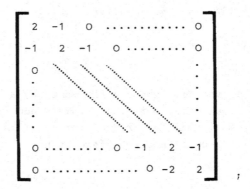

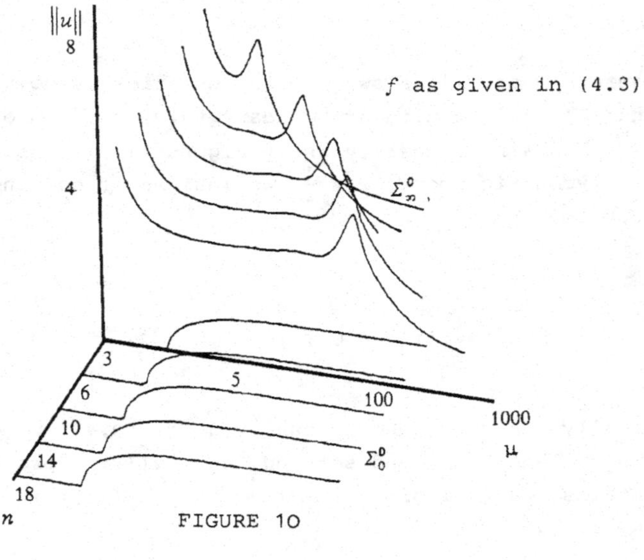

FIGURE 10

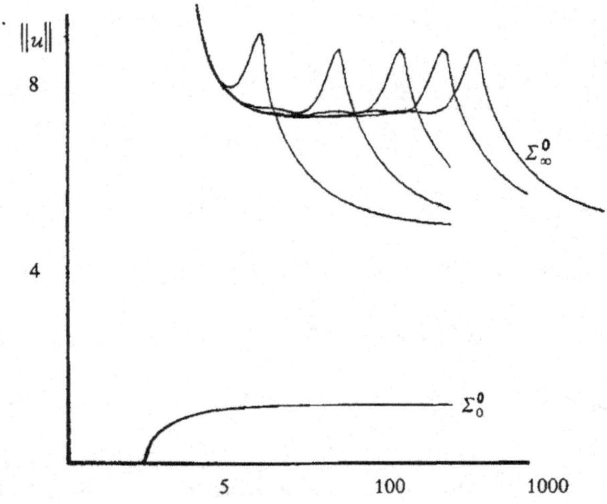

FIGURE 11

The solution branches Σ_0^0 and Σ_∞^0 are computed for various choices of the number of interior mesh points (Figure 10) and Figure 11 shows a projection of these curves onto the $(\lambda, ||u||_\infty)$ plane indicating that Σ_0^0 is very stable with respect to n whereas only the 'first part' of Σ_∞^0 enjoys such a stability.

We next investigate (3.3) in some more detail and show that the solution structure of this equation is indeed much more complicated than the solution structure of the corresponding infinite dimensional problem.

THEOREM 3.2. *Let f satisfy the conditions of theorem 3.1, and assume that f changes sign at each zero s_i, $i = 0,\ldots, 2m-1$. Then for all large μ (3.3) has at least $(2m)^n$ solutions in $\overset{o}{R}{}_+^n = \text{int } R_+^n$. These solutions may be found close to the zeros of F which are contained in $\overset{o}{R}_+$. Furthermore these solutions $(\mu >> 1)$ have the following properties:*

(i) *For each $i = 0,\ldots, 2m-1$ there are $(i+1)^n - i^n$ solutions with norm close to s_i.*

(ii) *There are $(i+1)^n - (i-1)^n$ solutions u such that*

$s_{i-1} < \|u\|_\infty < s_i$, *for* i *even and* $2 \le i \le 2m-2$.

(iii) *There is at least one solution* u *such that*

$0 < \|u\|_\infty < s_0$.

(iv) *There are no solutions* u *such that* $s_i < \|u\|_\infty < s_{i+1}$

for i even.

(v) *There are* $(2m)^n - (2m-1)^n$ *solutions* u *such that*

$\|u\|_\infty > s_\infty$.

Proof. The idea of the proof is a simple one. We write (3.3) equivalently as

$$\frac{1}{\mu}Au = F(u) \quad , \tag{3.5}$$

and view (3.5) for μ large as a perturbation of the uncoupled equation

$$0 = F(u) \ . \tag{3.6}$$

On the other hand

$\overset{\circ}{R}{}^n_+ \cap F^{-1}(0) = \{z\colon z_i = s_j,\ 1 \le i \le n,\ 0 \le j \le 2m-1\}$.

Furthermore if $z \in \overset{\circ}{R}{}^n_+ \cap F^{-1}(0)$, then for $r > 0$ sufficiently small $\deg_B(-F, B_r, 0) = \pm 1$ (since F is an uncoupled system), where B_r is an open ball of radius r centered at z. Hence for all μ sufficiently large

$$\deg_B(\tfrac{1}{\mu}A-F,\ B_r,\ 0) = \deg_B(-F,\ B_r, 0) \neq 0$$

implying that (3.5) and hence (3.3) has a solution in B_r. We thus merely need to count the elements in $\overset{\circ}{R}{}^n_+ \cap F^{-1}(0)$ and use earlier arguments to deduce the result.

We remark at this point, that if f is continuously differentiable the above degree theoretic argument may simply be replaced by an application of the implicit function theorem which then also guarantees the unique solvability of (3.3) for large μ near the elements of $F^{-1}(0)$, and such solutions will depend continuously upon μ. Using connectedness and continuation arguments one may, in the general case, also deduce the existence of continua of such solutions.

We also remark that since A is an M-matrix ([Sc]) it has the property that

$$A^{-1}(R^n_+ \setminus \{0\}) \subset \overset{\circ}{R}{}_+^n,$$

it therefore follows that all nontrivial R^n_+ solutions of (3.3) already lie in $\overset{\circ}{R}{}^n_+$. If we do not assume that A is an M-matrix the existence of $(2m)^n$ solutions of (3.3) may still be concluded, but of course the structural conclusions of theorem 3.2 no longer can be drawn.

If we consider the problem (3.4) with

$$(3.7) \qquad f(s) = \begin{cases} \frac{1}{2}\sin 2s \ , & 0 \le s \le 2\pi \\ s - 2\pi \ , & s > 2\pi \end{cases}$$

and apply the numerical procedures used to obtain Figures 10 and 11 we obtain the following solution picture ($n = 6$) Figure 12.

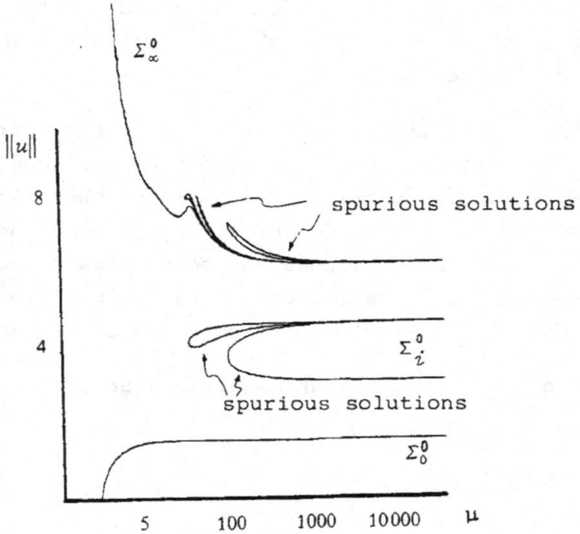

FIGURE 12

Since the phase portrait of the differential equation defined by f is as given in Figure 13, it follows that all solution branches lying between Σ_0^0 and Σ_∞^0 must be spurious.

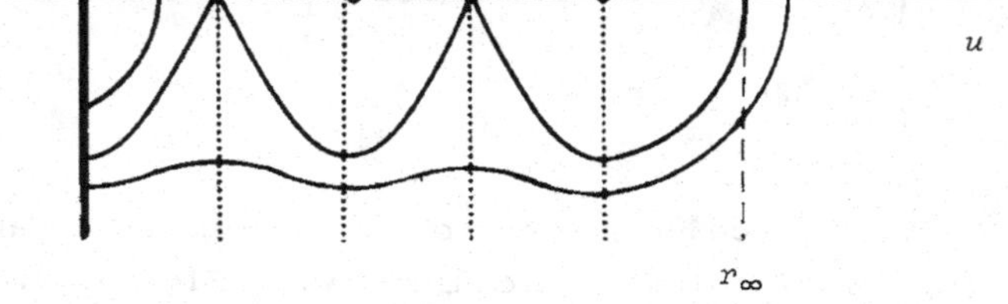

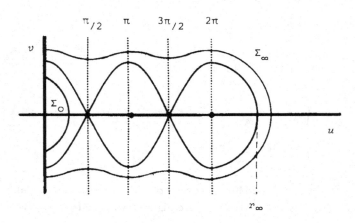

FIGURE 13

I.e. in this case there will be at least 3^n-1 solutions u, $\frac{\pi}{2} < ||u||_\infty < \frac{3\pi}{2}$ which are spurious. Furthermore the solutions u with $||u||_\infty > 2\pi$ (there are at least $4^n - 3^n$ of these) must have the property that $||u||_\infty \to 2\pi$ as $\mu \to \infty$, whereas the solutions of (3.4) which belong to Σ_∞ are such that $||u||_\infty > r_\infty$, where

$$\int_{\frac{3\pi}{2}}^{r_\infty} f(s)ds = 0 . \quad \text{(see Figure 13)}.$$

This phenomenon (described in more detail later) indicates that the numerical solutions on Σ_∞^0 become (for μ large) spurious.

If one uses only 2 interior mesh points and the same nonlinearity one obtains 16 nontrivial solutions which are all given in Figure 14.

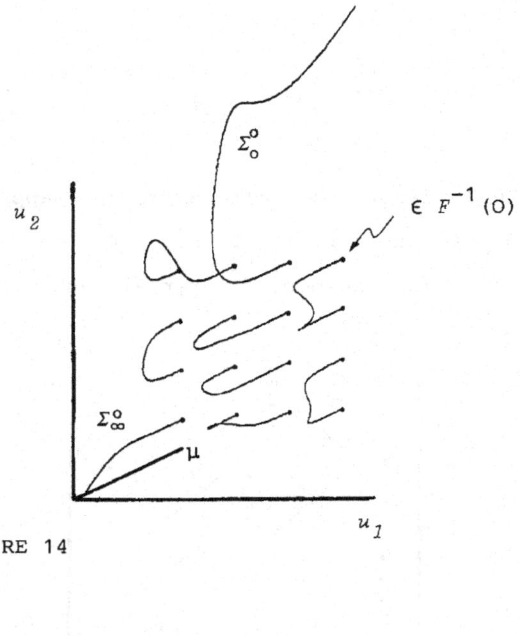

FIGURE 14

4. Characterization of Spurious Solutions

In our numerical studies of (1.1) we are able to distinguish between three structurally different types of spurious solutions. One should expect that the following characterization will carry over to the case of partial differential equations, however, our discussion is dependent upon an analysis of phase portraits associated with (1.1), and we therefore shall restrict ourselves to the case of ordinary differential equations.

As remarked earlier, it follows easily from a phase plane analysis that

$$(4.1) \qquad u'' + \lambda f(u) = 0 \quad , \qquad u(0) = 0 = u(1)$$

and

$$(4.2) \qquad u'' + \lambda f(u) = 0 \quad , \qquad u(0) = 0 = u'(\tfrac{1}{2})$$

are equivalent with respect to positive solutions.

We let

$$\Phi_i(u,\mu) = F(u) - \frac{1}{\mu}A^i u \qquad (i=1,2)$$

where $\Phi_1(u,\mu) = 0$ represents a finite difference approximation for (4.1) and $\Phi_2(u,\mu) = 0$ one for (4.2). E.g. A^1 and A^2 may be represented by $n \times n$ (n = number of internal mesh points) matrices

$$A^1 = \begin{bmatrix} 2 & -1 & 0 & \cdots & \cdots & 0 \\ -1 & 2 & -1 & 0 & & \vdots \\ 0 & \ddots & \ddots & \ddots & & \vdots \\ \vdots & & \ddots & \ddots & \ddots & 0 \\ \vdots & & & \ddots & \ddots & -1 \\ 0 & \cdots & \cdots & 0 & -1 & 2 \end{bmatrix}$$

$$A^2 = \begin{bmatrix} 2 & -1 & 0 & \cdots & & & 0 \\ -1 & 2 & -1 & 0 & & & 0 \\ 0 & \ddots & \ddots & \ddots & & & \vdots \\ \vdots & & & & & & 0 \\ 0 & & & 0 & -1 & 2 & -1 \\ 0 & \cdots & & & 0 & -2 & 2 \end{bmatrix}$$

Thus, from a numerical point of view, we have the freedom of choosing A^1 or A^2 and also the number of mesh points to be <u>even</u> or to be <u>odd</u>. We shall see, that with each different choice one obtains an essentially different qualitative picture of $\Phi_i^{-1}(0)$. For simplicity we again restrict ourselves to continuously differentiable f and assume that f has k simple positive zeros and also that $m_\infty^1 = m_\infty^2$, i.e. that f is asymptotically linear. In that case arguments due to Crandall and Rabinowitz [CR] imply that $\Phi_i^{-1}(0)$ will contain a locally unique bifurcation from $(\frac{\pi^2}{m_0},0)$ and a unique curve bifurcating from ∞ at $\frac{\pi^2}{m_\infty}$. Generically one may describe $\Phi_i^{-1}(0)$ as follows: Let $\bar{\varepsilon} \in \mathbb{R}^n$ be small and a regular value for Φ_i. Then

$$\Phi_i^{-1}(\bar{\varepsilon}) = \Sigma_0^{\bar{\varepsilon}} \cup \Sigma_\infty^{\bar{\varepsilon}} \cup \Sigma_i^{\bar{\varepsilon}} \cup S_j^{\bar{\varepsilon}} \qquad \begin{cases} i=1,\ldots,r \\ j=1,\ldots,s, \end{cases}$$

where all 1-manifolds $\Sigma^{\bar{\varepsilon}}$ are unbounded, i.e. $\Sigma^{\bar{\varepsilon}} \cong \mathbb{R}$ and $S^{\bar{\varepsilon}} \cong S^1$ (unit circle). Moreover $s \geq 0$ and $r = \frac{1}{2}(k)^n - 2$. Finally $\Sigma_0^{\bar{\varepsilon}}$ (respectively $\Sigma_\infty^{\bar{\varepsilon}}$) approximates the branch bifurcating from 0 (respectively ∞) of the infinite dimensional problem and

$$\lim_{\mu\to\infty} \{(\mu,u) \in \Sigma_0^{\bar{\varepsilon}}\} = \{z_0^{\bar{\varepsilon}}\} \subset F^{-1}(\bar{\varepsilon})$$

$$\lim_{\mu\to\infty} \{(\mu,u) \in \Sigma_\infty^{\bar{\varepsilon}}\} = \{z_\infty^{\bar{\varepsilon}}\} \subset F^{-1}(\bar{\varepsilon})$$

$$\lim_{\mu\to\infty} \{(\mu,u) \in \Sigma_i^{\bar{\varepsilon}}\} = \{z_{i_1}^{\bar{\varepsilon}}, z_{i_2}^{\bar{\varepsilon}}\} \subset F^{-1}(\bar{\varepsilon}).$$

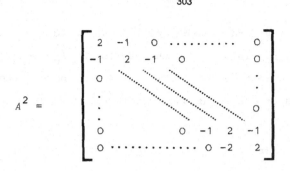

Letting $\bar{\varepsilon} \to 0$ we may obtain singularities in $\Phi_i^{-1}(0)$ which may be due to bifurcations. In any case the manifolds $\Sigma_0^{\bar{\varepsilon}}$ and $\Sigma_\infty^{\bar{\varepsilon}}$ approximate solution continua of (4.1) and (4.2). Whereas the manifolds $\Sigma_i^{\bar{\varepsilon}}$ may be spurious solutions or not depending upon the behavior of

$$G(u) := \int_0^u f(s)\,ds ,$$

(see e.g. Figures 6 and 13).

Let us proceed to classify the spurious solutions into three types.

TYPE I (Isolated Continua)

As $\bar{\varepsilon} \to 0$ some of the $\Sigma_i^{\bar{\varepsilon}}$ may keep their topological type, i.e. there exists an object Σ_i^0 such that

$$\text{and} \quad \begin{cases} \Sigma_i^0 \cong R \\ \\ \lim\limits_{\mu\to\infty} \{(\mu,u) \in \Sigma_i^0\} = \{z_{i_1}, z_{i_2}\} \subset F^{-1}(0) . \end{cases}$$

Then $\Sigma_i^0 \subset R_+ \times B(s_{2j}) \backslash \bar{B}(s_{2j-1}) := B_j$, some j, where $f^{-1}(0) = \{0, s_0, \ldots, s_\infty\}$. If $G(u)$ is such that (4.1) does not allow positive solutions in B_j (see e.g. Figure 13 or Figure 15), then Σ_i^0 will be spurious. E.g., assume that $G(u) < 0$ for all $u \in (s_1, s_\infty)$.

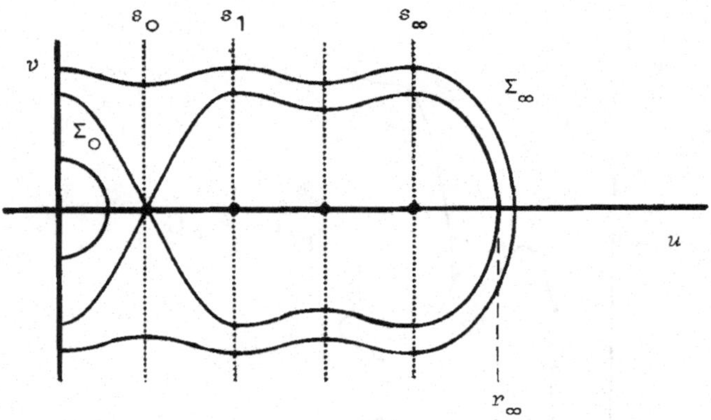

FIGURE 15

Figure 14 shows seven continua of this type. To see the asymptotic behavior with $\mu \to \infty$ of these continua we consider Figures 16 and 17.

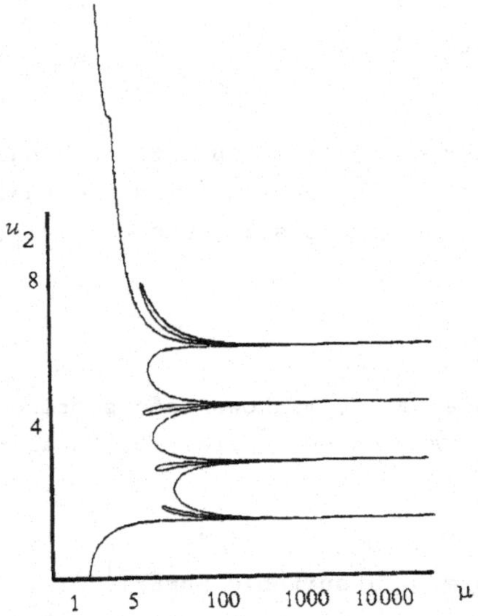

FIGURE 16

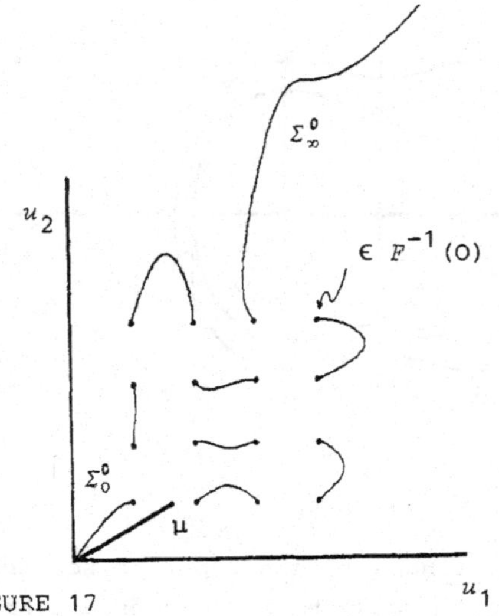

FIGURE 17

REMARK: If we change f (e.g. we consider a one-parameter family f_α, $f_0 = f$) such that

$$G_\alpha(u) = \int_0^u f_\alpha(s)\,ds$$

eventually allows solutions of (4.1) in B_j, then we shall have a family $\Sigma_j^0(\alpha)$ which for $\alpha = 0$ is spurious and eventually, for some α, be a family of approximate solutions of (4.1).

TYPE II (Transition)

As the phase plane analysis of (4.1) shows (see section 2) the solution branches of (4.1) must respect certain critical norm levels, which are given by

$$r_j := \int_{s^*}^{r_j} f(s)\,ds = 0, \text{ a suitable } s^* \in f^{-1}(0) .$$

For example Figure 7 shows that the solutions in Σ_∞ are bounded below in norm by r_∞. Theorem 3.2, however, shows e.g. that the numerical branch Σ_∞^0 corresponding to Σ_∞ must eventually dip below r_∞ in norm and approach (in norm) s_∞, i.e. it will undergo a transition from acceptable numerical solutions into spurious ones. Numerically this transition manifests itself by a 'cusp' like phenomenon in Σ_∞^0 (see e.g. Figure 11). A similar phenomenon may be observed in those Σ_i^0 which are not spurious. This explains the singularity in Figure 18.

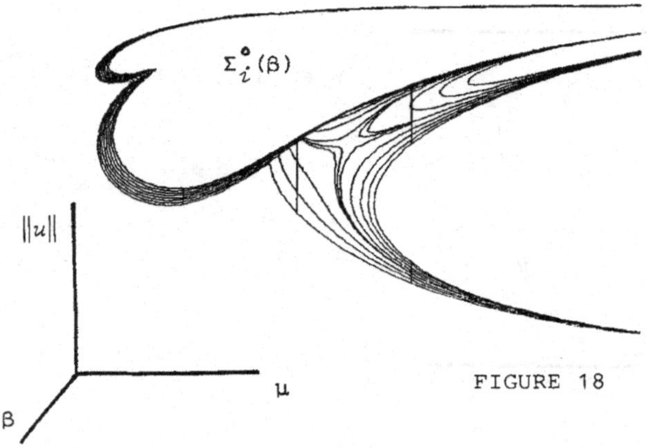

f_β as given in Figure 34 with $\beta = 1 - \alpha$

FIGURE 18

If $\mu^* = \mu^*(n)$ denotes the parameter value which locates the relative maximum in Σ_∞^0 in the $(\mu, ||\cdot||_\infty)$-plane, then as is shown in section 5, $\mu^*(n) \to \infty$, i.e. the transition will disappear as n increases. Figures 19 and 20 demonstrate this disappearance for 2 different non-linearities. In Figure 19 f is given by (3.7) and in Figure 20

$$f(s) = \begin{cases} \frac{1}{2}\sin(2s) & , 0 \le s \le \frac{\pi}{2} \\ -\frac{3}{2}\sin(\frac{2}{3}(s - \frac{\pi}{2})) & , \frac{\pi}{2} \le s \le 2\pi \\ s - 2\pi & , 2\pi \le s. \end{cases} \tag{4.3}$$

In both cases we not only plot the $||\cdot||_\infty$ of Σ_∞^0 versus μ but also the $||\cdot||_\infty$ of the second difference quotient.

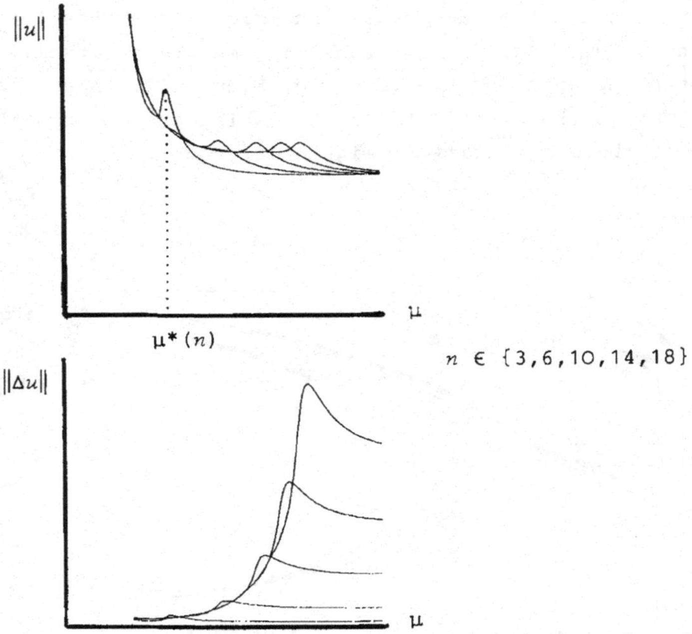

FIGURE 19

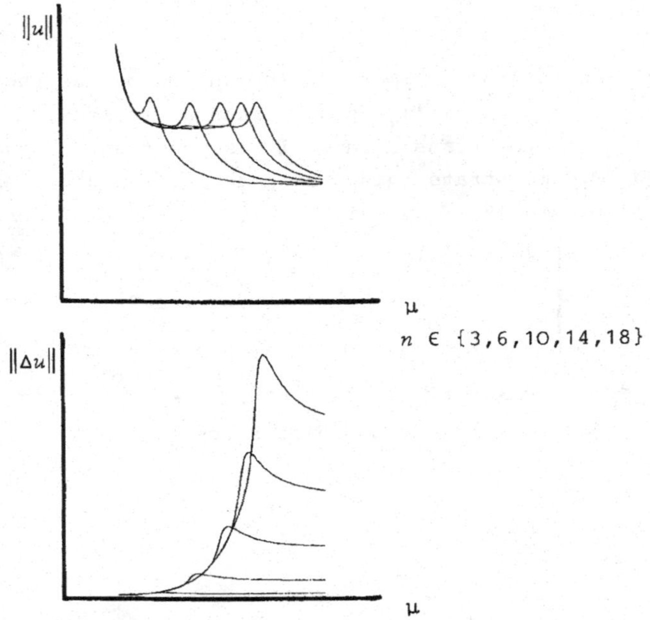

FIGURE 20

Spurious solutions of type I or type II will occur irregardless of the choice of the matrix A, i.e. whether symmetry of solutions is taken into consideration or not. Let

$$J = \begin{bmatrix} 0 \cdots 0 & 1 \\ \vdots \quad \cdot^{\cdot^{\cdot}} & \vdots \\ 1 & 0 \cdots 0 \end{bmatrix}$$

Then one observes that if $\Phi_1(z) = 0$ also $\Phi_1(Jz) = 0$; this observation leads to solutions of type III.

TYPE III (Nonsymmetric Continua)

Let $z \in F^{-1}(0)$ be such that $z \neq Jz$ (i.e. z is nonsymmetric) and consider Φ_1. Then there will exist branches in $\Phi_1^{-1}(0)$ which tend, as $\mu \to \infty$, to z and Jz. At least locally, for $\mu \sim \infty$, these branches will consist of nonsymmetric solutions and therefore will be spurious with respect to (4.1). Experimentally we found that there are entire branches of nonsymmetric solutions which are secondary bifurcations from symmetric acceptable solutions. The nature of the secondary bifurcation behavior in turn highly depends upon the number of zeros of f, and surprisingly also on whether n is even or odd.

Figures 21 and 22 show secondary bifurcations from Σ_∞^0 into nonsymmetric solutions in case n is even and

$$(4.4) \qquad f(s) = \begin{cases} \frac{1}{2}\sin(2s) & , \ 0 \le s \le \pi \\ s - \pi & , \ \pi \le s \end{cases}$$

And Figures 23 and 24 show the analogous phenomenon when n is odd. In both cases we have chosen as a measure of deviation from symmetry

$$\operatorname{asym}(z) := \|z - Jz\|_\infty$$

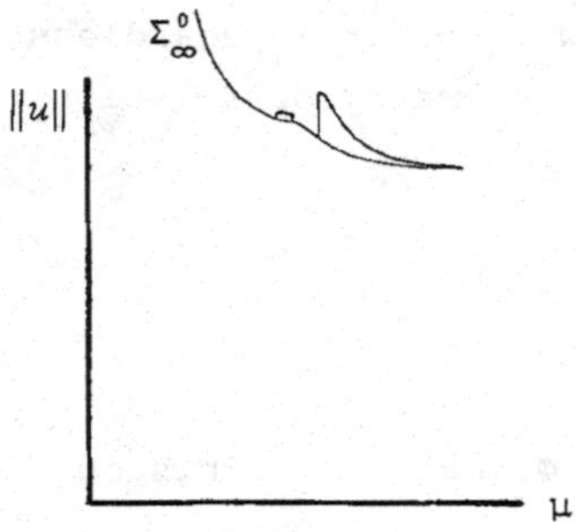

FIGURE 21

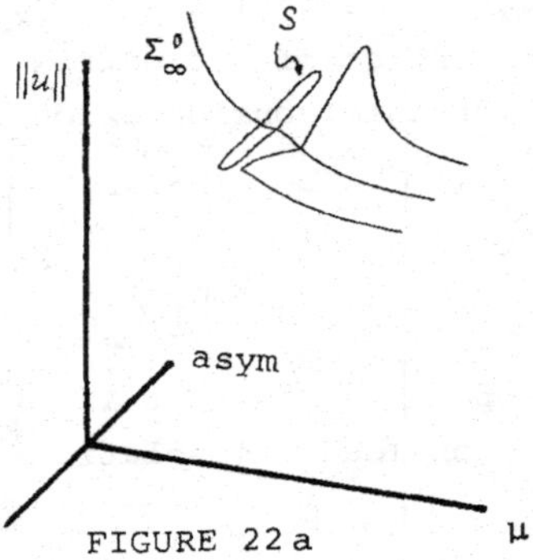

FIGURE 22a

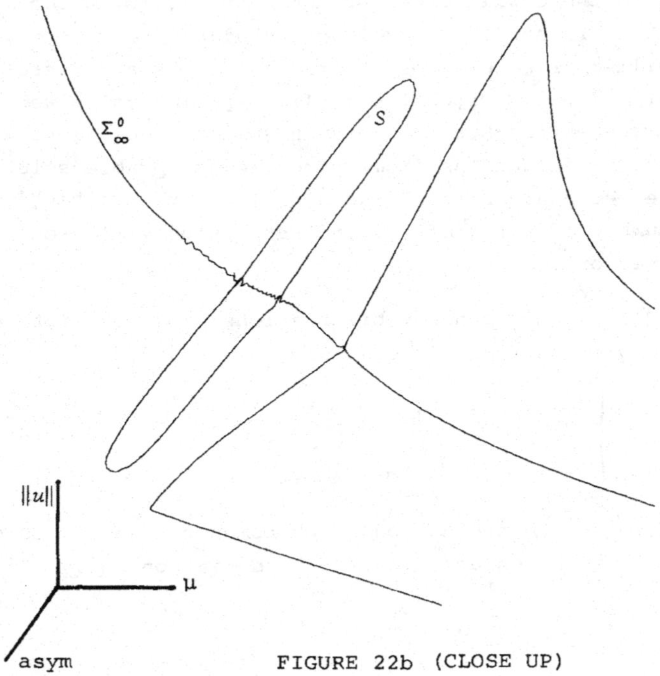

FIGURE 22b (CLOSE UP)

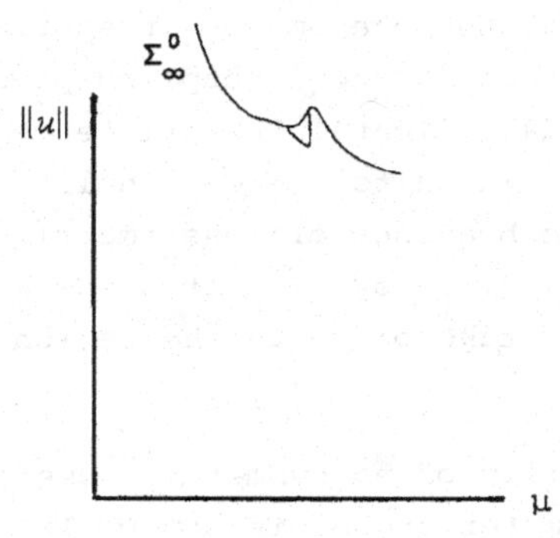

FIGURE 23

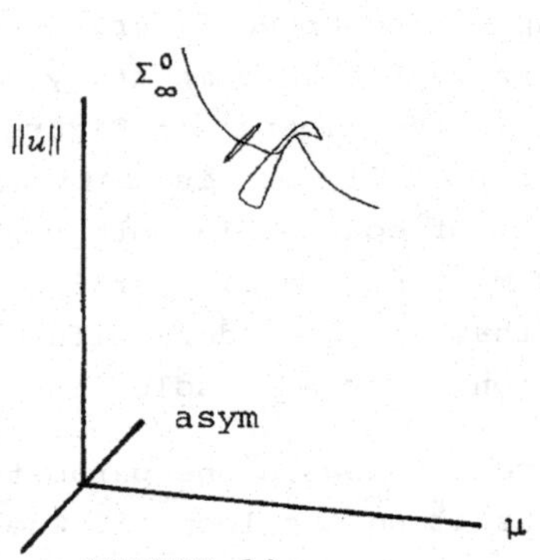

FIGURE 24a

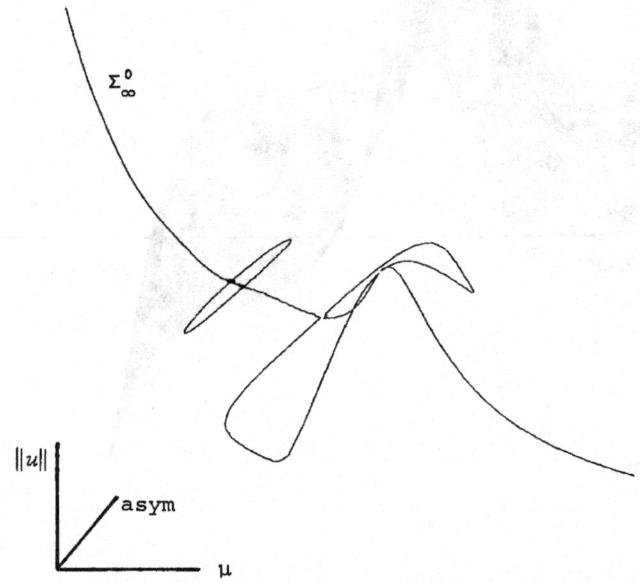

FIGURE 24b (CLOSE UP)

REMARK The above figures show that the phenomenon of spurious solutions cannot be completely understood by the decoupling process $\mu \to \infty$; as the 'loops' in Figures 21-24 indicate. The existence and the number of 'loops' is intimately related to the cardinality of $f^{-1}(0)$. Of course, in both cases the branches of nonsymmetric solutions must have a bilateral symmetry given by J. It seems worth noting that the secondary bifurcations take place in the region of transition of type II solutions.

Figure 25 shows a one-parameter family of nonsymmetric numerical solutions along the loop S. The parameter chosen is arclenght (n=20).

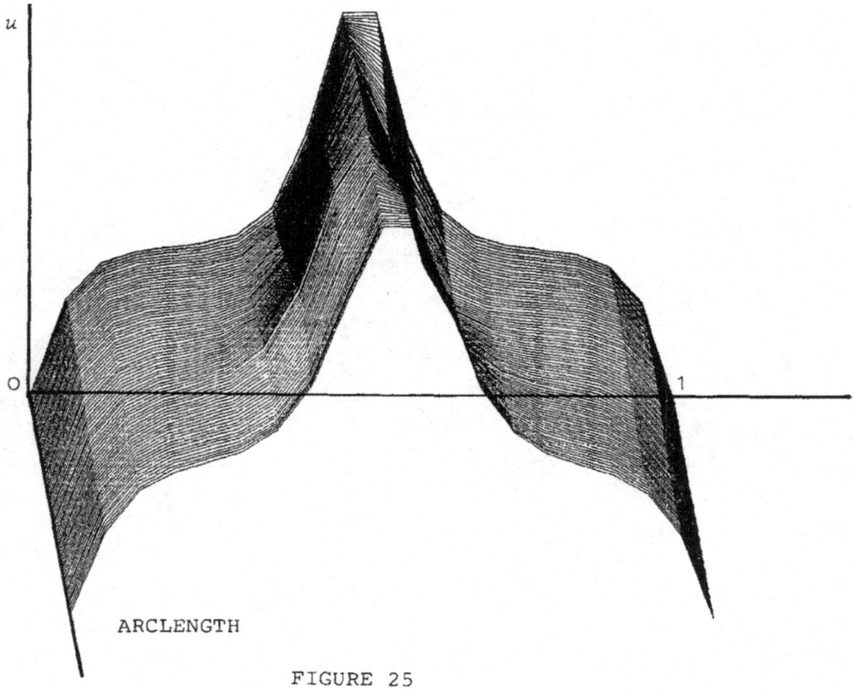

FIGURE 25

The analysis performed for $\mu \to \infty$ may, of course, equally well be considered by letting $\mu \to -\infty$. This corresponds to the case where F is replaced by $-F$ and letting $\mu \to \infty$. Depending upon the nature of f, more specifically of $G(u) = \int_0^u f(s)ds$, there may be no positive solutions at all or there may be some. Typical examples are given by Figures 26 and 27.

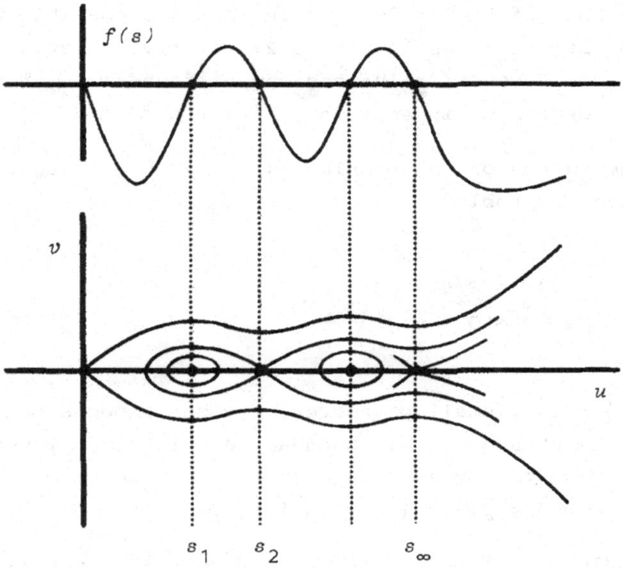

FIGURE 26

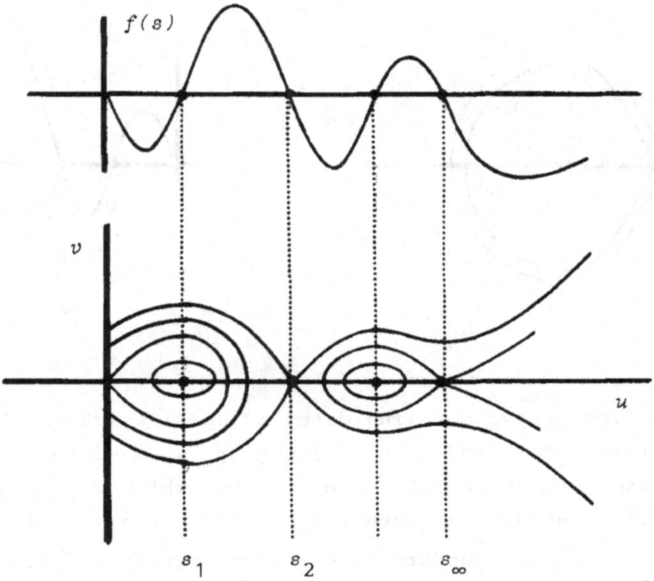

FIGURE 27

Thus in the case of Figure 26 no solutions are possible whereas for nonlinearities f as in Figure 27 positive solutions will exist. In either case, however, solutions may be found for large μ by the analysis developed in section 3 for equation (3.3).

For large values of μ problem (4.1) may be viewed as a singular perturbation problem

$$(4.5) \qquad \begin{cases} \varepsilon u'' + f(u) = 0 \\ u(0) = 0 = u(1) \end{cases}$$

where $\varepsilon = \frac{1}{\mu}$ is a small parameter, and one expects that the computed branches should exhibit boundary and interior layer effects. This may be verified by studying the time map for the associated system in the phase plane (see e.g.[OM]).

For example if the nonlinearity is such that a separatrix in the phase plane exists as indicated in Figure 28

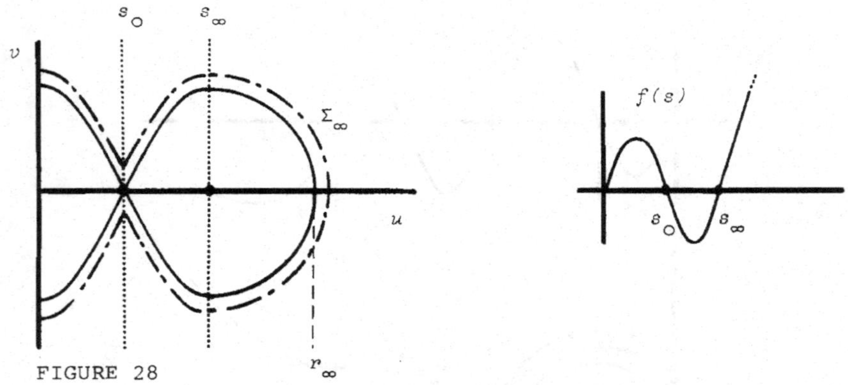

FIGURE 28

then a solution trajectory (indicated by dots) which is close to the separatrix will spend a long time near s_0 and not much time anywhere else, one hence obtains a solution which exhibits a boundary layer at either end and an interior layer at $x = 1/2$. Figure 29 shows this interior, boundary layer effect. There a family of Σ_∞-solutions is shown in a neighborhood of the transition to type II solutions.

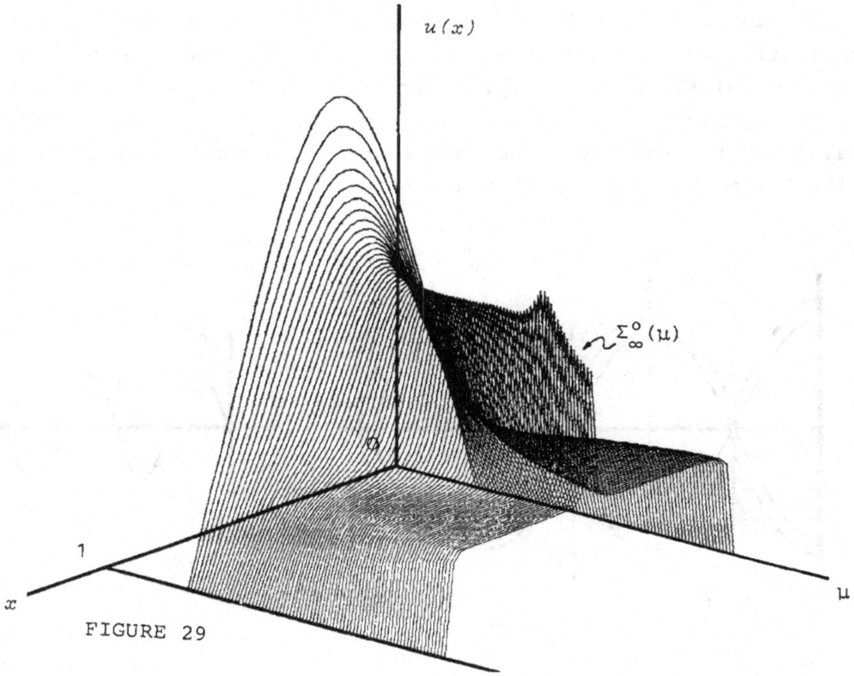

FIGURE 29

The solutions were computed using $n = 19$ interior mesh points. The various components of the vector u (as a function of μ) then show the following asymptotic behavior.

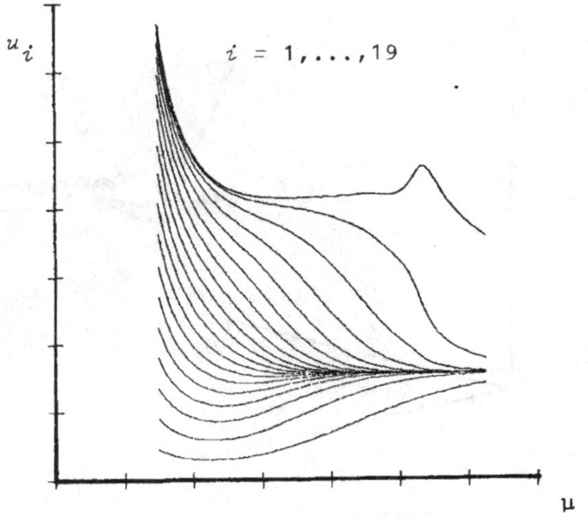

FIGURE 30

If, on the other hand one has a nonlinearity f which defines a separatrix in the phase plane as in Figure 31, then one expects to find solutions having 3 interior layers. This is again substantiated by Figures 32 and 33, where in Figure 33 the asymptotic behavior of the various components of the solution vector (as functions of μ) has been exhibited.

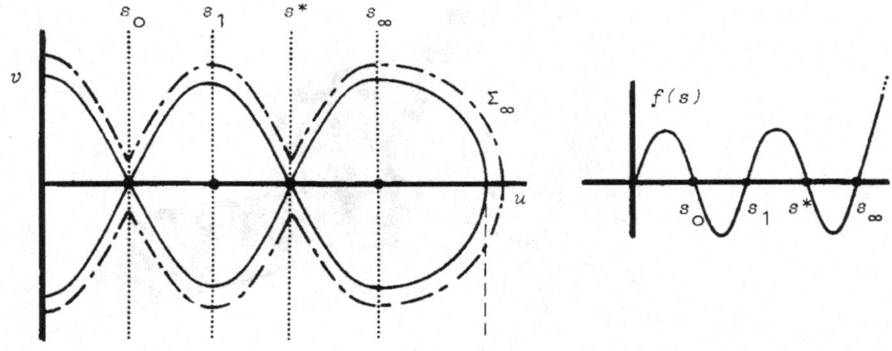

FIGURE 31

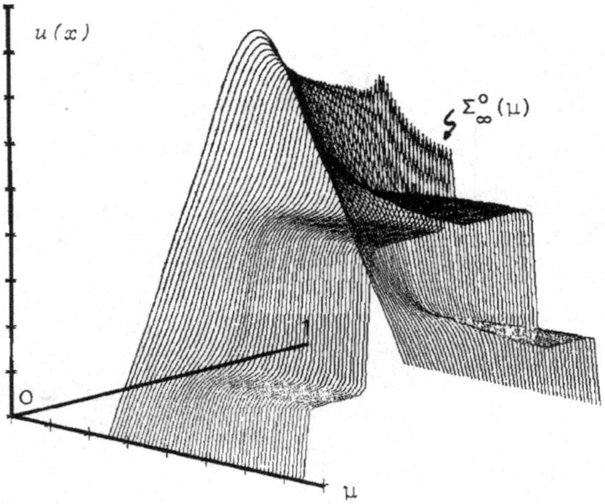

FIGURE 32a

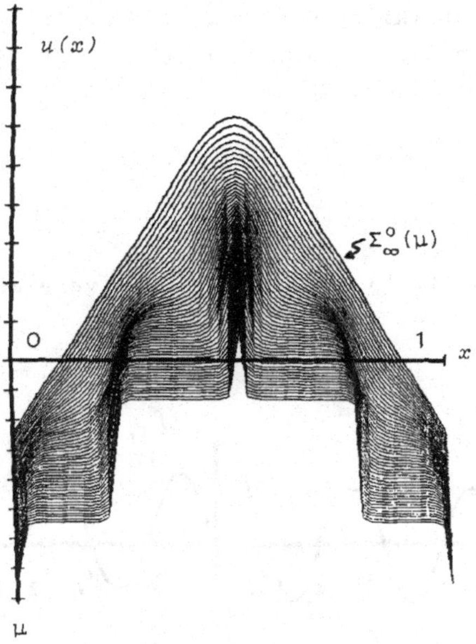

FIGURE 32b

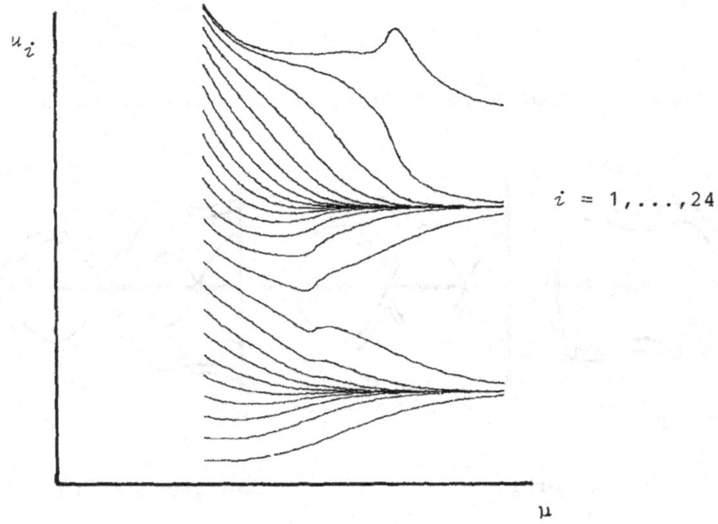

FIGURE 33

REMARK: There is another striking phenomenon hidden in the simple equation (4.1). Introducing a second parameter $\alpha \in R$ in (4.1) to obtain (4.6)

$$(4.6) \qquad \begin{cases} u'' + \lambda f_\alpha(u) = 0 \\ u(0) = 0 = u(1) \end{cases}$$

we look at positive solutions for $\lambda \sim \infty$ and let α vary as indicated in Figure 34.

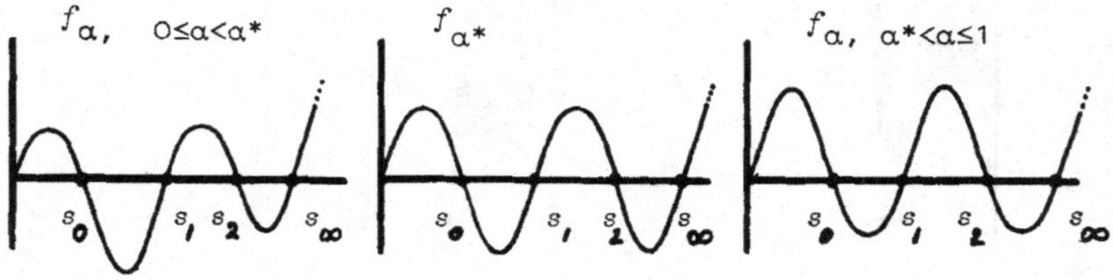

FIGURE 34

$$\int_{s_0}^{s_2} f_\alpha(s)ds < 0 \qquad \int_{s_0}^{s_2} f_\alpha(s)ds = 0 \qquad \int_{s_0}^{s_2} f_\alpha(s)ds > 0$$

The essential parts of phase portraits are given in Figure 35.

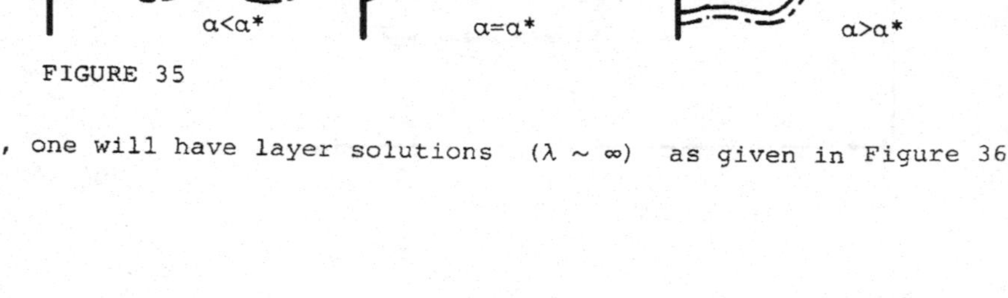

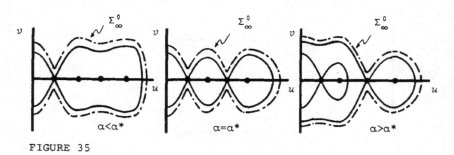

FIGURE 35

Thus, one will have layer solutions $(\lambda \sim \infty)$ as given in Figure 36.

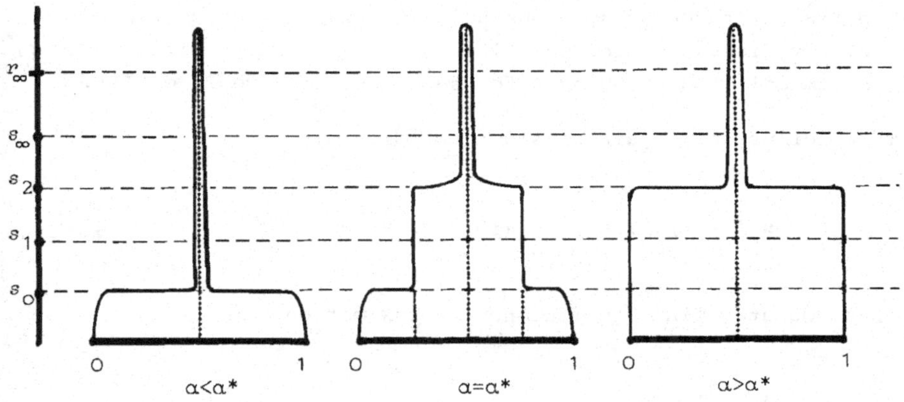

FIGURE 36

Obviously, $\alpha = \alpha^*$ is a critical situation in the 1-parameter family ($\lambda \sim \infty$, fixed) given by (4.6). Of course the singularity is due to breaking off the 'unstable' saddle-to-saddle connection in the phase portrait for f_{α^*}. Figure 36 thus depicts an interesting critical change in the positive solutions of (4.6) as α crosses α^*.

All numerical computations and computer plots are subject to the boundary conditions $u(0) = u(\pi) = 0$.

5. Dependence of Spurious Solutions on the Mesh Size

In this final section we once more consider problem (1.1) and finite dimensional approximations thereof and consider the dependence of spurious solutions upon the mesh size, and we shall show that for a given value of λ all spurious solutions will disappear provided the mesh size is made sufficiently fine. This type of behavior is very much dependent on the nature of the nonlinearity f (in our case it follows because f is independent of u') as examples of Gaines [G] have demonstrated, i.e. examples exist for which for a given λ spurious solutions exist for all sizes of the mesh.

The disappearance of spurious solutions is, in our context, most easily discussed, for solutions of type I. There will, however, also be immediate consequences for solutions of type II or III.

Thus let us once again consider the problem

$$(5.1) \qquad u'' + \lambda f(u) = 0 \ , \quad u(0) = 0 = u(1)$$

and an associated finite difference approximation

$$(5.2) \qquad Au = \lambda h^2 F(u) \ ,$$

where $u = (u_1, \dots, u_n)$, $u_0 = 0 = u_{n+1}$.

For $i = 1, \dots, n$ let

$$\nabla_{u_i} = u_i - u_{i-1} \ .$$

and $\nabla_u = (\nabla_{u_1}, \dots, \nabla u_n)$

According to [G] we have the following convergence lemma.

LEMMA 5.1. *Fix* λ *and assume that there exists* $h_0 > 0$, M, N *such that for all* h, $0 < h < h_0$, *the associated system* (5.2) *has a solution* u_h *such that* $||u_h||_\infty \le M$ *and* $||\nabla u_h||_\infty \le hN$. *Then there exists a solution of* (4.1) *which may be obtained as a limit of a sequence of polygonal curves which are determined by* $\{u_h\}$.

The following lemma, in turn provides sufficient conditions for the existence of N, once M is given.

LEMMA 5.2. *Let* $M > 0$ *be given. Then there exists a constant* $N = N(M,\lambda)$ *such that if* u_h *is a solution of* (5.2) *with* $||u_h||_\infty \le M$ *it follows that* $||\nabla u_h||_\infty \le hN$.

This lemma is true in much more general cases as has been observed in [G] and [M].

As an immediate corollary we obtain the following

COROLLARY 5.3. *Assume there exist* $h_o > 0$ *and* $M > 0$ *such that for* $\lambda = \lambda^*$ (5.2) *has a solution* u_h *for every* $h \in (0,h_o)$ *with* $||u_h||_\infty \le M$. *Then* (5.1) *has a solution which may be obtained as a limit of a sequence of polygonal curves obtained from* $\{u_h\}$.

Let us now assume that the phase portrait determined by f is such that all solutions (λ,u) of (5.1) must satisfy either $||u||_\infty < s_o$ or $||u||_\infty > s_\infty$ (see e.g. Figure 13), and let $\Sigma_i^o(h)$ denote a solution continuum of (5.2) such that for $(\lambda,u_h) \in \Sigma_i^o(h)$ it is the case that

$$(5.3) \qquad s_i < ||u_h||_\infty < s_{i+1}, \qquad i \text{ odd}.$$

Such continua exist, as has been established in section 3.

Let

$$(5.4) \qquad \bar{\lambda}(h) = \inf\{\lambda: \text{there exists a solution } (\lambda,u) \in \Sigma_i^o(h)\}.$$

The result below now follows from what has been established above.

THEOREM 5.4. *Let* $\Sigma_i^\circ(h)$ *denote a branch of spurious solutions of type I and let* $\overline{\lambda}(h)$ *be defined by* (5.4). Then

$$\lim_{h\to 0} \overline{\lambda}(h) = \infty , \tag{5.5}$$

i.e. $\Sigma_i^\circ(h)$ *will disappear as* $h \to 0$.

Proof Assume there exists λ^* such that for all small h there exists a solution u_h of (5.2) for $\lambda = \lambda^*$, with $(\lambda^*, u_h) \in \Sigma_i^\circ(h)_\infty$. It then follows from Corollary 5.3 that for $\lambda = \lambda^*$ (5.1) will have a solution u which may be obtained as a limit of polygonal arcs from $\{u_h\}$. This however will imply because of (5.3) that u satisfies (5.3) also, contradicting our assumption made above.

If we consider $\Sigma_\infty^\circ(h)$ we found that for large λ, $(\lambda,u) \in \Sigma_\infty^\circ(h)$ implied that $||u_h||_\infty < r_\infty$. Letting

$$\overline{\lambda}(h) = \inf \{\lambda : (\lambda,u) \in \Sigma_\infty^\circ\ (h) \text{ and } ||u||_\infty < r_\infty\}$$

we find by arguments similar to the ones used above that

$$\lim_{h\to 0} \overline{\lambda}(h) = \infty .$$

Considering other cases one may conclude in general that spurious solutions either disappear or become nonspurious as $h \to 0$.

References

[Al] E. Allgower, On a discretization of $y'' + \lambda y^k = 0$, Proc. Conf. Roy. Irish Acad., New York-London, 1975

[AH] A. Ambrosetti and P. Hess, Positive solutions of asymptotically linear elliptic eigenvalue problems, J. Math. Anal. Appl. 73 (1980), 411-422

[An] N. Angelstorf, Spezielle periodische Lösungen einiger autonomer zeitverzögerter Differentialgleichungen mit Symmetrien, Dissertation, Universität Bremen, 1980

[BD] W.-J. Beyn and E.J. Doedel, Stability and multiplicity of solutions to discretizations of nonlinear ordinary differential equations, (to appear in SIAM J. Sci. Stat. Comp.)

[B] E. Bohl, On the bifurcation diagram of discrete analogues for ordinary bifurcation problems, Math. Meth. Appl. Sci. 1 (1979), 566-571

[BB] K.J. Brown and H. Budin, On the existence of positive solutions for a class of semilinear elliptic boundary value problems, SIAM J. Math. Anal. 10 (1979), 875-883

[C] L. Collatz, The numerical treatment of differential equations, Springer Verlag, Berlin 1960

[CR] M. Crandall and P. Rabinowitz, Bifurcation from simple eigenvalues, J. Funct. Anal. 8 (1971), 321-340

[G] R. Gaines, Difference equations associated with boundary value problems for second order nonlinear ordinary differential equations, SIAM J. Numer. Anal. 11 (1974),411-434

[GT] D. Gilbarg and N.S. Trudinger, Elliptic Partial Differential Equations of Second Order, Springer, Heidelberg-New York, 1977

[JPS] H. Jürgens, H.O. Peitgen and D. Saupe, Topological perturbations in the numerical study of nonlinear eigenvalue and bifurcation problems, Proc. Conf. Analysis and Computation of Fixed Points, Academic Press, New York-London, 1980, 139-181

[IK] E. Isaacson and H. Keller, Analysis of Numerical Methods, John Wiley, New York, 1966

[M] P. Morse, Boundary Value Problems For Nonlinear Difference Equations, Ph. D. Thesis, University of Utah, Salt Lake City, 1980

[OM] R.E. O'Malley, JR., Phaseplane solutions to some singular perturbation problems, J. Math. Anal. Appl. 54 (1976), 449-466

[P] S.I. Pohozaev, Eigenfunctions of the equation $\Delta u + \lambda f(u) = 0$, Soviet Math. Dokl., 6 (1965), 1408-1411

[PP] H.O. Peitgen and M. Prüfer, The Leray Schauder continuation method is a constructive element in the numerical study of nonlinear eigenvalue and bifurcation problems, Proc. Conf. Functional Differential Equations and Approximation of Fixed Points, Springer Lecture Notes in Math. 730 (1980), 326-409

[PSS] H.O. Peitgen, D. Saupe and K. Schmitt, Nonlinear elliptic boundary value problems versus their finite difference approximations: numerically irrelevant solutions, J. reine angew. Mathematik 322 (1981), 74-117

[PS_1] H.O. Peitgen and K. Schmitt, Perturbations topologiques globales des problèmes non linéaires aux valeurs propres, C.R. Acad. Sc. Paris 291 (1980), 271-274

[PS_2] H.O. Peitgen and K. Schmitt, Global topological perturbations of nonlinear elliptic eigenvalue problems, (to appear) Report Nr. 33, Forschungsschwerpunkt "Dynamische Systeme", Universität Bremen 1981

[R] P. Rabinowitz, Some aspects of nonlinear eigenvalue problems, Rocky Mountain J. Math. 3 (1973), 162-202

[Sa] D. Saupe, On accelerating PL continuation algorithms by predictor corrector methods, (to appear in Math.Prog., Report Nr. 22, Forschungsschwerpunkt "Dynamische Systeme", Universität Bremen, 1981)

[Sc] J. Schröder, M-matrices and generalizations using an operator theory approach, SIAM Review 20 (1978), 213-244

[SA] H. Spreuer and E. Adams, Pathalogische Beispiele von Differenzenverfahren bei nichtlinearen gewöhnlichen Randwertaufgaben, ZAMM 57 (1977), T 304-T 305

[SS] A.B. Stephens and G.R. Shubin, Multiple solutions and bifurcation of finite difference approximations to some steady state problems of fluid dynamics, preprint of Naval Surface Weapons Center, 1981

[St] C.A. Stuart, Concave solutions of singular non-linear differential equations, Math. Z. 136 (1974), 117-135

CHANGE OF STRUCTURE AND CHAOS FOR SOLUTIONS OF $\dot{x}(t) = -f(x(t-1))$

BY

H. PETERS

Forschungsschwerpunkt "Dynamische Systeme"
Fachbereich Mathematik
Universität Bremen
D-2800 Bremen 33

CHANGE OF STRUCTURE AND CHAOS FOR SOLUTIONS OF $\dot{x}(t) = -f(x(t-1))$

Hubert Peters *

I. INTRODUCTION

In this paper we study the differential-delay equation

$$(1.1) \qquad \dot{x}(t) = -\alpha . f(x(t-1))$$

with

$$(1.2) \qquad f(x) = \text{sign} \sin (\pi x)$$

The motivation for this equation comes from a result proven by T. Furumochi in [2] . He observes that for certain parameters $\alpha, \delta \in R^+$ there exist "periodic solutions of the second kind" for the equation

$$(1.3) \qquad \dot{x}(t) = \delta - \sin x(t-\alpha)$$

i.e. solutions $x \in C^1(R^+)$ such that $x(t+p) = x(t) + C$ for all $t > 0$ and certain constants p and C.

If we think of equation

$$(1.4) \qquad \dot{x}(t) = -\alpha . \sin x(t-1) \quad =: \quad -\alpha . g(x(t-1))$$

as a model of equation (1.3) there is one important difference to the class of equations mainly discussed in the literature (cf. [4]): the nonlinearity $g(x) = \sin x$ does not satisfy the global feedback condition $x.g(x) > 0$. The nontrivial zeros $z_k = k\pi$ $(k \in Z-\{0\})$ of the sine-function correspond to nontrivial stationary solutions $x_k \equiv k\pi$ $(k \in Z-\{0\})$ of the delay equation (1.4).

---

*This work has been supported by Deutsche Forschungsgemeinschaft, research project "Multiple Bifurkation" at the University of Bremen

It follows from a result in [6] that in the case of equation (1.4) for each $\alpha > \pi/2$ there exists a "special periodic solution" x_α whose norm $|x_\alpha|$ is bounded by the smallest positive zero π of the nonlinearity g . (x_α is called a "special periodic solution" iff $x_\alpha(t+2) = -x_\alpha(t)$, $x_\alpha(-t) = -x_\alpha(t)$ for all $t \in R$, and $x_\alpha(t) > 0$ for $t \in (0,2)$).

Defining $y_\alpha^{\pm}(t) := x_\alpha(3t) \pm \pi$ one yields two further periodic solutions of period 4/3 oscillating around the stationary solutions $+\pi$ resp. $-\pi$.

Heuristically one would expect interferences between the "slowly" oscillating periodic solution x_α and the "rapidly" oscillating periodic solutions $y_\alpha^{\pm}$. In fact numerical computations show complicated behaviour of solutions of equation (1.4).

The special features of the nonlinearity $g = \sin$ therefore lead to the following questions:

<u>PROBLEM 1</u> : Is there a transition from periodic solutions to periodic solutions of the second kind depending on the parameter α ?

<u>PROBLEM 2</u> : Do the "humps" in the nonlinearity and interferences between "slowly" and "rapidly" oscillating solutions cause "chaos" in equation (1.1) ?

Up to now there seems to be no way to answer these questions for the "smooth" equation (1.4). We therefore discuss as a model problem equation (1.1) with nonlinearity (1.2). The choice of a step function as model-nonlinearity is motivated by the fact that we were able to prove "chaotic behaviour" in the sense of Li and Yorke [3] for the equation $\dot{x}(t) = -f_\alpha(x(t-1))$ where f_α is a suitably chosen step function which models the "hump"-nonlinearity $\tilde{f}_\alpha(x) = \alpha.x(1+x^8)^{-1}$ [7].

The following proposition shows that the model equation (1.1) reflects the known results for the smooth equation (1.4).

(1.5) PROPOSITION

(1.5.1) For each $\alpha > 0$ there exists a nontrivial special periodic solution x_α of equation (1.1) such that

$$|x_\alpha| = \sup_{t \in R} |x_\alpha(t)| \leqq 1$$

(1.5.2) For $\alpha = 2$ there exists a periodic solution of the second kind $\tilde{x}$ of equation (1.1).

Proof:

(1) For $0 < \alpha < 1$ define $\varphi_\alpha \in C[0,1]$ by $\varphi_\alpha(t) := \alpha . t$, $t \in [0,1]$. For $\alpha \geqq 1$ define $\varphi_\alpha \in C[0,1]$ by

$$\varphi_\alpha(t) := \begin{cases} 0 & ; \quad 0 \leq t \leq \frac{\alpha-1}{2\alpha} \\ \alpha t - \frac{\alpha-1}{2} & ; \quad \frac{\alpha-1}{2\alpha} \leq t \leq \frac{\alpha+1}{2\alpha} \\ 1 & ; \quad \frac{\alpha+1}{2\alpha} \leq t \leq 1 \end{cases}$$

It is easily seen that φ_α is segment of a periodic solution $x_\alpha = x^{\varphi_\alpha}$ with the described properties.

(2) Define $\psi \in C[0,1]$ by $\psi(t) = 2.t$, $t \in [0,1]$. The trajectory $\tilde{x} := x^\psi$ is a periodic solution of the second kind: $\tilde{x}(t+2) = \tilde{x}(t) + 2$.

qed.

(1.6) Remark

Let x_α be the nontrivial special periodic solution of equation (1.1). As in the case of a smooth nonlinearity we then have that $y_\alpha^{\pm}$ with $y_\alpha^{\pm}(t) := x_\alpha(3t) \pm 1$ is a periodic solution of period 4/3 .

II. RESULTS

Even in our model case we cannot give a complete answer to the questions given in the introduction. To provide at least partial results a restriction on special parameters ($1 \leq \alpha \leq 2$) and selected segments ("spikes", "simple" initial function) seems to be useful.

Before we state the main theorems some notations are in order:

We call the function $\Phi_\alpha \in C[0,1]$ defined by $\Phi_\alpha(t) = \alpha . t$ $(t \in [0,1])$ the "simple" initial function.

$x^\psi \in C(R^+)$ denotes the trajectory belonging to equation (1.1) with initial function $\psi \in C[0,1]$; $x_t^\psi \in C[0,1]$ is a segment of the trajectory x^ψ defined by $x_t^\psi(s) := x^\psi(t+s)$, $s \in [0,1]$.

Given an initial function $\psi \in C[0,1]$ denote by $z_\psi > 1$ the first zero of the trajectory x^ψ . The shift-operator which maps the initial function ψ on the segment $-x_{z_\psi}^\psi$ is denoted by S_α , i.e. $S_\alpha(\psi) = -x_{z_\psi}^\psi$. If z_ψ does not exist, define $S_\alpha(\psi) := 0$.

We call a solution x^ψ slowly oscillating iff two subsequent zeros z_i, z_{i+1} of the trajectory x^ψ are separated by a distance greater than 1 . A solution x^ψ is said to be slowly oscillating with respect to a level N (N $\in$ Z) iff x - N is slowly oscillating. A solution which is not slowly oscillating (with respect to N) is said to be rapidly oscillating (with respect to N).

2.1 THEOREM ("Transition from periodic solutions to periodic solutions of the second kind")

Let $\{\alpha_n\}_{n\in N}$, $\{\beta_n\}_{n\in N}$ with

$$1 = \alpha_0 < \alpha_1 < \ldots < \alpha^* = \frac{3}{2} < \ldots < \beta_1 < \beta_0 = 2$$

be recursively defined by

$$\alpha_{n+1} = \frac{4\alpha_n}{2\alpha_n + 1} \qquad \beta_{n+1} = \frac{4\beta_n}{2\beta_n + 1}$$

$$\alpha_0 = 1 \qquad \beta_0 = 2$$

(a) For $\alpha = \alpha_n$, the simple initial function Φ_α is a segment of an <u>(n+1)-spike-solution</u>, i.e. of a periodic solution $x^{(n+1)}$ of equation (1.1) with the following properties:

(a_1) $x^{(n+1)}$ is slowly oscillating with respect to $x \equiv 0$, rapidly oscillating with respect to $x \equiv +1$ and $x \equiv -1$.

(a_2) Let $z > 1$ be the smallest positive zero of $x^{(n+1)}$. There exist n+1 points $1 = t_0 < t_1 < \ldots < t_n < z$ where $x^{(n+1)}$ attains a local maximum $M_i := x^{(n+1)}(t_i)$. The sequence $\{M_i\}$ is monotonically decreasing, $M_0 = \alpha$, and $M_n = 1$.

(a_3) $x^{(n+1)}(z+t) = - x^{(n+1)}(t)$, i.e. Φ_α is a fixed point of the shift operator S_α .

(b) For $\alpha = \alpha^* = \frac{3}{2}$, the simple initial function Φ_α is an initial function for the rapidly oscillating periodic solution of period 4/3 belonging to the stationary solution $x \equiv +1$.

(c) For $\alpha = \beta_n$, the simple initial function Φ_α is a segment of an <u>(n+1)-spike-solution of the second kind</u> , i.e. of a solution $\tilde{x}^{(n+1)}$ of the second kind of equation (1.1) with the following properties:

(c_1) $\tilde{x}^{(n+1)}$ is slowly oscillating with respect to the stationary solutions $x \equiv 2k$ ($k \in Z$), rapidly oscillating with respect to the stationary solutions $x \equiv 2k+1$ ($k \in Z$).

(c_2) Let $\tilde{z} := \min \{ z > 1 \mid \tilde{x}^{(n+1)}(z) = 2 \}$. There exist n points $1 < \tilde{t}_1 < \tilde{t}_2 < \ldots < \tilde{t}_n < \tilde{z}$ where $\tilde{x}^{(n+1)}$ attains a local minimum $m_i := \tilde{x}^{(n+1)}(t_i)$. The sequence $\{m_i\}$ is monotonically increasing, $m_1 = \alpha - 1$, and $m_n = 1$.

(c_3) $\tilde{x}^{(n+1)}(\tilde{z} + t) = \tilde{x}^{(n+1)}(t) + 2$ for all $t > 0$.

(d) For $\alpha = \alpha_n$, the simple initial function Φ_α is a fixed point of the shift-operator S_α. For α near α_n this fixed point can be continued, more precisely:
There exist ε_n^-, $\varepsilon_n^+ > 0$ and a map

$$\psi : I_n := (\alpha_n - \varepsilon_n^- , \alpha_n + \varepsilon_n^+) \to C[0,1]$$
$$\alpha \to \psi(\alpha) =: \psi_\alpha$$

such that $S_\alpha \psi_\alpha = \psi_\alpha$ for all $\alpha \in I_n$. Moreover $\psi_\alpha = \Phi_\alpha$ for $\alpha \in I_n$ and $\alpha \le \alpha_n$.
With respect to S_α^2, the fixed point index of the fixed point ψ_α is $+1$ for $\alpha < \alpha_n$, $\alpha \in I_n$, and -1 for $\alpha > \alpha_n$, $\alpha \in I_n$. In particular, fixed points of S_α^2 bifurcate from fixed points of S_α at $\alpha = \alpha_n$.

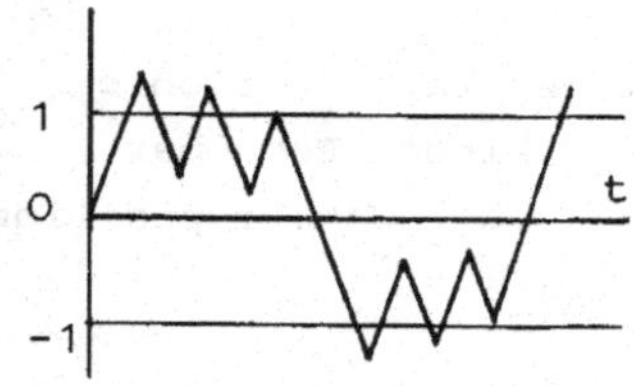

$\alpha = \alpha_2$

3-spike-solution

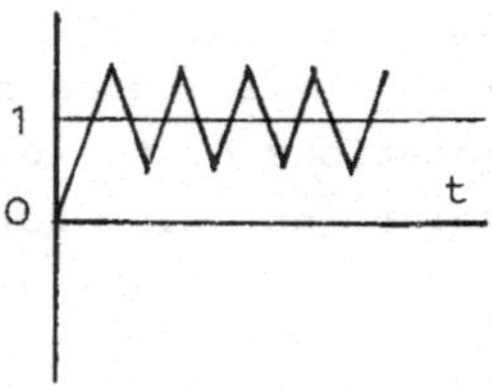

$\alpha = \alpha^*$

solution of period 4/3

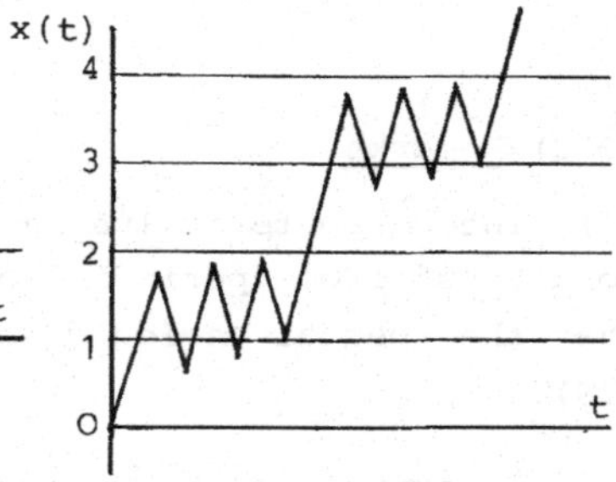

$\alpha = \beta_2$

3-spike-solution of the second kind

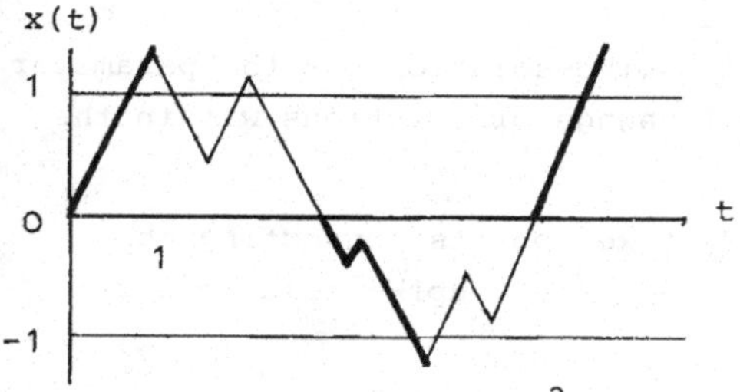

$\Phi_\alpha \to -S_\alpha(\Phi_\alpha) \to S_\alpha^2(\Phi_\alpha)$

$S_\alpha^2(\Phi_\alpha) = \Phi_\alpha$

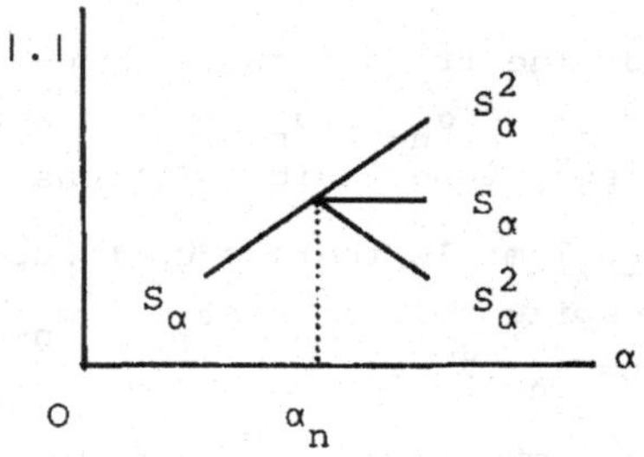

Bifurcation at $\alpha = \alpha_n$

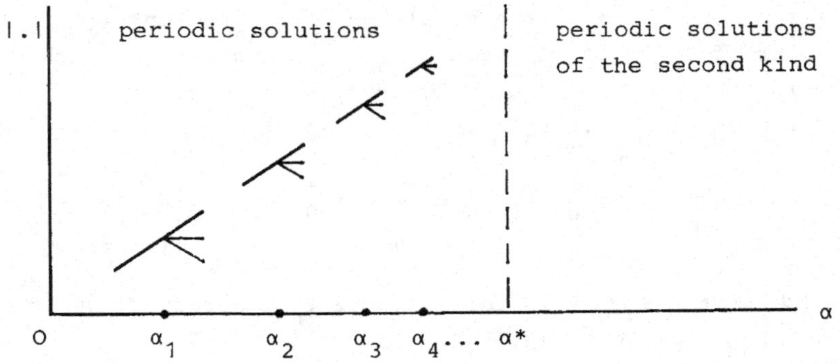

Partial Bifurcation Diagram

(2.2) Remarks

(1) The parameter value $\alpha = \alpha^* = \frac{3}{2}$ does not separate "periodic solutions" from "periodic solutions of the second kind". Remember that the special periodic solutions exist for all $\alpha > 0$ (Proposition 1.5).

(2) Interpreting the initial functions Φ_α that yield periodic solutions of the second kind as fixed points of a suitable shift-operator Σ_α (to be defined) one can also find parameters $\tilde{\beta}_n > \beta_n$, such that fixed points of Σ_α^2 bifurcate from fixed points of Σ_α at $\alpha = \tilde{\beta}_n$.

(3) The transition from n- to (n+1)-spike-solutions in the parameter range (α_{n-1}, α_n) is a structural change of solutions within the class of periodic solutions.

Problem: Is there a continuum of S_α-fixed points connecting the n-spike solutions at $\alpha = \alpha_{n-1}$ with the (n+1)-spike solutions at $\alpha = \alpha_n$?

(4) The simple initial functions Φ_α lead to different types of

solutions depending on the parameter α . Roughly speaking one can think of the transition from periodic solutions to periodic solutions of the second kind as "picking up" and "putting off" spikes. Nevertheless there is a sudden change from "periodic" to "periodic of the second kind" at $\alpha = \alpha^* = \frac{3}{2}$.

Problem: Do there exist parameters α such that Φ_α belongs to an aperiodic solution ?

(5) The parameter values α_n satisfy the relationship

$$\lim_{n\to\infty} \frac{\alpha_n - \alpha_{n-1}}{\alpha_{n+1} - \alpha_n} = c$$

where $c = 4$. Observe that c is different from Feigenbaum's number which occurs in cascading bifurcations [1] .

2.3 THEOREM (" Complex dynamic behaviour ")

Let $\alpha \in (1,2)$. Then there exist infinitely many periodic as well as aperiodic solutions of equation (1.1).

More precisely, for $\alpha \in (\frac{7}{5}, 2)$ there are at least three different types of periodic and aperiodic solutions which can be distinguished by their amplitudes and oscillation properties:

Type 1 : For every $\alpha \in (1,2)$ there exist infinitely many periodic and aperiodic solutions x_α such that

$$|x_\alpha| = \sup_{t\in R} |x_\alpha(t)| < \frac{1}{7}(3\alpha + 4)$$

and x_α is rapidly oscillating with respect to 0 .

Type 2 : For every $\alpha \in (\frac{15}{11}, 2)$ there exist infinitely many periodic and aperiodic solutions x_α which are slowly oscillating with respect to 0 .

<u>Type 3</u> : For every $\alpha \in (\frac{7}{5}, 2)$ there exist infinitely many periodic and aperiodic solutions x_α such that

$$|x_\alpha| = \sup_{t \in R} |x_\alpha(t)| > \frac{1}{7}(3\alpha + 4)$$

and x_α is rapidly oscillating with respect to O .

(2.4) <u>Remark</u> .

We use the phrase "infinitely many periodic and aperiodic solutions" as an abbreviation for the conclusions of the Li-Yorke theorem. In fact, the proof of theorem 2.3 will show that the dynamics of equation (1.1) can be described (at least partially) by a one-dimensional map h_α . This map has periodic and aperiodic points which correspond to periodic and aperiodic solutions of equation (1.1). However, it is difficult to translate the "lim inf" and "lim sup" relations which hold for the map h_α (cf. the Li-Yorke theorem [3]) into the framework of solutions of equation (1.1).

<u>Type 1</u>

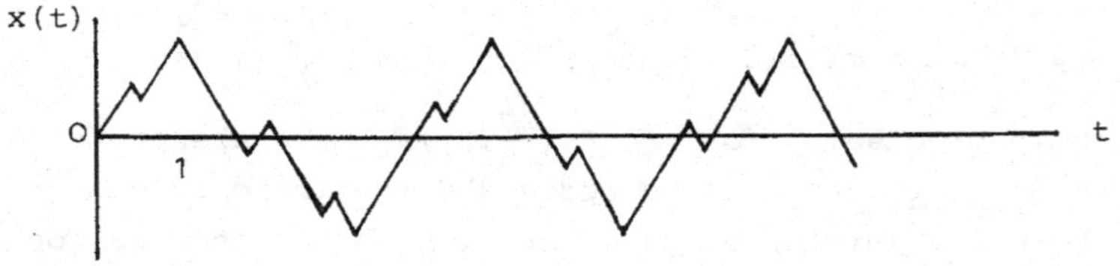

<u>Type 2</u>

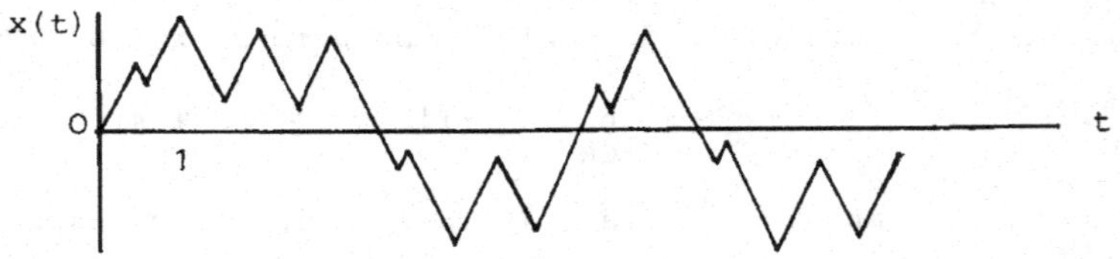

<u>Type 3</u>

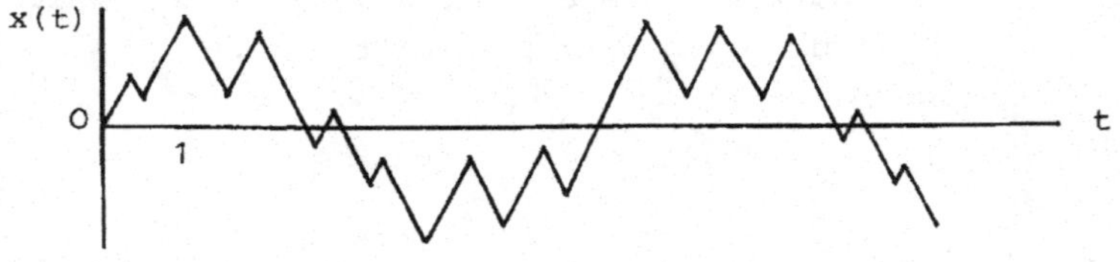

III. PROOF OF THEOREM 2.1

The solution x^{Φ_α} of equation (1.1) corresponding to the initial function Φ_α is piecewise linear and can be computed explicitly for certain parameter values α .

Part 1 : Construction of the (n+1)-spike-solution

If $\alpha > 1$ then there exists exactly one $z_0 \in [0,1]$ such that $\Phi_\alpha(z_0) = 1$. The segment $x^{\Phi_\alpha}_{z_0}$ is in the set

$$M_\alpha \quad := \quad \{ \varphi \in C[0,1] \mid \varphi(0)=1 \ ; \quad \dot{\varphi}(t) = \begin{cases} +\alpha \ , & t\in[0,y) \\ -\alpha \ , & t\in[y,1] \end{cases} \ ; \quad 0 \le y \le \frac{\alpha-1}{\alpha}\}$$

of spikes with derivative $\pm\alpha$. Since $\alpha \le 2$, for each $\varphi \in M_\alpha$ there is a point $z_1 \in (0,1]$ such that $\varphi(z_1) = 1$. Define a shift operator σ_α on M_α as follows:

$$\sigma_\alpha \ : \ M_\alpha \ \to \ M_\alpha$$
$$\varphi \ \to \ \begin{cases} x^\varphi_{z_2} & , \text{ if there exists a } z_2 > z_1 \text{ such that } x^\varphi(z_2) = 1 \\ 0 & \text{else} \end{cases}$$

Using y as a coordinate, the set M_α is homeomorphic to the interval $I_\alpha := [0, \frac{\alpha-1}{\alpha}]$. The shift operator σ_α induces a mapping $\tilde{\sigma}_\alpha : I_\alpha \to I_\alpha$. For $y \in [\frac{1}{4}, \frac{\alpha-1}{\alpha}]$ ($\neq \emptyset \Leftrightarrow \alpha \ge \frac{4}{3}$) one computes easily $\tilde{\sigma}_\alpha(y) = 4y - 1$. Denote the inverse of $\tilde{\sigma}_\alpha$ by $\tilde{\rho}_\alpha$: $\tilde{\rho}_\alpha(x) = \frac{1}{4}x + \frac{1}{4}$.

The values M_i at the maxima are given by $M_i = \alpha . \tilde{\sigma}^i_\alpha(y) + 1$. The condition $M_n = 1$ for the (n+1)-spike-solution is a condition on y , namely $\tilde{\sigma}^n_\alpha(y) = 0$ or $y = \tilde{\rho}^n_\alpha(0) =: y_n$.
This gives for y_n the recurrence formula

$$\begin{cases} y_{n+1} = \frac{1}{4} y_n + \frac{1}{4} \\ y_0 = 0 \end{cases}$$

Since y is related to α by $y = \frac{\alpha-1}{\alpha}$ one obtains for α_n the recurrence-formula

$$\begin{cases} \alpha_{n+1} = \dfrac{4\alpha_n}{2\alpha_n + 1} \\ \alpha_0 = 1 \end{cases}$$

<u>Part 2</u> : For $\alpha = \alpha^* = \frac{3}{2}$, $y = \frac{\alpha-1}{\alpha} = \frac{1}{3}$ is a fixed point of the map $\tilde{\sigma}_\alpha$ which corresponds to a fixed point of the shift operator σ_α .

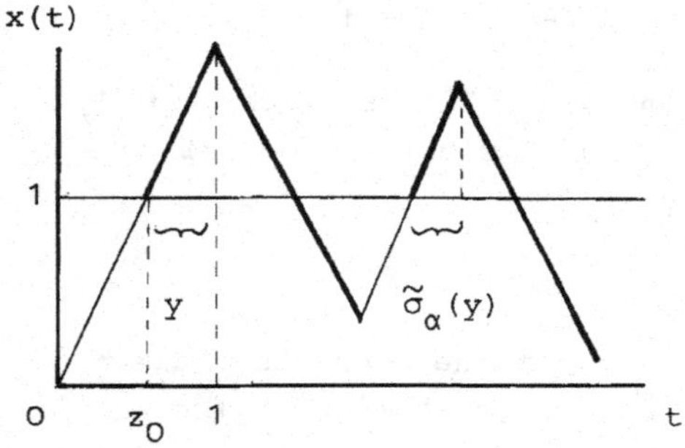

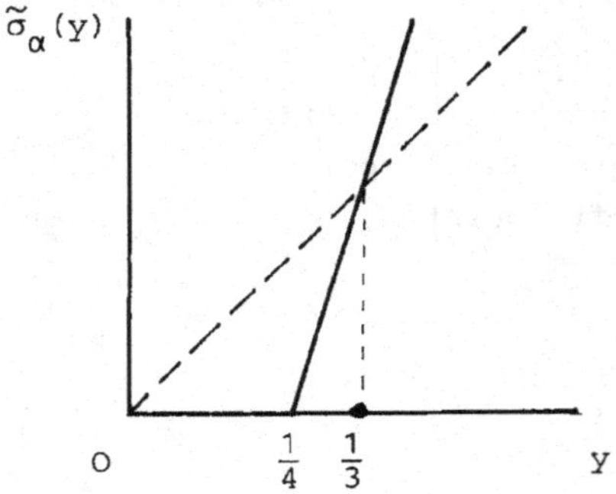

<u>Part 3</u> : Construction of the (n+1)-spike-solution of the second kind

The sequence of the m_i is given by $m_i = 1 - \alpha.(1 - 2\tilde{\sigma}^i_\alpha(y))$. m_n is equal to 1 if and only if $1 - 2\tilde{\sigma}^n_\alpha(y) = 0$. This is true iff $\tilde{\sigma}^n_\alpha(y) = \frac{1}{2}$. That means $y = \tilde{\rho}^n_\alpha(\frac{1}{2}) =: \tilde{y}_n$. Since $y = \frac{\alpha-1}{\alpha}$, one obtains for β_n the same recurrence formula as for α_n but with different initial value $\beta_0 = 2$.

<u>Part 4</u> : Bifurcation at $\alpha = \alpha_n$

Define

$$Q_\alpha := \left\{ \psi \in C[0,1] \;\middle|\; \begin{array}{ll} \dot{\psi} = +\alpha & \text{on } [0, \frac{\alpha-1}{\alpha} - y) \\ \dot{\psi} = -\alpha & \text{on } [\frac{\alpha-1}{\alpha} - y, \frac{\alpha-1}{\alpha}) \\ \dot{\psi} = +\alpha & \text{on } [\frac{\alpha-1}{\alpha}, 1] \end{array} ;\ \psi(0)=0 ;\ 0 \le y \le \frac{\alpha-1}{\alpha} \right\}$$

Observe that $\Phi_\alpha \in Q_\alpha$ ($y = 0$), and that Q_α is homeomorphic to the interval $I_\alpha := [\,0, \frac{\alpha-1}{\alpha}\,]$ via the coordinate y.

Let us even assume that $0 \leq y \leq \frac{\alpha-1}{2\alpha}$. In this case there exists exactly one $t_0 \in (0,1]$ such that $\psi(t_0) = 1$. Denote the coordinate of the "spike" $x^{\psi}_{t_0} \in M_\alpha$ by Y . One computes $Y = \frac{\alpha-1}{\alpha} - 2y$.

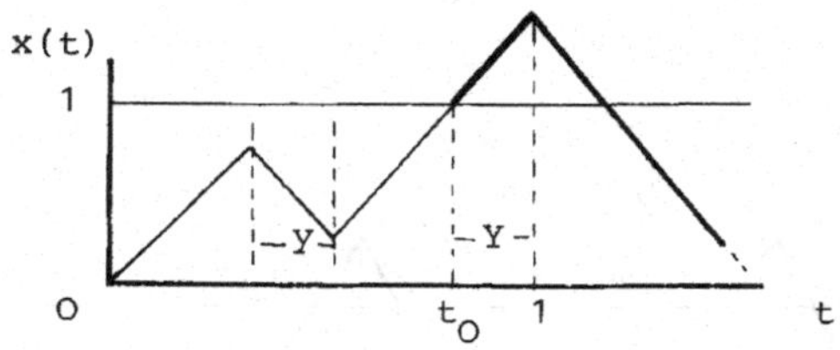

3.1 <u>LEMMA</u> Let $\tilde{\sigma}_\alpha$ defined by $\tilde{\sigma}_\alpha(Y) = 4Y - 1$ be the map which is induced by the shift operator σ_α on the interval $[\,\frac{1}{4}, \frac{\alpha-1}{\alpha}\,] \subset I_\alpha$. Let $Y = \frac{\alpha-1}{\alpha} - 2y$, y small and positive. Then the following conclusions hold:

If $\alpha \leq \alpha_n$ then $\tilde{\sigma}^n_\alpha(Y) \leq 0$

If $\alpha > \alpha_n$ then $\tilde{\sigma}^n_\alpha(Y) \leq 0$ for $\frac{\alpha - \alpha_n}{2\alpha\alpha_n} \leq y$

and $\tilde{\sigma}^n_\alpha(Y) \geq 0$ for $0 \leq y \leq \frac{\alpha - \alpha_n}{2\alpha\alpha_n}$

<u>Proof</u>: We already know from part 1 that $\tilde{\sigma}^n_\alpha(\frac{\alpha_n - 1}{\alpha_n}) = 0$. Since $\tilde{\sigma}_\alpha$ is monotonic, $\tilde{\sigma}^n_\alpha(Y) \geq 0$ (resp. ≤ 0) for $Y \geq \frac{\alpha_n - 1}{\alpha_n}$ (resp. $\leq \frac{\alpha_n - 1}{\alpha_n}$).

If $\alpha \leq \alpha_n$, $Y = \frac{\alpha-1}{\alpha} - 2y \leq \frac{\alpha_n - 1}{\alpha_n}$, therefore $\tilde{\sigma}^n_\alpha(Y) \leq 0$.

If $\alpha > \alpha_n$, one obtains

$$Y = \frac{\alpha-1}{\alpha} - 2y \geq \frac{\alpha_n - 1}{\alpha_n} \quad \Leftrightarrow \quad 0 \leq y \leq \frac{\alpha-\alpha_n}{2\alpha\alpha_n}$$

<u>qed</u>.

Therefore if y is positive and small there are two possibilities for $M_n = \alpha.\tilde{\sigma}_\alpha(Y) + 1$:

(a) either $M_n \geq 1$ $\quad$ ($\Leftrightarrow$ $\alpha > \alpha_n$ and $0 \leq y \leq \frac{\alpha-\alpha_n}{2\alpha\alpha_n}$)

(b) or $M_n < 1$ $\quad$ ($\Leftrightarrow$ $\alpha > \alpha_n$ and $y > \frac{\alpha-\alpha_n}{2\alpha\alpha_n}$

or $\alpha \leq \alpha_n$)

Case a : $M_n \geq 1$

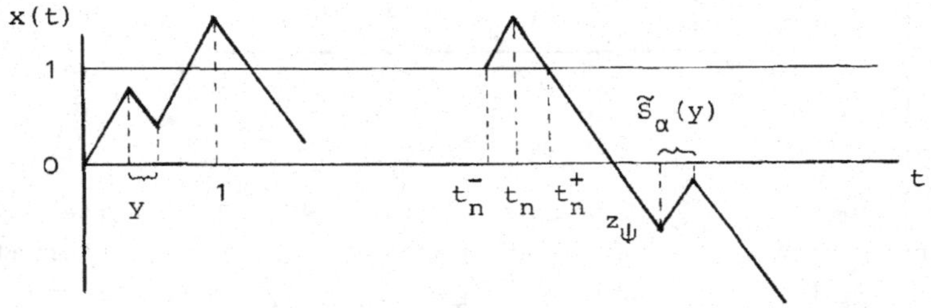

Let t_n be the point where the $(n+1)^{st}$ spike attains its maximum, $x^\psi(t_n) = M_n$. Define $t_n^- := t_n - \tilde{\sigma}_\alpha^n(Y)$, $t_n^+ := t_n + \tilde{\sigma}_\alpha^n(Y)$. Obviously we have that $x^\psi(t_n^-) = x^\psi(t_n^+) = 1$, and $z_\psi = t_n^+ + \frac{1}{\alpha}$ is the first positive zero of the trajectory x^ψ. If y is small $\tilde{\sigma}_\alpha^n(Y)$ is small too, and the segment $x_{z_\psi}^\psi$ has constant derivative $-\alpha$ on the interval $(z_\psi, t_n^- + 1)$ and constant derivative $+\alpha$ on $(t_n^- + 1, t_n^+ + 1)$. Since $t_n^+ + 1 - z_\psi = \frac{\alpha-1}{\alpha}$, it follows that $-x_{z_\psi}^\psi$ is element of Q_α with coordinate $t_n^+ - t_n^- = 2\tilde{\sigma}_\alpha^n(Y)$. Thus, the induced one-dimensional shift-operator $\tilde{S}_\alpha$ is given by

$$y \rightarrow \tilde{S}_\alpha(y) = 2\tilde{\sigma}_\alpha^n\left(\frac{\alpha-1}{\alpha} - 2y \right)$$

for y small and positive.

Case b : $M_n < 1$

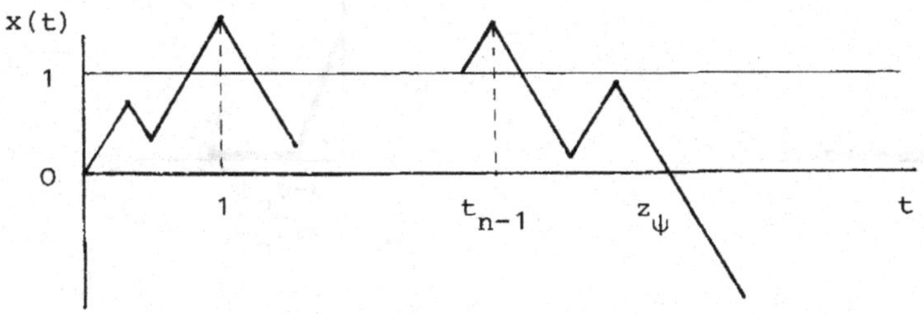

Let t_{n-1} be the unique point where the nth spike attains its maximum M_{n-1}. Define $t^-_{n-1} := t_{n-1} - \tilde{\sigma}^{n-1}_\alpha(Y)$, $t^+_{n-1} := t_{n-1} + \tilde{\sigma}^{n-1}_\alpha(Y)$. On the interval (t^+_{n-1}, z_ψ) which contains the interval $(t^+_{n-1}, t^+_{n-1}+1)$ of length 1 , the trajectory has only values between 0 and 1 . Therefore the segment $x^\psi_{z_\psi}$ following the zero z_ψ has constant derivative $-\alpha$. The induced one-dimensional shift operator $\tilde{S}_\alpha$ is therefore given by $\tilde{S}_\alpha(y) = 0$.

For α near α_n , y small and positive, we obtain the induced one dimensional shift operator

$$\tilde{S}_\alpha : [0,\varepsilon_n] \to [0,\varepsilon_n]$$

$$y \to \begin{cases} 0 & \text{for } \alpha < \alpha_n \\ 0 & \text{for } \alpha \geq \alpha_n \ ; \ \frac{\alpha-\alpha_n}{2\alpha\alpha_n} \leq y \leq \varepsilon_n \\ 2\tilde{\sigma}^n_\alpha(\frac{\alpha-1}{\alpha} - 2y) & ; \quad 0 \leq y \leq \frac{\alpha-\alpha_n}{2\alpha\alpha_n} \end{cases}$$

ε_n is chosen greater than $\frac{\alpha-\alpha_n}{2\alpha\alpha_n}$, but so small that $\tilde{S}_\alpha$ maps $[0,\varepsilon_n]$ into itself and such that the number of spikes produced by Φ_α is the same as the number of spikes produced by Φ_{α_n} .

One can give an explicit formula for ε_n . Also by an explicit computation one can find the numbers ε^-_n and ε^+_n occuring in the statement of the theorem. We omit this for reasons of length.

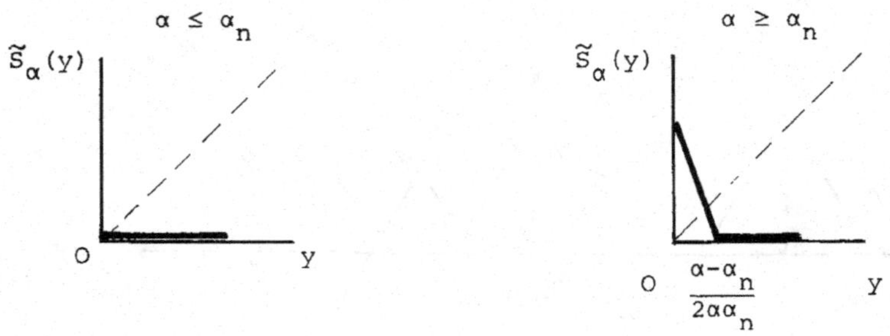

For $\alpha > \alpha_n$, $y = 0$ is no longer a fixed point of $\tilde{S}_\alpha$. If we denote the fixed point of $\tilde{S}_\alpha$ for $\alpha > \alpha_n$ by y^* , then the derivative of $\tilde{S}_\alpha$ at y^* is less than -1. With respect to $\tilde{S}^2_\alpha$, y^* has therefore the fixed point index -1. The bifurcating $\tilde{S}^2_\alpha$- fixed point is $y = 0$ (resp. $\tilde{S}_\alpha(0)$). Moreover we see from the one dimensional map that for $\alpha \le \alpha_n$ the $\tilde{S}_\alpha$ -fixed point is given by $y = 0$, i.e. the map S_α-fixed point is the simple initial function Φ_α .

QED.

IV. PROOF OF THEOREM 2.3

The proof of theorem 2.1 indicates that the shift operator S_α is not the appropriate tool to describe the dynamics of a trajectory. Moreover the computation of the induced one dimensional map $\tilde{S}_\alpha$ becomes difficult because there is in general a large time interval between two subsequent zeros.

A better choice would be, to take $x \equiv 1$ as reference level (as we have already done in defining the operator σ_α) and to consider the "spikes" $\varphi \in M_\alpha$. The proof of theorem 2.1 shows that under certain conditions on the coordinate y , segments of the trajectory come back into M_α after a short time. But in that case it also could happen, that the trajectory runs away from $x \equiv 1$ and oscillates many times around $x \equiv 0$ or $x \equiv -1$.

To avoid oscillations around $x \equiv -1$, one could force back the trajectory by choosing a different nonlinearity f with $f(x) < 0$ for $x < 0$ ("strict feedback for negative arguments"). Of course that would not be an appropriate model for equation (1.3).

Instead of this we take also $x \equiv -1$ as a reference level and define a map H_α which takes into account <u>both</u> reference levels $x \equiv +1$ and $x \equiv -1$ and which is no longer a shift operator in the classical sense. Nevertheless fixed points of H_α^p will correspond to periodic solutions of equation (1.1), because the nonlinearity f is odd.

After these remarks we start with the proof. First, define the set M_α^+ of "positive" spikes :

$$M_\alpha^+ := \{ \psi \in C[0,1] \mid \begin{array}{ll} \dot\psi = +\alpha & \text{on } [0,y) \\ \dot\psi = -\alpha & \text{on } [y,1] \end{array} ;\ \psi(0)=1 ;\ 0 \le y \le \frac{\alpha-1}{\alpha} \}$$

By $M_\alpha^- := \{ \psi \mid -\psi \in M_\alpha^+ \}$ we denote the set of "negative" spikes.

4.1 <u>LEMMA</u> Let $\alpha \le 2$. For every $\psi \in M_\alpha^+$ (resp. $\psi \in M_\alpha^-$) there exists $T_0(\psi) := \min \{ T > 1 \mid x_T^\psi \in M_\alpha^+ \cup M_\alpha^- \}$

<u>Proof</u>: The proof will follow from Lemma 4.6 .

4.2 <u>Remark</u>

For $\alpha > 2$ the lemma will not be true in general. For example the initial function ψ defined by $\psi(0) = 1$; $\dot\psi = 5/2$ on $[0,1/5)$; $\dot\psi = -5/2$ on $[1/5,1]$ yields a (rapidly oscillating) periodic solution of period $4/5$ around $x \equiv 0$. That means: $\psi \in M_\alpha^+$, but $x_T^\psi \notin M_\alpha^+ \cup M_\alpha^-$ for all $T > 1$.

Now define
$$H_\alpha : M_\alpha^+ \cup M_\alpha^- \to M_\alpha^+ \cup M_\alpha^-$$
$$\psi \to x^\psi_{T_0(\psi)}$$

4.3 Remark

H_α is not a shift operator in the classical sense, since there is no fixed reference level. Therefore H_α is not continuous:

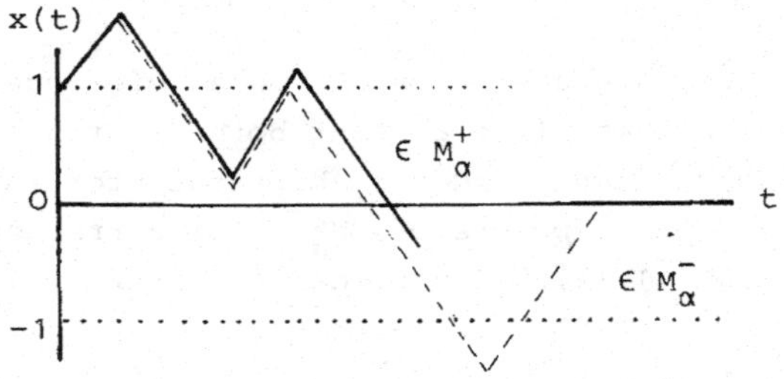

4.4 LEMMA Let $\varphi \in M_\alpha^+ \cup M_\alpha^-$ and $p \in N$. If $H_\alpha^p(\varphi) \equiv \varphi$ or $H_\alpha^p(\varphi) \equiv -\varphi$, then x^φ is a periodic solution of equation (1.1).

Proof: $H_\alpha^p(\varphi) \equiv \pm \varphi$ means: there exists a sequence $T_0(\varphi), \ldots T_p(\varphi)$ such that $x^\varphi_{T_i(\varphi)} \in M_\alpha^+ \cup M_\alpha^-$ and $x^\varphi_{T_p(\varphi)} \equiv \pm \varphi$.

Since equation (1.1) is autonomous and f is odd, x^φ is a periodic solution.

qed.

Because of the previous lemma, spikes in M_α^+ as well as spikes in M_α^- which reproduce a starting spike provide periodic solutions. We therefore identify "positive" and "negative" spikes: define φ and ψ to be equivalent if $\varphi \equiv \psi$ or $\varphi \equiv -\psi$. Denote by $\mathcal{M}_\alpha := M_\alpha^+ \cup M_\alpha^- / \sim$ the quotient space and by $\mathcal{H}_\alpha$ the induced map:

$$\mathcal{H}_\alpha : \mathcal{M}_\alpha \to \mathcal{M}_\alpha$$
$$[\varphi] \to \mathcal{H}_\alpha[\varphi] := [H_\alpha \varphi]$$

$\mathcal{M}_\alpha$ is homeomorphic to the interval $[0,\frac{\alpha-1}{\alpha}] =: I_\alpha$ using y as coordinate. Therefore we study the one dimensional map $h_\alpha : I_\alpha \to I_\alpha$ induced by $\mathcal{H}_\alpha$.

4.5 Remark

h_α has a simple geometric meaning: given two subsequent segments $x_{t_1}, x_{t_2} \in M_\alpha^+ \cup M_\alpha^-$, h_α maps the coordinate y of x_{t_1} onto the coordinate, say Y, of x_{t_2}, neglecting the order of derivatives $+\alpha$ / $-\alpha$ of the segments.

4.6 LEMMA The map $h_\alpha : I_\alpha \to I_\alpha$ is given by

$$y \to h_\alpha(y) = \begin{cases} -4y + \frac{\alpha-1}{\alpha} & ; \; y \in [0, \frac{1}{4}\frac{\alpha-1}{\alpha}] \\ \frac{\alpha-1}{\alpha} & ; \; y \in (\frac{1}{4}\frac{\alpha-1}{\alpha}, \frac{3}{8}\frac{\alpha-1}{\alpha}) \\ 8y - 3\frac{\alpha-1}{\alpha} & ; \; y \in [\frac{3}{8}\frac{\alpha-1}{\alpha}, \frac{1}{2}\frac{\alpha-1}{\alpha}] \\ \frac{\alpha-1}{\alpha} & ; \; y \in (\frac{1}{2}\frac{\alpha-1}{\alpha}, \frac{1}{4}) \\ 4y - 1 & ; \; y \in [\frac{1}{4}, \frac{\alpha-1}{\alpha}] \end{cases}$$

Proof: The proof proceeds by elementary computation. Just for illustration we compute h_α on the interval $(\frac{1}{4}\frac{\alpha-1}{\alpha}, \frac{3}{8}\frac{\alpha-1}{\alpha})$:

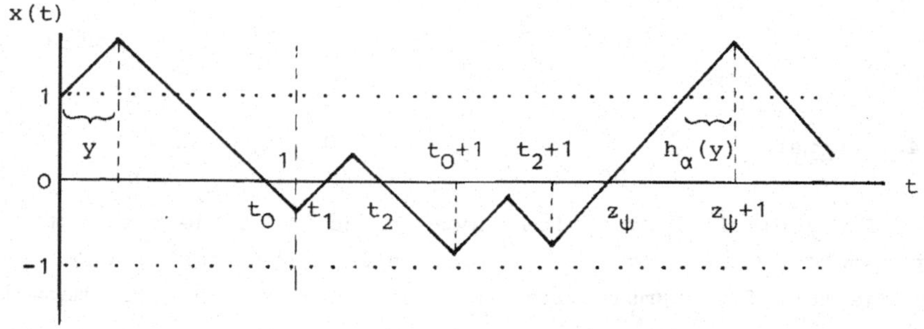

First observe that $t_0 = 2y + \frac{1}{\alpha} \leq 1$, since $y \leq \frac{\alpha-1}{2\alpha}$

$$x(1) = 1 - \alpha(1-2y) = 1 - \alpha + 2\alpha y \begin{cases} \leq 0 & , \text{ since } y \leq \frac{\alpha-1}{2\alpha} \\ \geq -1 & , \text{ since } y \geq \frac{\alpha-2}{2\alpha} \end{cases} \quad (\alpha < 2 \,!)$$

$$x(1+2y) = x(1) + 2\alpha y = 1 - \alpha + 4\alpha y \quad \begin{matrix} \geq 0 & \text{, since} & y \geq \frac{\alpha-1}{4\alpha} \\ \leq 1 & \text{, since} & y \leq \frac{1}{4} \end{matrix} \qquad (*)$$

Therefore there exists exactly one $t_1 \in (1, 1+2y)$ such that $x(t_1) = 0$. Because of symmetry we have that $t_1 = 1 + (1-t_0)$.

$$x(t_0+1) = x(1+2y+\frac{1}{\alpha}) = x(1+2y) - 1 \quad \begin{matrix} \geq -1 \\ \leq 0 \end{matrix} \quad \text{because of } (*)$$

Therefore there exists exactly one $t_2 \in (1+2y, 1+2y+\frac{1}{\alpha})$ such that $x(t_2) = 0$. Because of symmetry we have that $t_2 = t_1 + 2(1+2y-t_1)$.

$$x(t_1+1) = x(1+2y+\frac{1}{\alpha}+t_1-t_0) = x(1+2y+\frac{1}{\alpha}) + \alpha(t_1-t_0) =$$
$$= \alpha - 2 < 0 \text{ , since } \alpha < 2 .$$

Finally we have that

$$x(t_2+1) = x(t_1+1) - \alpha(t_2-t_1) = 3\alpha - 4 - 8y \geq -1 \text{ , since } y \leq \frac{3}{8}\frac{\alpha-1}{\alpha} .$$

Since the trajectory has only values between -1 and 0 on the interval (t_2,t_2+1), a segment with constant derivative $+\alpha$ follows leading to a spike $x_{T_0}^{\psi} \in M_{\alpha}^{+}$ with coordinate $\frac{\alpha-1}{\alpha}$.

Similar computations prove the other parts of the lemma.

<u>qed</u>.

4.7 <u>Remark</u>

The explicit computation demonstrates that the restriction on the parameter α ($1 < \alpha < 2$) is essential. This restriction keeps the trajectory for example from oscillating around zero too many times.

Now we can draw the graph of the map h_{α} :

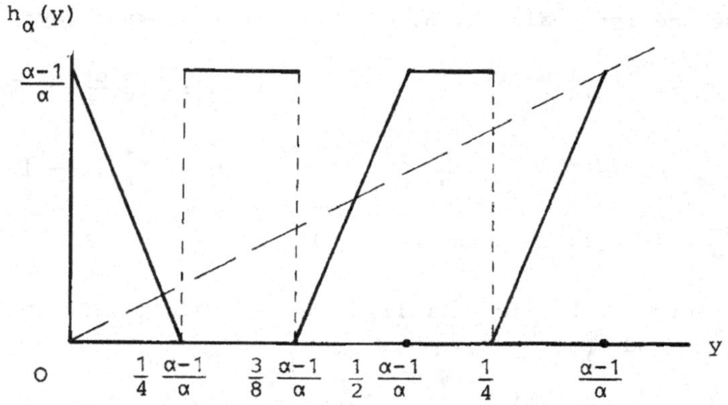

It is clear from the construction of h_α that periodic (aperiodic) points of h_α correspond to periodic (aperiodic) solutions of equation (1.1). Therefore the proof of the theorem will be complete if we show that h_α is chaotic in the sense of Li-Yorke [3]. Unfortunately the map h_α is not continuous. Thus the Li-Yorke theorem is not applicable. But H.W. Siegberg has observed [8] that the crucial assumption of the Li-Yorke theorem can be expressed as follows *) :

4.8 LEMMA Let $A,B \subset R$ be disjoint closed intervals, h continuous on A and B . If the covering property

$$\text{(P)} \qquad \begin{cases} h(A) \supset A \cup B \\ h(B) \supset A \end{cases}$$

is satisfied, then all conclusions of the Li-Yorke theorem hold. In particular there exist infinitely many periodic and aperiodic points of h .

*) K. Schmitt has pointed out to us that a similar "chaos producing" result is contained in the paper by Lasota-Ważewska, Mathematical models of the red cell system, Mat. Stosowana 6, 1976, 25-40

Now define intervals A, B, C, D in the following way:

$$A := [0, \frac{1}{4}\frac{\alpha-1}{\alpha}] \qquad B := [\frac{3}{8}\frac{\alpha-1}{\alpha}, \frac{3}{7}\frac{\alpha-1}{\alpha}]$$

$$C := [\frac{3}{7}\frac{\alpha-1}{\alpha}, \frac{1}{2}\frac{\alpha-1}{\alpha}] \qquad D := [\frac{1}{4}, \frac{\alpha-1}{\alpha}]$$

($\frac{3}{7}\frac{\alpha-1}{\alpha}$ is a fixed point of h_α)

Obviously (P) is satisfied for all $\alpha \in (1,2)$ in the case of the sets A and B :

$$h_\alpha(A) = [0, \frac{\alpha-1}{\alpha}] \supset A \cup B$$

$$h_\alpha(B) = [0, \frac{\alpha-1}{\alpha}] \supset A \cup B$$

This gives infinitely many periodic and aperiodic solutions of type 1.

With respect to A and D we have:

$$h_\alpha(A) = [0, \frac{\alpha-1}{\alpha}] \supset A \cup D .$$

But

$$h_\alpha(D) = [0, \frac{3\alpha-4}{\alpha}] \supset A = [0, \frac{1}{4}\frac{\alpha-1}{\alpha}]$$

if and only if $\alpha > \frac{15}{11}$. Therefore we get solutions of type 2 generated by h_α in A and D for $\alpha > 15/11$.

Finally we have

$$h_\alpha(C) = [\frac{3}{7}\frac{\alpha-1}{\alpha}, \frac{\alpha-1}{\alpha}] \supset C \cup D .$$

But

$$h_\alpha(D) = [0, \frac{3\alpha-4}{\alpha}] \supset C = [\frac{3}{7}\frac{\alpha-1}{\alpha}, \frac{1}{2}\frac{\alpha-1}{\alpha}]$$

if and only if $\alpha > 7/5$ which gives solutions of type 3.

Re-interpreting periodic respectively aperiodic points of h_α as segments in $M_\alpha^+ \cup M_\alpha^-$ one gets the assertion on the behaviour of the amplitudes. The fact that functions corresponding to points in B and C lead to rapidly oscillating segments (compare Lemma 4.6) finishes the proof.

QED.

V. FINAL REMARKS AND OPEN PROBLEMS

5.1 Multiple bifurcation at $\alpha = 1$

One could ask whether there exist even more periodic solutions which do not have segments in $M_\alpha^+ \cup M_\alpha^-$. Indeed this is the case. A local analysis of the shift operator S_α near $\alpha = 1$, $\alpha > 1$, on initial functions ψ_α defined by

$$\dot{\psi}_\alpha = \begin{cases} +\alpha & \text{on } [0,x) \\ -\alpha & \text{on } [x,x+y) \\ +\alpha & \text{on } [x+y,1] \end{cases} \quad ; \quad \psi_\alpha(0) = 0$$

with $x, y \geq 0$ and $x+y \leq 1$ gives four distinct periodic solutions bifurcating from the special periodic solution. The initial functions are fixed points of the shift operator S_α. Representing these fixed points by their coordinates (x,y), the four fixed points are given by

$$(x,y)_1 = \frac{\alpha-1}{\alpha}(1, 0) \quad ; \quad (x,y)_2 = \frac{\alpha-1}{\alpha}\left(\frac{4}{7}, \frac{3}{7}\right)$$

$$(x,y)_3 = \frac{\alpha-1}{\alpha}\left(\frac{1}{7}, \frac{6}{7}\right) \quad ; \quad (x,y)_4 = \frac{\alpha-1}{\alpha}\left(\frac{1}{3}, \frac{2}{5}\right)$$

The fixed points $(x,y)_2$ and $(x,y)_3$ correspond to fixed points of h_α, $(x,y)_1$ corresponds to a fixed point of h_α^2. However $(x,y)_4$ does not correspond to a fixed point or a periodic point of h_α. No segment of the trajectory belonging to $(x,y)_4$ reaches M_α^+ or M_α^-.

Periodic solution belonging to

$(x,y)_1$:

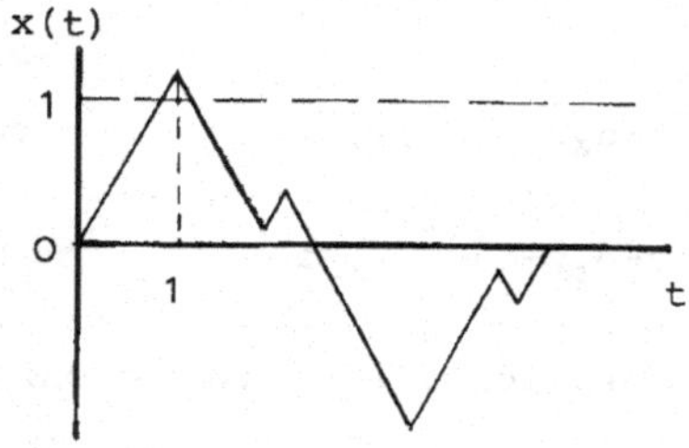

$(x,y)_2$:

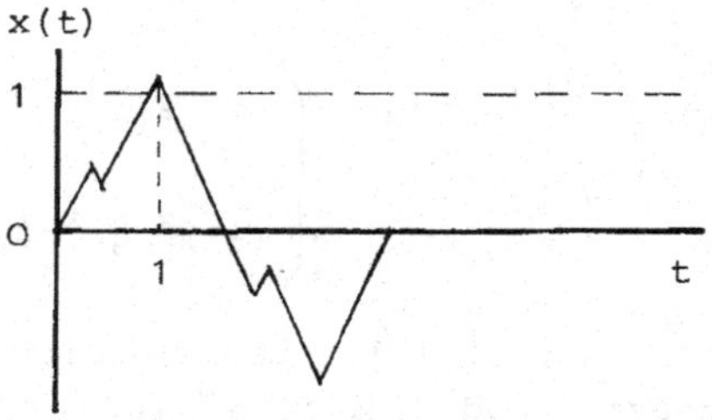

$(x,y)_3$:

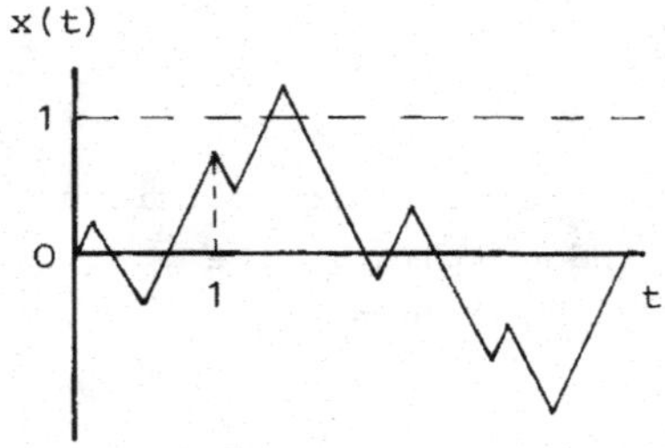

$(x,y)_4$:

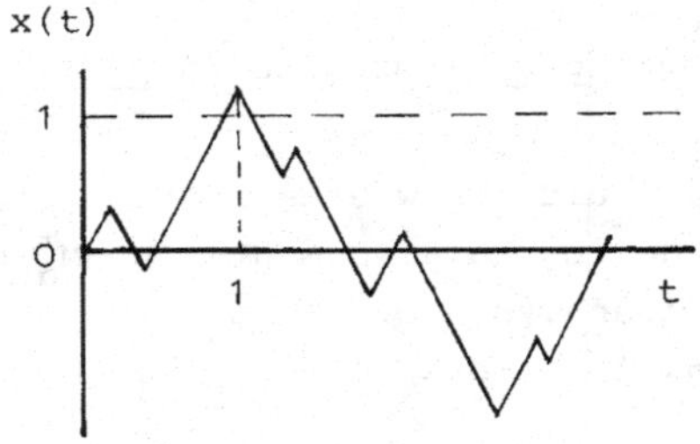

5.2 Chaos by the covering property

We have used Lemma 4.8 to provide several types of chaotic trajectories. Implicitely we have also used the fact that one can conclude on chaotic behaviour in knowing the underlying map only partially.

5.3 Approximation of step functions by continuous functions

The chaos result is not an artifact which is caused by the discontinuities of the nonlinear function (1.2). Replace f by a continuous approximating function f_δ of the following type:

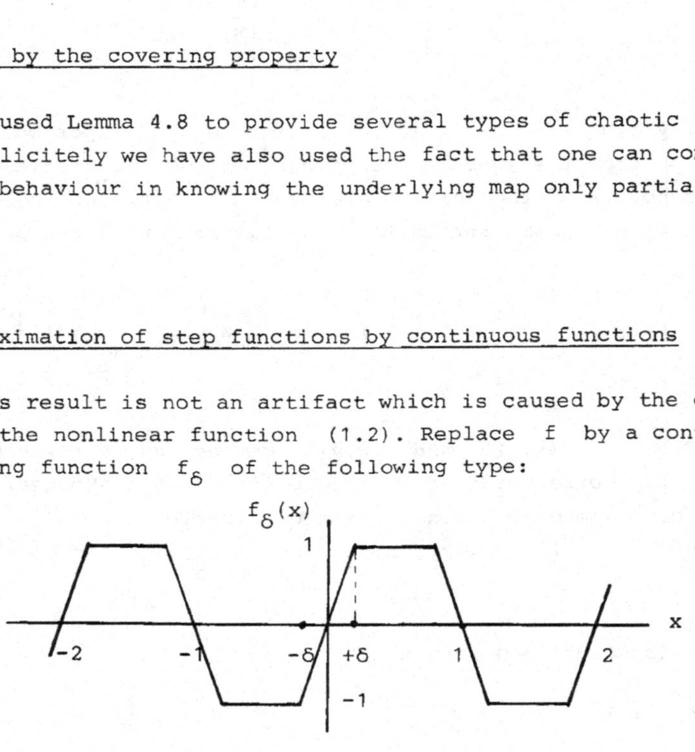

i.e.

$$f_\delta(x) = \begin{cases} f(x) & \text{on } [z+\delta, z+\delta-1], \quad z \in Z \\ (-1)^z \, g(\frac{x-z}{\delta}) & \text{on } [z-\delta, z+\delta], \quad z \in Z \end{cases}$$

where $g \in C[-1,1]$ is an arbitrary odd function which is monotonic increasing and satisfies $g(1) = 1$.

In general, the (differentiable) trajectories will become complicated. However, if α is suitably chosen and δ is small enough, then the smooth trajectory is almost the same as in the discontinuous case, the "peaks" being replaced by smooth "δ-caps". In particular the one dimensional map h_α is the same. In order to guarantee that the δ-caps fall into the region where f_δ is constant, the coordinate y must be restricted on subintervals of the intervals given in Lemma 4.6 depending on α and δ. Since for δ small enough the covering property is still satisfied, there is also chaos in the equation with the continuous nonlinearity f_δ.

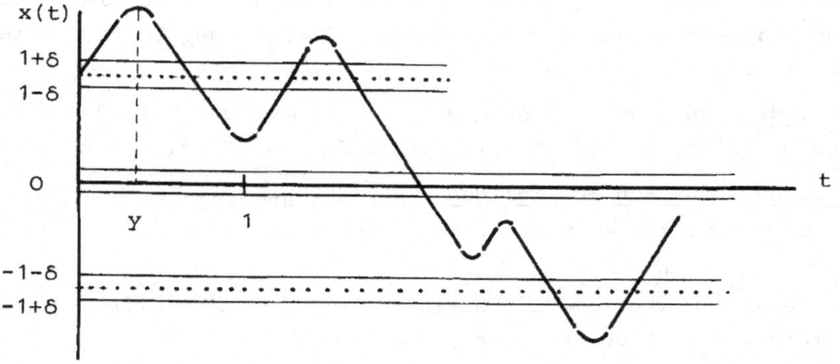

5.4 Problems

1) Extend the local diagrams near α_n to a global bifurcation diagram!

2) What happens for $\alpha > 2$? The example following Lemma 4.1 indicates that there should also be interferences with periodic solutions oscillating around 0.

ACKNOWLEDGMENT

I would like to thank my colleague H.W. Siegberg for stimulating discussions on chaos and Prof. H.-O. Peitgen for correcting some errors in an earlier version of this paper.

VI. REFERENCES

[1] M.J. Feigenbaum, Quantitative universality for a class of nonlinear transformations, J.Stat.Phys. 19, 1978, 25-52

[2] T. Furumochi, Existence of periodic solutions of one-dimensional differential-delay equations, Tôhoku Math.J. 30, 1978, 13-35

[3] T.Y. Li - J.A. Yorke, Period three implies chaos, Am.Math.Monthly 82, 1975, 985-992

[4] R.D. Nussbaum, Periodic solutions of nonlinear autonomous functional differential equations, in "Functional Differential Equations and Approximation of Fixed Points", Springer Lecture Notes in Math. 730, 1979, 283-326

[5] R.D. Nussbaum, Uniqueness and nonuniqueness for periodic solutions of $x'(t) = -g(x(t-1))$, J.Diff.Eq. 34, 1979, 25-54

[6] R.D. Nussbaum - H.-O. Peitgen, Spurious and special periodic solutions of $\dot{x}(t)=-\lambda f(x(t-1))$, to appear

[7] H. Peters, Comportement chaotique d'une équation différentielle retardée, C.R.Acad.Sci. Paris 290, 1980, 1119-1122

[8] H.W. Siegberg, PhD-thesis, Bremen 1981

Fachbereich Mathematik
Forschungsschwerpunkt "Dynamische Systeme", Universität Bremen
Postfach 33 04 40
D-2800 Bremen 33 , F.R.G.

Note added in proof :

H.O. Walther gave a talk in Bremen on a chaos result he proved for the equation $\dot{x}(t) = f(x(t-1))$ where f is a smoothed-out step function with two zeros at -1 and 0. Moreover f satisfies a strict feedback condition for $x > 0$, i.e. $x.f(x) < 0$ for $x > 0$. Therefore the existence of infinitely many periodic and aperiodic solutions can be proved with the aid of one fixed shift operator having $x \equiv -1$ as reference level. (H.O. Walther, Homoclinic solution and chaos in $\dot{x}(t) = f(x(t-1))$, preprint)

CHAOTIC MAPPINGS ON S^1
PERIODS ONE, TWO, THREE IMPLY CHAOS ON S^1

BY

H.-W. SIEGBERG

Forschungsschwerpunkt "Dynamische Systeme"
Fachbereich Mathematik
Universität Bremen
D-2800 Bremen 33

CHAOTIC MAPPINGS ON S^1

Periods one, two, three imply chaos on S^1

Hans-Willi Siegberg*

A fascinating result of Li and Yorke [8] resp. Šarkovskii [12], [13] states that even simple continuous maps $f : [a,b] \to [a,b]$ may develop a very complex behavior under iteration, see also [7], [11].

The essential feature of their "Period three implies chaos" theorem is the following: Let $A_1, A_2 \subset R$ be closed intervals such that $A_1 \cap A_2$ is empty or a single point, say $\{p\} = A_1 \cap A_2$. Let $f : A_1 \cup A_2 \to R$ be continuous such that $f(A_1) \supset A_1 \cup A_2$ and $f(A_2) \supset A_1$, i.e. f induces a transition matrix

$$A = ((\ a_{ij}\ )) = \begin{bmatrix} 1 & 1 \\ 1 & 0 \end{bmatrix} \quad ; \quad a_{ij} = \begin{cases} 1\ , & f(A_i) \supset A_j \\ 0\ , & \text{otherwise} \end{cases}$$

Then f has points of all periods,
and f has an uncountable number of aperiodic ("chaotic") points. In particular, if $f : [a,b] \to [a,b]$ is continuous with a point of period three, say $a = f^3(a) < f(a) < f^2(a)$, choose $A_1 = [f(a), f^2(a)]$, $A_2 = [a, f(a)]$.

(In the formulation above the Li-Yorke-Šarkovskii result generalizes easily to higher dimensions, provided A_1, A_2 are closed balls, $A_1 \cap A_2 = \emptyset$, $f|A_i$ is injective, $i = 1,2$, and $f|A_1$ satisfies appropriate expansion properties, see e.g. [10].)

---

*Research supported by the 'Deutsche Forschungsgemeinschaft' (SFB 72 Universität Bonn; DFG-Projekt "Multiple Bifurkation" Universität Bremen)

A first glance on the circle indicates that there cannot be a "Period three implies chaos" theorem for mappings $f : S^1 \to S^1$: a rotation $f : S^1 \to S^1$ through the angle $2\pi/3$ is not chaotic in the sense of Li and Yorke. Because of the topology of S^1 a point of period three does not force necessarily a transition scheme A, which generates points of all periods.

In this note we present a Li-Yorke-Šarkovskii type result for mappings $f : S^1 \to S^1$ which detects the periods generating chaotic behavior.

<u>Theorem</u>: Let $f : S^1 \to S^1$ be a continuous mapping with points of period one (= fixed point), two and three. Then

- for all $n \in N$ there is a point $x \in S^1$ of period n, i.e. $f^n(x) = x$, $f^m(x) \neq x$ for $m < n$;
- there exists an uncountable set $S \subset S^1$ (containing no periodic points) such that

  i) for every $p,q \in S$, $p \neq q$,

  $$\limsup_{n\to\infty} d(\ f^n(p), f^n(q)\ ) > 0$$

  $$\liminf_{n\to\infty} d(\ f^n(p), f^n(q)\ ) = 0$$

  ii) for every periodic point $p \in S^1$ and for every $q \in S$,

  $$\limsup_{n\to\infty} d(\ f^n(p), f^n(q)\ ) > 0$$

  (d = natural metric on S^1)

The theorem can be proved by lifting f to the real line and applying the same arguments as used by Li and Yorke in [8]. Instead of this (nasty) proof, in which one has to distinguish lots of cases, we give a more elegant proof which uses elements of a paper of Bowen and Franks [3] (where interval maps are discussed). Utilizing elementary singular homology theory and relative Lefschetz numbers we show how certain periodic points generate transition schemes which provide chaotic behavior. Moreover, following Bowen and Franks [3] we use these transition schemes to give (crude) estimates for the topological entropy of a chaotic map.

After the results of this note were proved and written up we became aware of similar results obtained independently by Block [1], and by Block, Guckenheimer, Misiurewicz and Young in [2].

I. PRELIMINARIES

In this section we present some preliminaries which are needed in the following, see e.g. [6] .

(1.1) Homology on S^1: Let $x_0,\dots,x_k \in S^1$ be mutually different points on S^1 which are indexed with respect to a fixed orientation of S^1. Denote by c_j the closed arc from x_j to x_{j+1} for $0 \leq j < k$ and by c_k the closed arc from x_k to x_0.

The arcs $c_0,\dots,c_k$ are cycles mod $T := \{x_0,\dots,x_k\}$, and, hence, they induce homology classes in $H_1(S^1,T)$ (integer coefficients). For simplicity the homology classes of $c_0,\dots,c_k$ are denoted by the same symbol.

$c_0,\dots,c_k$ are generators of $H_1(S^1,T) = Z^{k+1}$, and, moreover, the following property holds:

If $c : [a,b] \to S^1$ is a path on S^1 with endpoints in T, and if

$$c = \sum_{j=0}^{k} a_j c_j \in H_1(S^1,T) \quad ; \quad a_j \in Z, \quad j = 0,\dots,k,$$

then there are $|a_j|$ closed sub-intervals of $[a,b]$ which are mapped onto c_j. If, moreover, $|a_j| > 1$, the sub-intervals of $[a,b]$ which are mapped onto c_j are mutually disjoint.

(1.2) Traces and Fixed Points: Let $x_0,\dots,x_k \in S^1$ be mutually different points on S^1 (indexed with respect to a fixed orientation of S^1), and let $f : S^1 \to S^1$ be a continuous map with $f(T) \subset T$ ($T := \{x_0,\dots,x_k\}$) .

Denote by

$$A(f,T) = ((a_{ij}))_{0 \leq i,j \leq k}$$

the matrix representation of $f_* : H_1(S^1,T) \to H_1(S^1,T)$ with respect to the basis $c_0,\dots,c_k$:

$$f(c_i) = \sum_{j=0}^{k} a_{ij}c_j \quad , \quad i = 0,\dots,k.$$

(1.2.1) Lemma: If $a_{ii} \neq 0$ then f has $|a_{ii}|$ fixed points in c_i.

The lemma follows immediately from the intermediate value theorem.

In view of the previous lemma the trace of the "absolute value" $A_+(f,T)$ of $A(f,T)$,

$$A_+(f,T) := ((|a_{ij}|))_{0 \leq i,j \leq k} ,$$

provides an estimation for the number of fixed points of f. In order to compute the trace of $A(f,T)$ or $A_+(f,T)$ the following relative Lefschetz formula is useful, see [4] .

(1.2.2) Relative Lefschetz Formula :

$$\Lambda(\ f:S^1 \to S^1\ ) = \Lambda(\ f:T \to T\ ) + \Lambda(\ f:(S^1,T) \to (S^1,T)\ )$$

Because $H_0(S^1,T)$ and $H_1(T)$ are trivial this formula implies

(1.2.3) i) $\mathrm{tr}\ A(f,T) = \deg f + F - 1$

ii) $\mathrm{tr}\ A_+(f,T) \geq |\deg f + F - 1|$,

where deg f is the topological degree ("winding number") of f , and F is the number of fixed points of f in T .

The following theorem demonstrates the importance of $A_+(f,T)$, see [3] :

(1.2.4) <u>Theorem</u>:

$$\operatorname{card}\{x \in S^1 \mid f^n(x) = x\} \geq \operatorname{tr}(A_+(f,T)^n) - \operatorname{card} T$$

for all $n \in N$.

<u>Proof</u>: The i-th diagonal element b_{ii} of $A_+(f,T)^n$ can be written as

$$b_{ii} = \sum_I a_I ,$$

where

$$a_I := |a_{ii_1}| \cdots |a_{i_j,i_{j+1}}||a_{i_{j+1},i_{j+2}}| \cdots |a_{i_{n-1},i_n}|$$

and $I := (i;i_1,\ldots,i_n)$ runs over all strings such that $i_n = i$ (i fixed).

To every string $I = (i;i_1,\ldots,i_m)$ (not necessary $i_m = i$) with $a_I \neq 0$ a subset $c_i(I) \subset c_i$ of c_i is associated recursively in the following way:

$m = 1$: If $|a_{ii_1}| = 1$, let $c_i(I)$ be a closed sub-arc of c_i such that $f(\operatorname{int} c_i(I)) = \operatorname{int} c_{i_1}$ and $f(c_i(I)) = c_{i_1}$.
If $|a_{ii_1}| > 1$, let $c_i(I)$ be $|a_{ii_1}|$ mutually disjoint closed sub-arcs of c_i such that each of these arcs is mapped onto c_{i_1} as described above.

$m > 1$: Let $I = (i;i_1,\ldots,i_m)$ and $J = (i_1;i_2,\ldots,i_m)$.
If $|a_{ii_1}| = 1$ let $c_i(I)$ be a closed sub-arc of c_i such that $f(\operatorname{int} c_i(I)) = \operatorname{int} c_{i_1}(J)$ and $f(c_i(I)) = c_{i_1}(J)$;
if $c_{i_1}(J)$ is not connected $c_i(I)$ splits into a

set of closed sub-arcs of c_i which are mapped onto the components of $c_{i_1}(J)$.

If $|a_{ii_1}| > 1$ the set $c_i(I)$ is defined analogously as in the case $m = 1$.

Thus, if $I = (i;i_1,\ldots,i_n)$ is a string such that $i_n = i$ and $a_I \neq 0$, then each of the a_I components of $c_i(I)$ contains a fixed point of f^n.

If $I = (i;i_1,\ldots,i_n)$ and $I' = (i;i_1',\ldots,i_n')$ are two different strings such that $a_I, a_{I'} \neq 0$, then there exists k, $1 \leq k < n$, such that $i_k \neq i_k'$. Thus, the sets $c_i(I)$ and $c_i(I')$ intersect at most in common boundary points, because

$$f^k(\text{ int } c_i(I)\ ) \cap f^k(\text{ int } c_i(I')\ ) = \emptyset \ .$$

However, all boundary points of $c_i(I)$ and $c_i(I')$ are mapped by f^n into the set T. Hence,

$$\text{card}\,\{x \in S^1 \mid f^n(x) = x\} \geq b_{oo}+\ldots+b_{nn} - \text{card}\, T \ . \quad \blacksquare$$

If some diagonal element of $A_+(f,T)$ is bigger than 1 the previous theorem implies that the fixed point set of f^n is strictly increasing when n tends to infinity. Moreover, in this situation the theorem can be improved in the following way.

For a continuous map $f : S^1 \to S^1$ and for any $n \in N$ define

$$\text{Per } f^n := \{x \in S^1 \mid f^n(x) = x,\ f^k(x) \neq x \text{ for } k < n\},$$

the set of periodic points with period n.

(1.2.5) <u>Corollary</u>: If there is a diagonal element of $A_+(f,T)$ which is bigger than 1, then

$$\text{Per } f^n \neq \emptyset \text{ for all } n \in N \ .$$

Proof: If $|a_{ii}| > 1$ then there are two disjoint sub-arcs $d_0, d_1 \subset c_i$ such that $f(\text{ int } d_j) = \text{int } c_i$, $f(d_j) = c_i$ for $j = 0, 1$.

Set $C := \text{cl}\{S^1 \setminus (d_0 \cup d_1)\}$.

Then $H_1(d_0 \cup d_1, \partial d_0 \cup \partial d_1)$ and $H_1(S^1, C)$ are both isomorphic to Z^2 with d_0 and d_1 as generators.
Let M be the absolute value of the matrix representation of

$$f_* : H_1(d_0 \cup d_1, \partial d_0 \cup \partial d_1) \to H_1(S^1, C)$$

with respect to d_0 and d_1:

$$M = ((a_{ij}))_{0 \le i,j \le 1} = \begin{bmatrix} 1 & 1 \\ 1 & 1 \end{bmatrix} .$$

Since $\text{tr } M^n = 2^n$, there exists an aperiodic string $I = (i; i_1, \ldots, i_n)$ with $i_n = i$, i.e. the sequence $i_1, \ldots, i_n$ is not periodic ($i, i_j \in \{0,1\}$ for $1 \le j \le n$), and, thus, f^n has a fixed point in $d_i(I)$.
Since d_0 and d_1 are disjoint this fixed point must be contained in $\text{Per } f^n$. ●

In order to compute or estimate the trace of the n-th iterate of a matrix A the following recursive formula is often useful.

(1.2.6) Lemma: Let A be a $(4,4)$-matrix. Then

$$a_n = \text{tr } A.a_{n-1} - S_2.a_{n-2} + S_3.a_{n-3} - \det A.a_{n-4} \quad , n > 4$$

where $a_n = \text{tr } A^n$, and S_2 and S_3 are the symmetric functions of the eigenvalues of A, i.e.

$$S_2 := l_1 l_2 + l_1 l_3 + l_1 l_4 + l_2 l_3 + l_2 l_4 + l_3 l_4$$
$$S_3 := l_1 l_2 l_3 + l_1 l_2 l_4 + l_1 l_3 l_4 + l_2 l_3 l_4$$

(l_1, l_2, l_3, l_4 eigenvalues of A)

The easy proof is omitted.

(1.2.7) Remark: Recall that the symmetric functions tr A, S_2, S_3, det A occur in the characteristic polynomial of A:

$$\det(A - X.Id) = X^4 - \operatorname{tr} A.X^3 + S_2.X^2 - S_3.X + \det A .$$

II. PERIODIC POINTS

The aim of this chapter is to apply the techniques of the previous chapter to the study of periodic points. It seems to be reasonable to distinguish several cases according to the topological degree of $f : S^1 \to S^1$.

(2.1) $|\deg f| \geq 2$

(2.1.1) Theorem: Let $f : S^1 \to S^1$ be a continuous map with $|\deg f| \geq 2$.
Then $\operatorname{Per} f^n \neq \emptyset$ for all $n \in N$, provided

- $|\deg f| \geq 3$, or
- $\deg f = 2$, or
- $\deg f = -2$ and $\operatorname{Per} f^2 \neq \emptyset$.

Proof: Since $\deg f \neq 1$, f has a fixed point x_o, and, since $|\deg f| \geq 2$, there exists a point $x_1 \in S^1$ with $f(x_1) = x_o$. Set $T := \{x_o, x_1\}$.
Then

$$A(f,T) = \begin{bmatrix} l & l \\ m & m \end{bmatrix} , \quad l, m \in Z$$

with $l + m = \deg f$.

In view of (1.2.5) we have only to consider the case where

$$A_+(f,T) = \begin{bmatrix} 1 & 1 \\ 1 & 1 \end{bmatrix} .$$

Let $n \geq 3$ and choose an aperiodic string $I = (i;i_1,\dots,i_n)$, $i_n = i$. (Observe that I produces (n-1) additional aperiodic

strings, namely $(i_1;i_2,\ldots,i_n,i_1)$ and so on.)
Let $x \in c_i(I)$ be a fixed point of f^n. If x is not contained in $\text{Per } f^n$ then there exists k, $1 \leq k < n$, such that $f^k(x) \in T = \{x_0, x_1\}$
This implies $x = x_0$, and thus $\text{Per } f^n \neq \emptyset$.

Now, let $n = 2$ and assume $\deg f = 2$. In this case the sets $c_0(0;1,0)$ and $c_1(1;0,1)$ can intersect at most in $\{x_1\}$ which implies that $\text{Per } f^2 \neq \emptyset$.
This conclusion fails in the situation $\deg f = -2$ as the following example demonstrates:
Let f be the PL-extension of the following vertex map

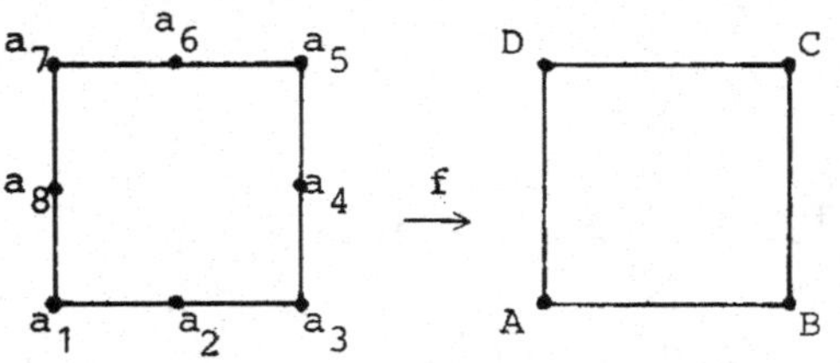

$$f(a_1) = f(a_5) = D$$
$$f(a_2) = f(a_6) = C$$
$$f(a_3) = f(a_7) = B$$
$$f(a_4) = f(a_8) = A$$

(2.1.2) <u>Remark</u>: There is an alternative proof of the theorem above using Nielsen-Wecken theory, see [5] .
On the circle S^1 the Nielsen number $N(f)$ of a continuous mapping $f : S^1 \to S^1$ coincides with the absolute value of the Lefschetz number of f, see [5, ch.VII,C]

$$N(f) = |\Lambda(f)| = | 1 - \deg f |.$$

Thus, $N(f^n) \to \infty$ for $n \to \infty$, and the theorem follows easily from the definition of the Nielsen number.

(2.2) $\boxed{\deg f = 0}$

Since any continuous proper map $g : R \to R$ may be extended to a continuous map on the circle we have to assume at least the existence of a point of period three. It turns out that this assumption is sufficient.

(2.2.1) <u>Theorem</u>: Let $f : S^1 \to S^1$ be continuous with $\deg f = 0$.
If $\operatorname{Per} f^3 \neq \emptyset$ then $\operatorname{Per} f^n \neq \emptyset$ for all $n \in N$.

<u>Proof</u>: Since $\operatorname{Per} f^3 \neq \emptyset$ there are points $x_0, x_1, x_2 \in S^1$ indexed with respect to a fixed orientation of S^1 such that $f(x_0) = x_1$, $f(x_1) = x_2$, $f(x_2) = x_0$. Set $T := \{x_0, x_1, x_2\}$. Then

$$A(f,T) = \begin{bmatrix} k & k+1 & k \\ l & l & l+1 \\ m+1 & m & m \end{bmatrix} \qquad k,\, l,\, m \in Z,$$

with $k + l + m = \deg f - 1 = -1$.

In view of (1.2.5) we have to consider only the two cases

i) $|k| + |l| + |m| = 1 = \operatorname{tr} A_+(f,T)$

ii) $|k| + |l| + |m| = 3 = \operatorname{tr} A_+(f,T)$

where $|k|, |l|, |m| \leq 1$.

The characteristic polynomials of $A_+(f,T)$ which can occur in these cases are

i) $-x^3 + x^2 + x$ (roots: 0, $1/2 + \sqrt{5}/2$, $1/2 - \sqrt{5}/2$)

ii) $-x^3 + 3x^2 - x$ (roots: 0, $3/2 + \sqrt{5}/2$, $3/2 - \sqrt{5}/2$)

The traces $a_n = \operatorname{tr} A_+(f,T)^n$ satisfy the following "Fibonacci" sequences

i) $a_n = a_{n-1} + a_{n-2}$, $a_1 = 1$ and $a_2 = 3$

ii) $a_n = 3a_{n-1} - a_{n-2}$, $a_1 = 3$ and $a_2 = 7$.

Since in any case $a_n + a_{n+1} \leq a_{n+2}$ there exists for any n an "aperiodic" *) string $I = (i; i_1, \ldots, i_n)$, $i_n = i$, $a_I \neq 0$. Thus, f^n has a fixed point in $c_i(I)$, and, if $n \notin 3N$ this fixed point must be contained in $\operatorname{Per} f^n$.
If $n \in 3N$, $n \geq 6$, then the number of aperiodic strings I is bigger than 6, and, hence, the theorem is proved. ●

*) Remark: If the elements of $A_+(f,T)$ are not contained in $\{0,1\}$ then there may exist strings $I = (i;i_1,\ldots,i_n)$, $a_I > 1$. Thus, b_{ii} (the i-th diagonal element of $A_+(f,T)$) does not describe the number of different strings $I = (i;i_1,\ldots,i_n)$ with $i_n = i$ and $a_I \neq 0$. However, in view of the definition of $c_i(I)$ (see (1.2.4) and also (1.2.5)) the string I can be interpreted and counted as a_I different strings $I^1,\ldots,I^p$, $p = a_I$ - each of these strings produces exactly one component of $c_i(I)$.
In this interpretation formally periodic strings I, say $I = (i;i_1,\ldots,i_n,i_1,\ldots,i_n)$, with $a_I > 1$ split into $a_I^2 - a_I$ aperiodic strings $(i;I^k,I^l)$.

(2.3) **deg f = -1**

(2.3.1) Theorem: Let $f : S^1 \to S^1$ be continuous with $\deg f = -1$.
If $\mathrm{Per}\, f^3 \neq \emptyset$ then $\mathrm{Per}\, f^n \neq \emptyset$ for all $n \in \mathbb{N}$.

Proof: Choose points $x_0, x_1, x_2 \in S^1$ indexed with respect to a fixed orientation of S^1 such that $f(x_0) = x_1$, $f(x_1) = x_2$, $f(x_2) = x_0$. Set $T := \{x_0, x_1, x_2\}$.
Then

$$A(f,T) = \begin{bmatrix} k & k+1 & k \\ l & l & l+1 \\ m+1 & m & m \end{bmatrix} \qquad k,\ l,\ m \in \mathbb{Z},$$

with $k + l + m = \deg f - 1 = -2$.

In view of (1.2.5) we have to consider only the case

$$|k| + |l| + |m| = 2 = \mathrm{tr}\, A_+(f,T)$$

where $|k|, |l|, |m| \leq 1$.

The characteristic polynomial of $A_+(f,T)$ which occurs in this case is

$$-x^3 + 2x^2 - 1 \quad (\text{ roots: } 1,\ 1/2 + \sqrt{5}/2,\ 1/2 - \sqrt{5}/2).$$

Thus, the traces $a_n = \text{tr } A_+(f,T)^n$ satisfy the following recursion

$$a_n = 2a_{n-1} - a_{n-3}\ ,\quad a_1 = 2,\ a_2 = 4,\ a_3 = 5\ .$$

Moreover, we see that $a_n + a_{n+1} = a_{n+2} + 1$.
Hence, for any $n \in N$ there exists an aperiodic string I , and the theorem follows as in (2.2.1). ●

(2.4) $\boxed{\deg f = 1}$

It turns out that the case $\deg f = 1$ is much more complex than the other cases, however the same arguments still work.

(2.4.1) Theorem: Let $f : S^1 \to S^1$ be continuous with $\deg f = 1$.
If f has a fixed point, and if $\text{Per } f^2, \text{Per } f^3 \neq \emptyset$ then $\text{Per } f^n \neq \emptyset$ for all $n \in N$.

Proof: Choose points $x_0, x_1, x_2, x_3 \in S^1$ which are indexed with respect to a fixed orientation of S^1 such that $f(x_0) = x_0,\ f(x_1) = x_2,\ f(x_2) = x_3,\ f(x_3) = x_1$.
Set $T := \{x_0, x_1, x_2, x_3\}$.
Then

$$A(f,T) = \begin{bmatrix} k+1 & k+1 & k & k \\ l & l & l+1 & l \\ m+1 & m & m & m+1 \\ p & p+1 & p+1 & p+1 \end{bmatrix} \qquad k, l, m, p \in Z$$

with $k+1 + l + m + p+1 = \deg f = 1$.
In view of (1.2.5) we have to consider the following cases:

i) $\text{tr } A_+(f,T) = 1$:
1) $k = 0,\ l = 0,\ m = 0,\ p = -1$
2) $k = -1,\ l = 0,\ m = 0,\ p = 0$
3) $k = -1,\ l = 0,\ m = 1,\ p = -1$
4) $k = -1,\ l = 1,\ m = 0,\ p = -1$

ii) tr $A_+(f,T) = 3$:

5) $k = 0, \quad l = 1, \quad m = -1, \quad p = -1$

6) $k = 0, \quad l = -1, \quad m = 1, \quad p = -1$

7) $k = -2, \quad l = 1, \quad m = 1, \quad p = -1$

8) $k = 0, \quad l = 1, \quad m = 0, \quad p = -2$

9) $k = 0, \quad l = -1, \quad m = 0, \quad p = 0$

10) $k = -2, \quad l = 1, \quad m = 0, \quad p = 0$

11) $k = 0, \quad l = 0, \quad m = 1, \quad p = -2$

12) $k = 0, \quad l = 0, \quad m = -1, \quad p = 0$

13) $k = -2, \quad l = 0, \quad m = 1, \quad p = 0$

14) $k = -1, \quad l = 1, \quad m = 1, \quad p = -2$

15) $k = -1, \quad l = 1, \quad m = -1, \quad p = 0$

16) $k = -1, \quad l = -1, \quad m = 1, \quad p = 0$

One computes the following characteristic polynomials and recursion formulas for the traces $a_n = \operatorname{tr} A_+(f,T)^n$:

1) $x^4 - x^3 - x - 1$;
$a_{n+1} = a_n + a_{n-2} + a_{n-3}$ with $a_1 = 1,\ a_2 = 1,\ a_3 = 4$
$a_4 = 8$

2) $x^4 - x^3 - 2x^2 - x - 1$;
$a_{n+1} = a_n + 2a_{n-1} + a_{n-2} + a_{n-3}$ with $a_1 = 1,\ a_2 = 5$
$a_3 = 10,\ a_4 = 25$

3) $x^4 - x^3 - 4x^2 - x + 1$;
$a_{n+1} = a_n + 4a_{n-1} + a_{n-2} - a_{n-3}$ with $a_1 = 1,\ a_2 = 9$
$a_3 = 16,\ a_4 = 49$

4) $x^4 - x^3 - 2x^2 + x + 1$;
$a_{n+1} = a_n + 2a_{n-1} - a_{n-2} - a_{n-3}$ with $a_1 = 1,\ a_2 = 5$
$a_3 = 4,\ a_4 = 9$

5) $x^4 - 3x^3 + x + 1$;

$a_{n+1} = 3a_n - a_{n-2} - a_{n-3}$ with $a_1 = 3,\ a_2 = 9$
$a_3 = 24,\ a_4 = 65$

6) $x^4 - 3x^3 + 2x^2 - x + 1$;

$a_{n+1} = 3a_n - 2a_{n-1} + a_{n-2} - a_{n-3}$ with $a_1 = 3,\ a_2 = 5$
$a_3 = 12,\ a_4 = 26$

7) $x^4 - 3x^3 - 6x^2 - x + 1$;

$a_{n+1} = 3a_n + 6a_{n-1} + a_{n-2} - a_{n-3}$ with $a_1 = 3,\ a_2 = 21$
$a_3 = 84,\ a_4 = 377$

8) $x^4 - 3x^3 - 3x - 1$;

$a_{n+1} = 3a_n + 3a_{n-2} + a_{n-3}$ with $a_1 = 3,\ a_2 = 9$
$a_3 = 36,\ a_4 = 121$

9) $x^4 - 3x^3 + 3x - 1$;

$a_{n+1} = 3a_n - 3a_{n-2} + a_{n-3}$ with $a_1 = 3,\ a_2 = 9$
$a_3 = 18,\ a_4 = 49$

10) $x^4 - 3x^3 - 2x^2 - x - 1$;

$a_{n+1} = 3a_n + 2a_{n-1} + a_{n-2} + a_{n-3}$ with $a_1 = 3,\ a_2 = 13$
$a_3 = 48,\ a_4 = 177$

11) $x^4 - 3x^3 - x - 1$;

$a_{n+1} = 3a_n + a_{n-2} + a_{n-3}$ with $a_1 = 3,\ a_2 = 9$
$a_3 = 30,\ a_4 = 97$

12) $x^4 - 3x^3 + 2x^2 + x - 1$;

$a_{n+1} = 3a_n - 2a_{n-1} - a_{n-2} + a_{n-3}$ with $a_1 = 3,\ a_2 = 5$
$a_3 = 6,\ a_4 = 9$

13) $x^4 - 3x^3 - 4x^2 - x - 1$;

$a_{n+1} = 3a_n + 4a_{n-1} + a_{n-2} + a_{n-3}$ with $a_1 = 3,\ a_2 = 17$
$a_3 = 66,\ a_4 = 273$

14) $x^4 - 3x^3 - 6x^2 - x + 1$;

$a_{n+1} = 3a_n + 6a_{n-1} + a_{n-2} - a_{n-3}$ with $a_1 = 3,\ a_2 = 21$
$a_3 = 84,\ a_4 = 377$

15) $x^4 - 3x^3 - x + 1$;

$$a_{n+1} = 3a_n + a_{n-2} - a_{n-3} \quad \text{with} \quad a_1 = 3,\ a_2 = 9,\ a_3 = 30,\ a_4 = 89$$

16) $x^4 - 3x^3 - 2x^2 + x + 1$;

$$a_{n+1} = 3a_n + 2a_{n-1} - a_{n-2} - a_{n-3} \quad \text{with} \quad a_1 = 3,\ a_2 = 13,\ a_3 = 42,\ a_4 = 145$$

It is not hard to prove (by induction and/or by comparison with "known" Finonacci sequences) that in any case the sequence $\{a_n\}$ tends sufficiently fast to infinity, such that for all $n \geq 4$ there is an aperiodic string $I = (i; i_1, \ldots, i_n)$ $i_n = i$, $a_I \neq 0$, providing a periodic point. The tedious proofs are omitted.

Clearly, case 1) indicates that f does not have a point of period two necessarily. A concrete example is the PL-extension of the following vertex map

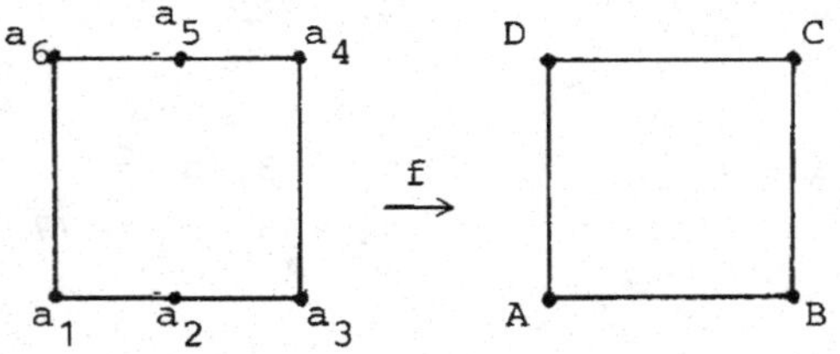

$f(a_1) = f(a_5) = A$
$f(a_2) = f(a_6) = B$
$f(a_3) = C$
$f(a_4) = D$ ●

III. APERIODIC POINTS AND ENTROPY

In this chapter we discuss the second aspect of chaotic mappings $f : S^1 \to S^1$ - the existence of aperiodic points and the topological entropy.

(3.1) <u>Theorem</u>: Let $f : S^1 \to S^1$ be continuous such that the conditions of (2.1.1), (2.2.1), (2.3.1) or of (2.4.1) are satisfied.
Then there exists an uncountable set $S \subset S^1$ (containing no periodic points) such that

i) for every $p, q \in S$, $p \neq q$,

$$\limsup_{n\to\infty} d(\ f^n(p), f^n(q)\ ) > 0$$

$$\liminf_{n\to\infty} d(\ f^n(p), f^n(q)\ ) = 0$$

ii) for every periodic point $p \in S^1$ and for every $q \in S$

$$\limsup_{n\to\infty} d(\ f^n(p), f^n(q)\ ) > 0\ ,$$

where d is the natural metric of S^1.

<u>Proof</u>: The proof follows the proof of Li and Yorke in [8] , see also [10] . ●

Moreover, it turns out that the matrix $A_+(f,T)$ resp. the traces of $A_+(f,T)^n$ can be used for an estimation of the topological entropy along the same lines as described by Bowen and Franks in [3].

(3.2) <u>Theorem</u>: Let $f : S^1 \to S^1$ be continuous such that the conditions of (2.1.1), (2.2.1), (2.3.1) or of (2.4.1) are satisfied.
Then the topological entropy $h(f)$ of f is bigger than zero, more precisely

$$h(f) \geq (\log 2)/3$$

<u>Proof</u>: The proof is essentially a recapitulation of the paper of Bowen and Franks [3] :

For fixed $N \geq 2$ and for any $k \geq 1$ there exists a (Nk, ε_N)-separated set Q_k for f (see [3]) such that

$$\text{card } Q_k \geq 3^{-k} \left(\text{tr } A_+(f,T)^{Nk} - \text{card } T \right)$$

($\varepsilon_N \to 0$ for $N \to \infty$).

Thus

$$h(f) \geq \limsup_{k\to\infty} (Nk)^{-1} \left(\log \text{tr } A_+(f,T)^{Nk} - k \log 3 \right).$$

In any case (except in case 4) in the proof of (2.4.1)) $A_+(f,T)^3$ has at least one diagonal element which is bigger than 1.

Thus, if $N \in 3\mathbb{N}$ then $\text{tr } A_+(f,T)^N \geq (\sqrt[3]{2})^N$, and hence,

$$h(f) \geq \limsup_{k\to\infty} (Nk)^{-1} \left(\log (\sqrt[3]{2})^{Nk} - k \log 3 \right)$$

$$\geq \log (\sqrt[3]{2}) - N^{-1} \log 3 \quad \text{for all } N \in 3\mathbb{N}.$$

Thus,

$$h(f) \geq (\log 2)/3.$$

(Since in case 4) in the proof of (2.4.1) $A_+(f,T)^2$ has a diagonal element which is bigger than 1, we conclude in this case $h(f) \geq (\log 2)/2 > (\log 2)/3$.) ●

(3.2.1) Remark: Clearly, the estimates of $h(f)$ can be sharpened by looking at the eigenvalues of $A_+(f,T)$.
For instance

in (2.1.1) : $h(f) \geq \log |\deg f|$, this estimate follows also from Mannings result in [9] ;

in (2.2.1)
and (2.3.1) : $h(f) \geq \log (1/2 + \sqrt{5}/2)$.

REFERENCES

[1] L. BLOCK : Periodic points of continuous mappings of the circle, Trans. AMS 260 (1980), 555 - 562

[2] L. BLOCK, J. GUCKENHEIMER, M. MISIUREWICZ, L-S. YOUNG : Periodic points and topological entropy of one dimensional maps, in ' Global Theory of Dynamical Systems ' Springer Lecture Notes in Mathematics, vol 819 (1980), 18 - 34

[3] R. BOWEN, J. FRANKS : The periodic points of maps of the disk and the interval, Topology 15 (1976), 337 - 342

[4] C. BOWSZYC : Fixed point theorems for the pairs of spaces, Bull. Acad. Polon. Sci. 16 (1968), 845 - 850

[5] R.F. BROWN : The Lefschetz Fixed Point Theorem, Scott, Foresman and Comp. Glenview, Illinois (1971)

[6] M. GREENBERG : Lectures on Algebraic Topology, New York, Benjamin, (1967)

[7] J. GUCKENHEIMER : Bifurcations of maps of the interval, Inventiones Math. 39 (1977), 165 - 178

[8] T-Y. LI, J.A. YORKE : Period three implies chaos, Amer. Math. Monthly 82 (1975), 985 - 992

[9] A. MANNING : Topological entropy and the first homology group, in ' Dynamical Systems - Warwick 1974 ' Springer Lecture Notes in Mathematics, vol 468 (1975), 185 - 190

[10] F.R. MAROTTO : Snap-back repellers imply chaos in R^n, J. Math. Anal. and Appl. 63 (1978), 199 - 223

[11] R.M. MAY : Biological populations with nonoverlapping generations: stable points, stable cycles, and chaos, Science 186 (1974), 645 - 647

[12] A.N. ŠARKOVSKII : Coexistence of cycles of a continuous map of a line into itself, Ukr. Mat. Z. 16 (1964), 61 - 71

[13] P. ŠTEFAN : A theorem of Sarkovskii on the existence of periodic orbits of continuous endomorphisms of the real line, Comm. Math. Phys. 54 (1977), 237 - 248

Fachbereich Mathematik
Forschungsschwerpunkt "Dynamische Systeme"
Universität Bremen
Bibliothekstraße
Postfach 330 440
2800 Bremen 33
West Germany

AN ALGORITHM FOR ULTRASONIC TOMOGRAPHY BASED ON INVERSION OF THE HELMHOLTZ EQUATION

BY

F. STENGER

Department of Mathematics
University of Utah
Salt Lake City, Utah 84112
USA

AN ALGORITHM FOR ULTRASONIC TOMOGRAPHY BASED ON INVERSION OF THE HELMHOLTZ EQUATION

$$\nabla^2 u + \frac{\omega^2}{c^2} u = 0$$

by

Frank Stenger*
Department of Mathematics
University of Utah
Salt Lake City, UT 84112

*Research supported by U. S. Army Research Contract No. DAAG-29-77-G-0139.

ABSTRACT

This paper describes a numerical method for reconstructing the function $f(\overline{r}) = \omega^2[c(r)^{-2}-c_0^{-2}]$, where $c(\overline{r})$ denotes the speed of sound in a bounded body, and c_0 denotes the speed of sound in the medium surrounding the body, for both the case of plane wave excitation, $e^{i(\overline{k}\cdot\overline{r}-\omega t)}$, and spherical wave excitation, $e^{ik|\overline{r}-\overline{r}_s|-i\omega t}/[4\pi|\overline{r}-\overline{r}_s|]$. It is assumed that the body is located in the interior of a cylinder of radius a , having the z axis as its axis of symmetry, that the ultrasonic sound pressure is measured on the surface of this cylinder at the points $(a\cos\theta_j\ , a\sin\theta_j\ , z_p)$, where $\theta_j = j\pi/(2N+1)$, $z_p = ph$, $p = 1,2,\ldots,2N+5$ We then describe the reconstruction of $f(x,y,z_p) = F_p(\rho,\theta)$ in the form

$$F_p(\rho,\theta) = \sum_{j=1}^{2N+1} \sum_{m=-2N}^{2N} F_{jm}\, S_j(\rho)e^{im\theta}$$

where the F_{jm} are complex numbers and the $S_j(\rho)$ are "Chapeau" splines on a nonequi-spaced mesh. If h and $a\pi/(2N+1)$ are of the order of $1/k^{1/3}$, where $k = \omega/c_0 = 2\pi/\lambda$, then the constructed solution F_ℓ satisfies $F_p(\rho,\theta) = f(\rho,\theta,zp) + O(1/k^2)$, where f denotes the exact solution to the Rytov approximation to the Helmholtz equation.

1. INTRODUCTION AND SUMMARY.

In the present paper it is assumed that due to sonic excitation, the pressure $p = p(\bar{r},t)$ within a body may be sufficiently accurately described by the wave equation

$$(1.1) \qquad (\nabla^2 - \frac{1}{c^2(\bar{r})}\frac{\partial^2}{\partial t^2})p = 0$$

where $\nabla = (\partial/\partial x, \partial/\partial y, \partial/\partial z)$ is the gradient operator, t denotes time, and $c(\bar{r})$ is the speed of sound at the point $\bar{r} = (x,y,z)$ in the body. If $\rho = \rho(\bar{r})$ denotes the density, and $\kappa = \kappa(\bar{r})$ the compressibility, then $c(\bar{r}) = [\rho(\bar{r})\kappa(r)]^{-1/2}$; the equation (1.1) results if we assume that the amplitude of the sound waves is small enough so that when excitation causes changes in ρ and κ of $\Delta\rho$ and $\Delta\kappa$, then $|\Delta\rho|^2$ and $|\Delta\kappa|^2$ can be ignored relative to $|\Delta\rho|$ and $|\Delta\kappa|$ [5].

In the present paper we assume the geometry of Fig. 1.1. For both the case of sonic plane wave excitation

(1.2) $$p_0 = e^{i\bar{k}\cdot\bar{r}-i\omega t}$$

or sonic spherical wave excitation

(1.3) $$p_0 = \frac{e^{ik|\bar{r}-\bar{r}_s|-i\omega t}}{4\pi|\bar{r}-\bar{r}_s|}$$

we describe a numerical-asymptotic method for reconstructing the function

(1.4) $$f(\bar{r}) = \omega^2\left(\frac{1}{c^2(\bar{r})} - \frac{1}{c_0^2}\right) .$$

In (1.2) $\bar{k}$ is the propogation vector of sound, $|\bar{k}| = k = \omega/c_0 = 2\pi/\lambda$, and where $\omega = (2\pi)\times(\text{frequency})$. In (1.3) $\bar{r}_s$ denotes the source of the spherical wave excitation. In (1.4) $c(\bar{r})$ denotes the speed of sound at the point $\bar{r}$ in the body, and c_0 denotes the speed of sound in the medium surrounding the body. Notice that $f = 0$ on the exterior of the body.

Assuming sound at a constant frequency ω , we take

(1.5) $$p(\bar{r},t) = u(\bar{r})e^{-i\omega t}$$

in (1.1), so that

(1.6) $$\nabla^2 u + \frac{\omega^2}{c^2(\bar{r})} u = 0 .$$

In Sec. 2 which follows, we employ the Rytov approximation on the solution u of (1.6) to reduce the inversion of (1.6) to the inversion of a more tractible integral equation for f, in terms of the measured pressure p.

In Sec. 3 we employ the asymptotic integration method of stationary phase to further simplify the problem of solving for f, reducing it to the solution of a sequence of one-dimensional line integrals of the form

$$(1.7) \qquad W_p(\bar{r}_s,\bar{r}_d) = \int_0^1 [f((1-t)\bar{r}_d+t\bar{r}_s,z_p') + \tfrac{1}{k}B(\bar{r}_s,\bar{r}_d,t)]\,|\bar{r}_s-\bar{r}_d|\,dt$$

where $\bar{r}_s$ and $\bar{r}_d$ are vectors in the plane $z = z_j$ such that in the absence of a body the waves travel in a direction from $\bar{r}_s$ to $\bar{r}_d$, and where $W_p(\bar{r}_s,\bar{r}_d)$ is measured at the point $(\bar{r}_d,z_p)$ on the cylinder. The first term in the integral (1.7) is the "Eikonal approximation" which is the basis on which existing machines operate ([3]). The first term in (1.7) is a simple linear combination of the va of f on a few neighboring paths parallel to the path of integration. The contribution B consists of the dominant diffraction effects. The solution of (1.7) thus yields $f(x,y,z_p)$ to within a relative error of order $1/k^2$.

In Sec. 4 we describe the numerical solution of (1.7) for f. To this end, we approximate f in the plane $z = z_p$ by F, where

$$(1.8) \qquad F(\rho,\theta) = \sum_{m=-2N}^{2N} \sum_{\ell=1}^{2N+1} F_{\ell m}\, S_\ell(\rho) e^{im\theta} \quad .$$

In (1.8) the $F_{\ell m}$'s are complex numbers and the $S_\ell(\rho)$ are "Chapeau" splines; these later are defined on a nonequi-spaced mesh which is conveniently chosen, based on the equi-spaced location of the vectors $\bar{r}_s$ and $\bar{r}_d$ on the circumference of the cylinder. This makes it possible later in Sec. 6 to solve for the numbers $F_{\ell m}$ in (1.8) by means of the solution of lower triangular linear matrix systems of the form

$$\text{(1.9)}\qquad \begin{bmatrix} J_{m11} & & & \\ J_{m12} & J_{m22} & & \\ J_{m13} & J_{m23} & J_{m33} & \\ & \cdots & & \\ J_{m1,2N+1} & J_{m2,2N+1} & \cdots & J_{m,2N+1,2N+1} \end{bmatrix} \begin{bmatrix} F_{1m} \\ F_{2m} \\ F_{3m} \\ \vdots \\ F_{2N+1,m} \end{bmatrix} = \begin{bmatrix} \tilde{W}_{1m} \\ \tilde{W}_{2m} \\ \vdots \\ \tilde{W}_{2N+1,m} \end{bmatrix}$$

where the $\tilde{W}_{jm}$ are easily obtained in terms of the data. The points $\bar{r}_s$ and $\bar{r}_d$ are located at discrete points on the cylinder, namely at $ae^{ij\Delta}$, $j = 0,1,\ldots,4N+1$, where $\Delta = \pi/(2N+1)$, $J_{mj\ell}$ is then defined (see Sec. 5) by the line integral

$$\text{(1.10)}\qquad J_{mj\ell} = \int_{P_\ell} S_j(\rho)e^{im\theta}ds$$

where P_ℓ is the line segment joining the points a and $ae^{i\ell\Delta}$. The $J_{mj\ell}$ are evaluated by Gaussian quadrature.

In Sec. 6 the results derived in earlier sections are combined in explicit algorithms, for reconstructing the function f in (1.4) in both the case of a plane wave source and the case of a spherical wave source. It is assumed that sonic excitation and data collection is thus carried out on each plane $z = z_p$, where $z_p = ph$, $p = 1,2,\ldots,2N+5$, and f is reconstructed at each height $z_p, p = 3,4,\ldots,2N+3$. Under the assumption that h and $a\Delta$ are of order $k^{-1/3}$, the resulting approximation F in (1.8) is accurate to within a relative error of order k^{-2}. The most popular frequency is 2.25×10^6 cycles/sec; working in cm. units, this implies $k^{-2} < 5\times10^{-6}$. In Sec. 7 we conclude with some results involving our experiences on the computer implementation of these algorithms.

The results of the present paper are an improvement over those in [9] in three ways. While we also used Chapeau splines in [9], these were equi-spaced and hence we were not able to solve for $F_{\ell m}$ by means of a lower triangular matrix. In addition, vertical diffraction was ignored in [9], whereas it is included in the present paper. Finally, a two dimensional problem was solved in [9] by means of a slowly convergent Fourier expansion of the kernel, whereas in the present paper we use asymptotics together with interpolatory approximation to reduce the solution for f to a sequence of one-dimensional problems.

The method of the present paper differs considerably from the method of [1] and [6], in that we do not use Fourier transforms in the present method of solution. As is clearly evident from [1] that the Fourier transform $\hat{f}(\bar{\lambda})$ of $f(\bar{r})$ while defined everywhere in theory can be reconstructed from data only for $|\bar{\lambda}| \leq 2k$. The theory of the Whittaker Cardinal Function [4] then tells us that f cannot be reproduced on a grid with mesh size smaller than $\lambda/4$. Thus the Fourier transform techniques have a relative error of at least $O(k^{-2})$ built into them at the outset. On the other hand, it would be easy to extend the approach of the present paper to produce a solution having a relative error of order k^{-3} , or even smaller.

2. ASYMPTOTIC APPROXIMATION OF THE SOLUTION OF THE HELMHOLTZ EQUATION.

Let us start with Eq. (1.6), and let f be defined by Eq. (1.4). We then obtain

$$\nabla^2 u + k^2 u = fu \tag{2.1}$$

where

$$k = \frac{\omega}{c_0} \quad . \tag{2.2}$$

Let us now apply a plane wave excitation (see Eq. (1.2))

$$u_0 = e^{i\bar{k}\cdot\bar{r}} \quad (|\bar{k}| = k) \quad . \tag{2.3}$$

Then, setting

$$u = e^{i\bar{k}\cdot r+W} \tag{2.4}$$

in (2.1)

$$\nabla^2 W + 2i\bar{k}\cdot\nabla W + (\nabla W)\cdot(\nabla W) = f \quad . \tag{2.5}$$

The Rytov approximation now predicts that the term $(\nabla W)\cdot(\nabla W)$ is negligible relative to the other terms. Dropping the terms and solving for W, yields

$$W(\bar{r};\bar{k}) = \iiint_{\mathbb{R}^3} f(\bar{R}) \frac{\exp[ik|\bar{r}-\bar{R}|-i\bar{k}\cdot(\bar{R}-\bar{r})]}{4\pi|\bar{r}-\bar{R}|} d\bar{R} \quad . \tag{2.6}$$

Similarly, for the case of the sperical wave source (1.3) we take

(2.7) $$u_0 = \frac{e^{ik|\bar{r}-\bar{r}_s|}}{4\pi|\bar{r}-\bar{r}_s|} ,$$

set $u = u_0 e^W$ to find that (see [1])

(2.8) $$W(\bar{r};\bar{r}_s) = -\frac{1}{(4\pi)^2} \iiint_{\mathbb{R}^3} f(\bar{R}) \frac{e^{ik\{|\bar{r}-R|+|\bar{r}_s-\bar{R}|\}}}{|\bar{r}-\bar{R}|\,|\bar{r}_s-\bar{R}|} d\bar{R} .$$

3. INCLUSION OF DIFFRACTION EFFECTS VIA THE METHOD OF STATIONARY PHASE.

3.1 The Case of a Plane Wave Source.

Let us assume in Eq. (2.6) that $\bar{k} = (k_x, k_y, 0)$, let us take $\bar{R} = (x,y,z)$, so that $d\bar{R} = dxdydz$, and let us carry out one integration with respect to the variable z . In order to achieve this within a relative error of order k^{-2} , we assume f to be known on the planes $z = z_p = ph$, $j = 1,2,3,\dots,2N+5$, we set

$$g(t) = f(x,y,ph+t) \quad , \qquad (3.1)$$

$$g_\ell = g(\ell h) \ , \ \ell = 0 \ , \ \pm 1 \ , \ \pm 2$$

and we use the approximation

$$\begin{aligned} g(t) = g_0 &+ \frac{t}{12h} \ (g_{-2} - 8g_{-1} + 8g_1 - g_2) \\ &+ \frac{t^2}{2h^2} \ [\delta^2 g_0 - \frac{1}{12} \delta^4 g_0] \\ &+ \frac{t^3}{12h^3} (-g_{-2} + 2g_{-1} - 2g_1 + g_2) \\ &+ \frac{t^4}{24h^4} \ \delta^4 g_0 \quad . \end{aligned} \qquad (3.2)$$

In Eq. (3.2) δ denotes the central difference operator, $\delta g(t) = g(t+\frac{1}{2}h)-g(t-\frac{1}{2}h)$; Eq. (3.2) therefore interpolates g at ℓh , $\ell = 0,\pm 1,\pm 2$, and the error of interpolation for $-2h \leq t \leq 2h$ is of order h^5 as $h \to 0$. Thus the error in the coefficient of t^2 compared with the exact value $g''(0)$ is of order h^3 .

Let us set

$$(3.3)\quad \begin{cases} \bar{r} = (a\cos\varphi\ ,\ a\sin\varphi\ ,\ z_j) = (\bar{a}, z_j) \\ \bar{R} = (\rho\cos\theta\ ,\ \rho\sin\theta\ ,\ z) = (\bar{\rho}, z) \\ \omega = |\bar{a}-\bar{\rho}| = \{a^2 - 2a\rho\cos(\theta-\varphi) + \rho^2\}^{1/2} \end{cases} .$$

Then

$$(3.4)\qquad |\bar{r}-\bar{R}| = \{\omega^2 + (z-z_p)^2\}^{1/2} = \{\omega^2 + t^2\}^{1/2} .$$

The integration with respect to z in (2.6) involves an integral of the form

$$(3.5)\quad \begin{cases} \Phi \equiv \displaystyle\int_{-\infty}^{\infty} g(t)\, \frac{e^{ik\{\alpha^2+t^2\}^{1/2}}}{\{\alpha^2+t^2\}^{1/2}}\, dt \\ \quad = e^{ik\alpha}\displaystyle\int_{-\infty}^{\infty} e^{ikt^2/(2\alpha)}\Big[g_0 + \frac{t^2}{2h^2}(\delta^2 g_0 - \frac{1}{12}\delta^4 g_0)\Big]\cdot\Big[1 - \frac{ikt^4}{8\alpha^3}\Big]\Big[\frac{1}{\omega} - \frac{t^2}{2\alpha^3} + O(k^{-2})\Big]dt \\ \quad = \Big(\dfrac{2\pi}{\alpha k}\Big)^{1/2} e^{ik\omega+i\pi/4}\Big[(1-\frac{i}{8\alpha k})g_0 + \frac{i\alpha}{2h^2k}(\delta^2 g_0 - \frac{1}{12}\delta^4 g_0) + O(k^{-2})\Big] \end{cases} .$$

That is, in the notation of (3.3),

$$(3.6)\quad \begin{aligned} W(\bar{a}, z_p; k) = \frac{e^{i\pi/4}}{2(2\pi k)^{1/2}} \iint_{\mathbb{R}^2} \frac{e^{ik|\bar{a}-\bar{\rho}|}}{|\bar{a}-\bar{\rho}|^{1/2}}&\Big[\{1 - \frac{i}{8k|\bar{a}-\bar{\rho}|}\}F_p(\bar{\rho}) \\ &+ \frac{i|\bar{a}-\bar{\rho}|}{2h^2k}\{(\delta_z^2 - \frac{1}{12}\delta_z^4)F_p(\bar{\rho})\} + O(k^{-2})\Big]d\bar{\rho} \end{aligned}$$

where

$$F_p(\bar{\rho}) = f(\bar{\rho}, z_p)$$

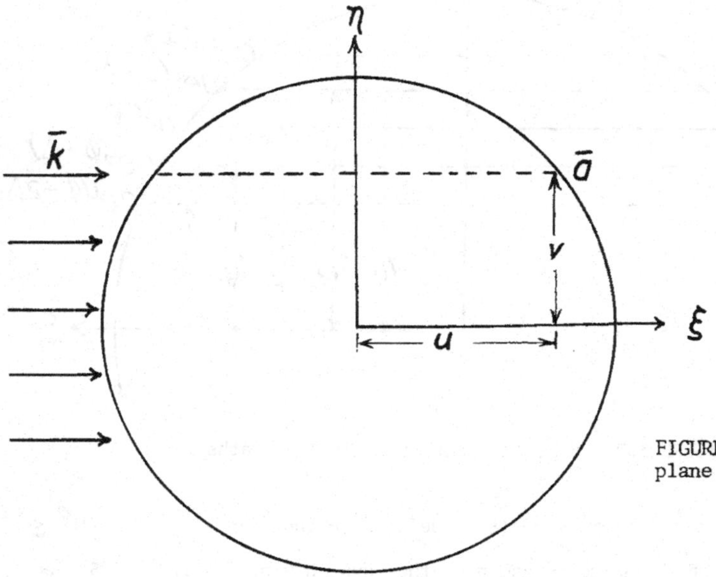

FIGURE 3.1. Incident plane wave detected at $\bar{a}$.

Let us next consider the situation in Fig. 3.1, in which the incident plane wave travels parallel to the ξ-axis and we measure W at the point $\bar{a}$ on the cylinder, such that the point $\bar{a}$ lies in the region $\{(\xi,\eta): \xi > 0\}^*$. In the coordinates of (ξ,η)

$$(3.7) \quad \begin{cases} \bar{a} = (u,v) \\ \bar{\rho} = (\xi,\eta) \\ \bar{a}-\bar{\rho} = (u-\xi,v-\eta) = (u-\xi,t); \; t = v-\eta \\ \bar{k}\,(\bar{a}-\bar{\rho}) = k(u-\xi) \end{cases}$$

*Here (ξ,η) is a rotation of (x,y) .

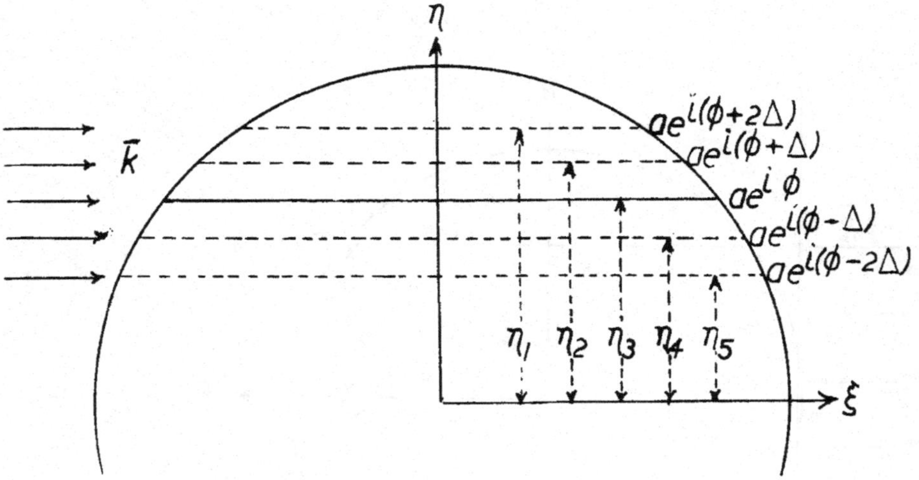

FIGURE 3.2. Five adjacent straight line paths.

Now consider a function $g = g(\xi,\eta)$ defined on the disc $\{(\xi,\eta)\colon \xi^2+\eta^2 \leq a^2\}$. Assuming the values of g to be known on the rays $\eta = \eta_\ell, \ell = 1,2,3,4,5$ (see Fig. 3.2), where $\eta_1 > \eta_2 > \eta_3 > \eta_4 > \eta_5$, and setting

$$g_\ell = g(\xi,\eta_\ell)$$

we can interpolate g as a function of η by means of the Newton polynomial

$$\begin{aligned} g(\xi,\eta) \cong g_1 &+ [1,2](\eta-\eta_1) + [1,2,3](\eta-\eta_1)(\eta-\eta_2) \\ &+ [1,2,3,4](\eta-\eta_1)(\eta-\eta_2)(\eta-\eta_3) \\ &+ [1,2,3,4,5](\eta-\eta_1)(\eta-\eta_2)(\eta-\eta_3)(\eta-\eta_4) \end{aligned} \tag{3.8}$$

where the numbers in brackets denote the usual divided differences, such that the right-hand side equals $g(\xi,\eta_\ell)$ when $\eta = \eta_\ell, \ell = 1,2,\ldots,5$. The error of interpolation in this formula is $O(d^5)$ where $d = \max_{(i=1,2,3,4)} \eta_i - \eta_{i+1}$. In the notation of (3.7), we take $\eta_3 = v$, so that the coefficient of t^2 in the Taylor series expansion in powers of t of the right-hand side of (3.8) is, B , i.e.

$$g(x,v-t) = g(x,v) + At + Bt^2 + \ldots \tag{3.9}$$

Then by differentiating the right-hand side of (3.8) with respect to η

$$\begin{aligned} B = {} & [1,2,3] + (2\eta_3-\eta_2-\eta_1)[1,2,3,4] \\ & + [(\eta_3-\eta_2)(\eta_3-\eta_4) + (\eta_3-\eta_1)(\eta_3-\eta_4) + (\eta_3-\eta_1)(\eta_3-\eta_2)][1,2,3,4,5] \end{aligned} \tag{3.10}$$

The error in this coefficient compared with the exact value $\frac{1}{2}g_{\eta\eta}(\xi,v)$ is of order d^3 for sufficiently smooth g .

We are thus in position to do another integration in (3.6). Having replaced $\bar{\rho}$ with a pair of orthogonal coordinates (ξ,η) where ξ is parallel and $\eta(\eta = v-t)$ is perpendicular to the propogation vector $\bar{k}$, we expand $|\bar{a}-\bar{\rho}|$ in (3.6) in powers of t , integrate termwise using the method of stationary phase, to get

$$W(\bar{a},z_p,\bar{k}) = \frac{i}{2k}\int_0^{2u} \{F_p(\bar{\rho}) + \frac{is}{k}[2B+C] + O(k^{-2})\}ds \tag{3.11}$$

In (3.11)

(3.12) $$\bar{\rho} = \bar{a} - \frac{\bar{k}}{k} s \quad ,$$

i.e., the path of integration is parallel to the propogation vector, starting from the detection point $\bar{a}$, and traversing the total length $2u$ of this path lying in the interior of the cylinder. B is expressed as in (3.10), and involves divided differences of f on adjacent rays parallel to the integration ray, which lie in the plane $z=z_p$ while

(3.13) $$C = \frac{1}{2h^2} (\delta_z^2 - \frac{1}{12} \delta_z^4) F_p(\bar{\rho})$$

and involves values of f on a adjacent rays parallel to the integration ray, which lie above and below the integration ray.

An algorithm for approximating F based on (3.11) is given in Sec. 5.

3.2 The Case of a Spherical Wave Source.

Let us now apply the method of stationary phase to Eq. (2.8). We again use the notation (3.1) and the approximation (3.2). In (2.8) we assume that the detection point $\bar{r} = (\bar{a}, z_p)$ and the source point $\bar{r}_s = (\bar{b}, z_p)$ are both located on the edge of the cylinder in the plane $z=z_p$, where

(3.14) $$\bar{a} = (a \cos\varphi,\ a \sin\varphi)$$
$$\bar{b} = (a \cos\alpha,\ a \sin\alpha) \quad .$$

Next, setting

$$(3.15)\qquad \begin{cases} \bar{R} = (\rho\cos\theta,\ \rho\sin\theta, z) = (\bar{\rho}, z) \\ \omega = |\bar{a}-\bar{\rho}| \\ \beta = |\bar{b}-\bar{\rho}| \end{cases}$$

and as in Sec. 3.1, performing one integration with respect to the variable z in (2.8) via use of the method of stationary phase, we get

$$w(\bar{a},z_p;\bar{b},z_p) = -i^{1/2}\left(\frac{2\pi}{k}\right)^{1/2} \iint_{\mathbb{R}^2} \exp\{ik(\omega+\beta)\}\omega^{-1/2}\beta^{-1/2}(\omega+\beta)^{-1/2} \cdot$$

$$(3.16)\qquad \cdot\left[\left\{1-\frac{i}{8k}\left(\frac{1}{\omega}+\frac{1}{\beta}+\frac{1}{\omega+\beta}\right)\right\}F_p(\bar{\rho})\right.$$

$$\left. + \frac{i}{k}\frac{\omega\beta}{\omega+\beta}\frac{1}{2h^2}\left(\delta_z^2-\frac{1}{12}\delta_z^4\right)F_p(\bar{\rho}) + O\left(\frac{1}{k^2}\right)\right]d\bar{\rho}\ .$$

Let us now consider a sonic point source emanating at $(\bar{b},z_p)$, and the result of this source being measured at $(\bar{a},z_p)$. We again introduce (ξ,η) coordinates, where ξ is parallel to the line segment joining $\bar{a}$ and $\bar{b}$, and η is perpendicular to this segment. (See Fig. 3.2.) We then again use the interpolation procedure of (3.8), and set

$$(3.17)\qquad \begin{cases} \bar{\rho} = (\xi,\eta) \\ \bar{a} = (u,v) \\ \bar{b} = (-u,v) \\ \bar{a}-\bar{\rho} = (u-\xi, v-\eta) = (u-\xi, t)\ ;\ t = v-\eta \\ \bar{b}-\bar{\rho} = (-u-\xi, v-\eta) = (-u-\xi, t) \end{cases} \quad .$$

Expanding F_p , ω and β in powers of t , substituting into (3.16) and integrating, using the method of stationary phase, we get

$$W(\bar{a},z_p;\bar{b},z_p) = \frac{-i}{16\pi k}\frac{e^{2iku}}{u}\int_0^{2u}[(1-\frac{7i}{32ku})F_p(\bar{\rho}) + \frac{is(2u-s)}{2ku}(B+C) + O(k^{-2})]ds$$

(3.18)

where

$$u = \frac{1}{2}|\bar{b}-\bar{a}|$$

(3.19)

$$\bar{\rho} = \bar{a} + \frac{\bar{b}-\bar{a}}{2u}s$$

and where B and C are defined as in (3.11).

An algorithm for approximating F based on (3.18) is described in Sec. 5.

4. APPROXIMATION OF F .

It is possible to reconstruct a three dimensional grey-scale view of f once f is known on closely spaced planes perpendicular to the z-axis. In this section we shall therefore describe a method of approximating $F(\bar{\rho})=F(\rho,\theta)=f(\bar{\rho},z_p)$ on the plane $z=z_p$ and in the interior of the disc

$$\{\bar{\rho} = \rho e^{i\theta} : 0 \leq \rho \leq a \ , \ -\pi \leq \theta \leq \pi\} \quad . \tag{4.1}$$

The approximation takes the form

$$F(\rho,\theta) = \sum_{m=-2N}^{2N} e^{im\theta} \sum_{j=1}^{2N+1} F_{jm} S_j(\rho) \tag{4.2}$$

where N is a suitably chosen positive integer and the $S_j(\rho)$ are "Chapeau" splines which we describe in what follows.

Let us set

$$\begin{cases} \Delta = \dfrac{\pi}{2N+1} \\ \rho_\ell = a \cos \dfrac{\ell\Delta}{2} \ , \ \ell = 0,1,\ldots,2N+1 \end{cases} \quad . \tag{4.3}$$

We then define $S_j(\rho)$ by

$$
(4.4)\quad
\begin{cases}
S_j(\rho) = \left\{\begin{array}{ll}
0 & \text{if } \rho \leqq \rho_{j+1} \\[2ex]
\dfrac{\rho-\rho_{j+1}}{\rho_j-\rho_{j+1}} & \text{if } \rho_{j+1} \leqq \rho \leqq \rho_j \\[2ex]
\dfrac{\rho_{j-1}-\rho}{\rho_{j-1}-\rho_j} & \text{if } \rho_j \leqq \rho \leqq \rho_{j-1} \\[2ex]
0 & \text{if } \rho \geqq \rho_{j-1}
\end{array}\right\} \quad j = 1,2,\ldots,2N \\[8ex]
S_{2N+1}(\rho) = \begin{cases}
\dfrac{\rho_{2N}-\rho}{\rho_{2N}} & \text{if } 0 \leqq \rho \leqq \rho_{2N} \\[2ex]
0 & \text{if } \rho > \rho_{2N}
\end{cases}
\end{cases}
$$

5. THE ELEMENTS OF THE TRIANGULAR MATRICES.

In this section we derive an algorithm for reconstructing f based on either (3.11) or (3.18).

Consider for the moment, the case of $N=5$ in Eq. (4.2). With reference to Fig. 5.1, let us assume that a detector is located at the point a on the horizontal axis, and that sound propogater along the dashed line segment joining $ae^{i6\Delta}$ and a. We consider the integration of $e^{im\theta} S_j(\rho)$ along this segment, from a to $ae^{i6\Delta}$. The "x"s which one meets along this segment are at distance ρ_j $(j = 1,2,\ldots,6,5,\ldots,1)$ from the origin. They are located at a distance r_{6j} from a, where $r_{6j} = u_6-\sqrt{\rho_j^2-\rho_6^2}$ for $j = 1,\ldots,6$, $r_{6j} = u_6 + \sqrt{\rho_{12-j}^2-\rho_6^2}$ for $j = 7,8,9,\ldots,11$, where u_ℓ denotes half of the length of the line segment from a to $ae^{i6\Delta}$. Thus this line segment cuts the supports of $S_1,S_2,\ldots,S_6$, and no others, since the remaining S_j are zero along this segment. If $j=6$, the support of $S_j(\rho)$ is cut only once; if $1 \leq j < 6$, it is cut twice.

Now let N be an arbitrary positive integer, and let us set

$$
(5.1)\qquad \begin{cases}
\Delta = \dfrac{\pi}{2N+1} \\[2ex]
\rho_j = a\cos\left(\dfrac{j\Delta}{2}\right) \\[2ex]
u_\ell = a\sin\dfrac{\ell\Delta}{2} \\[2ex]
r_{j\ell} = \begin{cases} u_\ell - \sqrt{\rho_j^2-\rho_\ell^2}\ , & j = 0,1,\ldots,\ell \\[1ex] u_\ell + \sqrt{\rho_{2\ell-j}^2-\rho_\ell^2}\ , & j = \ell+1,\ldots,2\ell \end{cases}
\end{cases} .
$$

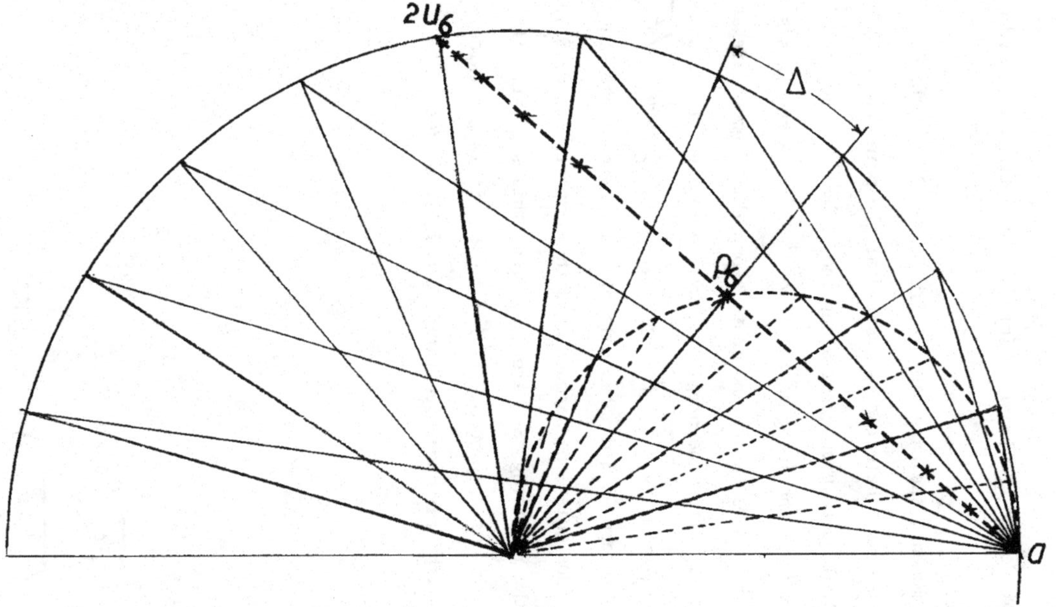

FIGURE 5.1. (N = 5 , $\Delta = \pi/11$ in figure).

Radii ρ_ℓ start at origin in direction $ae^{i\ell\Delta/2}$ and end on circle with center at $\frac{a}{2}$ and radius $a/2$.

Points '×' on line segment from a to $ae^{i\ell\Delta}$ at distance $r_{j\ell}$ from a .

Length of line segment from a to $ae^{i\ell\Delta}$ is $2u_\ell$.

Inspection of Eqs. (3.11), (3.18) and (4.2) shows that we shall require the evaluation of three different integrals, thereby handling both the case of a plane wave source and a spherical wave source simultaneously. To this end, we consider the integrals

$$J^q_{mj\ell} = \int_0^{2u_\ell} r^q e^{im\theta} S_j(\rho)dr \tag{5.2}$$

where $q = 0,1$ or 2, and where the integration is along the line segment of length $2u_\ell$, starting at a and ending at $ae^{i\ell\Delta}$. In view of the symmetric spacing of the points $r_{j\ell}$ about the center point of this line segment, we obtain

$$\text{(5.3)} \quad \begin{cases} J^q_{m\ell\ell} = \displaystyle\int_{r_{\ell-1,\ell}}^{r_{\ell\ell}} S_\ell(|a + ire^{i\ell\Delta/2}|) \cdot \\ \quad \cdot \left[r^q \left(\dfrac{a + ire^{i\ell\Delta/2}}{a - ire^{-i\ell\Delta/2}}\right)^{m/2} + (2u_\ell - r)^q\, e^{i\ell m\Delta} \left(\dfrac{a - ire^{-i\ell\Delta/2}}{a + ire^{i\ell\Delta/2}}\right)^{m/2}\right] dr , \\ \quad \ell = 1,2,\ldots,2N+1 \\ J^q_{mj\ell} = \left(\displaystyle\int_{r_{j-1,\ell}}^{r_{j\ell}} + \int_{r_{j\ell}}^{r_{j,\ell+1}}\right) S_j(|a + ire^{i\ell\Delta/2}|) \cdot \\ \quad \cdot \left[r^q \left(\dfrac{a + ire^{i\ell\Delta/2}}{a - ire^{-i\ell\Delta/2}}\right)^{m/2} + (2u_\ell - r)^q\, e^{i\ell m\Delta} \left(\dfrac{a - ire^{-\ell\Delta/2}}{a + ire^{i\ell\Delta/2}}\right)^{m/2}\right] dr , \\ \quad j = 1,2,\ldots,\ell-1 \; ; \; \ell = 1,2,\ldots,2N+1 \quad . \\ J^q_{mj\ell} = 0 \text{ if } 1 \leq \ell < j \leq 2N+1 \quad . \end{cases}$$

Next, we consider the integration of $r^q e^{im\theta}\ S_j(\rho)$ along the segment from $\bar{\rho} = ae^{i\sigma\Delta}$ to $\bar{\rho} = ae^{i(\sigma+\ell)\Delta}$, where $1 \leq \ell \leq 2N+1$. We denote this path by $P_{\sigma\ell}$. Then we have

$$\int_{P_{\sigma\ell}} r^q\, e^{im\theta}\, S_j(\rho)dr = e^{im\sigma\Delta}\, J^q_{mj\ell} \quad . \tag{5.4}$$

Therefore, if F is given by (4.2), Eq. (5.4) yields

$$\int_{P_{\sigma\ell}} r^q F(\rho,\theta)dr = \sum_{m=-2N}^{2N} e^{im\sigma\Delta} \sum_{j=1}^{\ell} F_{jm}\, J^q_{mj\ell} \quad . \tag{5.5}$$

This formula is the basic intermediate formula which illustrates the triangular nature of the equations as a result of our special mesh spacing of the $S_j(\rho)$. We note, in view of (5.3) that

$$J^q_{-m,j,\ell} = \overline{J^q_{m,j,\ell}}$$

where the bar denotes the complex conjugate; consequently it is necessary to compute $J^q_{mj\ell}$ only for $m = 0,1,\ldots,2N$; $j = 1,2,\ldots,\ell$; $\ell = 1,2,\ldots,2N+1$.

In (5.3) the integral for $J_{mj\ell}$ is split into two parts in order to be able to obtain a smooth integrand, so that each integral can be accurately evaluated by means of Gauss-Legendre quadrature. We note also that if $\ell = 2N+1$, the ratio $(a-ire^{-i\ell\Delta/2})/(a+ire^{i\ell\Delta/2})$ reduces to 1 .

6. DERIVATION OF AN ALGORITHM FOR f .

6.1 The Basic Equations and Algorithms.

Let us multiply both sides of (3.11) by $-2ki$, let us assume that the sound pressure is measured at $(ae^{i\sigma\Delta},z_p)$ and that $\bar{k}$ propogates in the direction from $(ae^{i(\sigma+\ell)\Delta},z_p)$ to $(ae^{i\sigma\Delta},z_p)$. Ignoring the $O(k^{-2})$ term, we write the resulting equation (3.11) in the form

$$W_p(\ell,\sigma) = \int_{P_{\sigma\ell}} \{F_p(\rho,\theta) + \frac{ir}{k}[2B_p + C_p]\}dr \tag{6.1}$$

where we have denoted $-2ik\, W\left\{ae^{i\sigma\Delta},z_p \;;\; k\,\frac{e^{i\sigma\Delta}-e^{i(\sigma+\ell)\Delta}}{|e^{i\sigma\Delta}-ei^{(\sigma+\ell)\Delta}|}\right\}$ by $W_p(\ell,\sigma)$.

Similarly, in (3.18) we take $\bar{a} = ae^{i\sigma\Delta}$, $\bar{b} = e^{i(\sigma+\ell)\Delta}$, we multiply both sides by $16\pi kiu_\ell \exp(-2iku_\ell)(1+\frac{7i}{32ku_\ell})$, set the resulting left-hand side equal to $U_p(\ell,\sigma)$ and ignore terms of order k^{-2} , to get

$$U_p(\ell,\sigma) = \int_{P_{\sigma\ell}} \{F_p(\rho,\theta) + \frac{ir(2u_\ell - r)}{2ku_\ell}[B_p + C_p]\}dr \quad . \tag{6.2}$$

Let $V_\ell(\sigma\Delta)$ denote either one of $W_p(\ell,\sigma)$ or $U_p(\ell,\sigma)$, and let us illustrate a method of producing an approximate solution to the equation

$$V_\ell(\sigma\Delta) = \int_{P_{\sigma\ell}} G(\rho,\theta)dr \tag{6.3}$$

for G given $V_\ell(\sigma\Delta)$, $\ell = 1,2,\ldots,2N+1$, $\sigma = 1,2,\ldots,4N+2$ where

(6.4) $$G(\rho,\theta) = \sum_{m=-2N}^{2N} e^{im\theta} \sum_{j=1}^{2N+1} G_{jm} S_j(\rho) \quad .$$

Applying Eq. (5.5) to (6.3), we get

(6.5) $$V_\ell(\sigma\Delta) = \sum_{m=-2N}^{2N} e^{im\sigma\Delta} \sum_{j=1}^{\ell} G_{jm} J^0_{mj\ell} \quad .$$

Given two functions of θ, g and γ defined and periodic with period 2π, we can define the discrete inner product

(6.6) $$(g,\gamma) = \frac{1}{4N+2} \sum_{\sigma=1}^{4N+2} g(\sigma\Delta)\overline{\gamma(\sigma\Delta)} \quad .$$

Thus, setting

(6.7) $$V_{\ell m} = \frac{1}{4N+2} \sum_{\sigma=1}^{4N+2} V_\ell(\sigma\Delta)e^{-im\sigma\Delta}$$

and applying the inner product (6.6) to both sides of (6.5), where $\gamma(\theta) = e^{im\theta}$, we get

(6.8) $$V_{\ell m} = \sum_{j=1}^{\ell} G_{jm} J^0_{mj\ell} \quad .$$

Taking $\ell=1$, we simply get G_{1m}:

(6.9) $$G_{1m} = V_{1m}/J^0_{m11}$$

since $J_{m11} \neq 0$ if $-2N \leq m \leq 2N$. The remaining $F_{\ell m}$ are obtained in succession

$$G_{\ell m} = [V_{\ell m} - \sum_{j=1}^{\ell-1} G_{jm} J^0_{mj\ell}]/J^0_{m\ell\ell} \quad \text{if} \quad J_{m\ell\ell} \neq 0$$

$$(6.10) \qquad = 0 \quad \text{if} \quad J^0_{m\ell\ell} = 0$$

$$(\ell = 2,3,\ldots,2N+1) \quad .$$

If N is sufficiently large, so that $a\Delta = O(k^{-1/3})$, the approximate solution G obtained in this way differs from the exact solution of (6.1) by a relative error of order k^{-2} at most. In this case, the solution $G=G_p$ obtained via (6.5)-(6.10) satisfies

$$(6.11) \qquad G(\rho,\theta) = F_p(\rho,\theta) + O(k^{-1}) \quad .$$

Since the terms B_p and C_p in (6.1) and (6.2) are simple linear combinations of F_p , we have

$$(6.12) \qquad \begin{cases} B(F_p) \equiv B_p = B(G_p) + O(\frac{1}{k}) \\ C(F_p) \equiv C_p = C(G_p) + O(\frac{1}{k}) \quad . \end{cases}$$

Hence if we solve the problems (compare (6.1), (6.2))

$$(6.13) \qquad \int_{P_{\sigma\ell}} \Phi_p(\rho,\theta)dr = W_p(\sigma+\ell,\sigma) - \frac{i}{k}\int_{P_{\sigma\ell}} r[2B(G_p) + C(G_p)]dr$$

$$(6.14) \qquad \int_{P_{\sigma\ell}} \Psi_p(\rho,\theta)dr = U_p(\sigma+\ell,\sigma) - \frac{i}{2ku_\ell}\int_{P_{\sigma\ell}} r(2u_\ell-r)[B(G_p) + C(G_p)]dr$$

for Φ_p and Ψ_p , then

$$F_p = \Phi_p + O(k^{-2}) \tag{6.15}$$

and also

$$F_p = \Psi_p + O(k^{-2}) \quad . \tag{6.16}$$

We are now in position to outline the algorithmic steps which enable us to determine F_p . We do this for both a plane wave and a spherical wave source. We assume that the cylinder has radius a , and that viewing takes place throughout in a vertical section of height b . Having chosen the integer N , we take $\Delta = \pi/(2N+1)$, $h = b/(2N+5)$. We arbitrarily take $z_0 = 0$, and then reconstruct images at heights jh , j = 3,4,...,2N+3 . There are essentially two main parts for both the plane wave and spherical wave case. The first part consists of the generation of the numbers $J^q_{mj\ell}$ required in these algorithms. The second part is the reconstructive part, which involves modification of the data so that the equation (6.1) and (6.2) apply, and the solution of these equations.

I. FIRST PART: GENERATION OF MATRIX ELEMENTS

1. Use Eqs. (5.3) to generate $J^q_{mj\ell}$, for $q = 0,1,2$; $m = 0,1,\ldots,2N$; $\ell = 1,2,\ldots,2N+1$. Evaluate these integrals via n-point Gauss-Legendre quadratures, $n \geq 3N$.

2. Store $J^0_{mj\ell}$ and $J^1_{mj\ell}$ in storage area P .

3. Set

(6.17) $$K_{mj\ell} = (2u_\ell J^1_{mj\ell} - J^2_{mj\ell})/(2u_\ell) \quad .$$

Store the $J^0_{mj\ell}$ and $K_{mj\ell}$ in storage area S .

II. SECOND PART: RECONSTRUCTION IN PLANE WAVE CASE

1. Read in $J^0_{mj\ell}$ and $J^1_{mj\ell}$ from storage area P .

2. Read the data $W(\ell,\sigma,p)\left\{=W\left(ae^{i\sigma\Delta}, z_p \; ; \; k\,\frac{e^{i\sigma\Delta}-e^{i(\sigma+\ell)\Delta}}{|e^{i\sigma\Delta}-e^{i(\sigma+\ell)\Delta}|}\right)\right\}$.

3. Multiply the data by $-2ki$.

4. Solve the equations

(6.18) $$W(\ell,\sigma,p) = \int_{P_{\sigma\ell}} F_p(\rho,\theta)dr \ , \ p = 1,2,\ldots,2N+5$$

for F_p , using (6.4) to (6.10). This involves the $J^0_{mj\ell}$.

5. Evaluate the numbers

(6.19) $$V(\ell,\sigma,p) = \int_{P_{\sigma\ell}} rF_p(\rho,\theta)dr = \sum_{m=-2N}^{2N} e^{im\sigma\Delta} \sum_{j=1}^{\ell} F_{jm}J^1_{mj\ell}$$

for $p = 1,2,\ldots,2N+5$, $\sigma = 1,2,\ldots,4N+2$, $\ell = 1,2,\ldots,2N+1$. This involves the use of the matrix elements $J^1_{mj\ell}$.

6. Use the numbers $V(\ell,\sigma,p)$ to form

(6.20) $$V(\ell,\sigma,p) \leftarrow \{2B(V(\cdot,\cdot,p)) + C(V(\cdot,\cdot,p)\}_{\ell,\sigma}$$

via (3.10) and (3.13), and then set

(6.21) $$W(\ell,\sigma,p) \leftarrow W(\ell,\sigma,p) - \frac{i}{k} V(\ell,\sigma,p) \quad .$$

Do this for $\ell = 1,2,\ldots,2N+1$, $\sigma = 1,2,\ldots,4N+2$, $p = 3,4,\ldots,2N+3$. We shall say more about the evaluation of $B(\cdot)$ and $C(\cdot)$ in Sec. 6.2.

7. Solve the equations

(6.22) $$W(\ell,\sigma,p) = \int_{P_{\ell\sigma}} F_p(\rho,\theta)dr$$

for F_p , $p = 3,4,\ldots,2N+3$.

8. At this point, the "surfaces" F_p are known at $2N+1$ planes; these can now be "grey scaled" and then viewed by known procedures [8].

III. SECOND PART: RECONSTRUCTION IN THE SPHERICAL WAVE CASE

1. Read in $J^0_{mj\ell}$ and $K_{mj\ell}$ from the storage area S .

2. Read the data $W(\ell,\sigma,p)(=W(ae^{i(\ell+\sigma)\Delta},z_p\ ;\ ae^{i\sigma\Delta},z_p)$.

3. Multiply the data by $16\pi kiu_\ell(1+\frac{7i}{32ku_\ell})\exp(-2iku_\ell)$. Here we recommend forming $\exp(-2iku_\ell)$ by first finding $\cos(2ku_\ell)$ and $\sin(2ku_\ell)$ in double precision (due to the relatively large magnitude of ku_ℓ) and then taking the results of these in single precision.

4. Solve the equations

(6.23) $$W(\ell,\sigma,p) = \int_{P_{\sigma\ell}} F_p(\rho,\theta)dr \ , \ p = 1,2,\ldots,2N+5$$

for F_p, using (6.4) to (6.10). This involves the $J^0_{mj\ell}$.

5. Evaluate the numbers

(6.24) $$V(\ell,\sigma,p) = \int_{P_{\sigma\ell}} \frac{r(2u_\ell - r)}{2u_\ell} F_p(\rho,\theta)dr = \sum_{m=-2N}^{2N} e^{im\sigma\Delta} \sum_{j=1}^{\ell} F_{jm} K_{mj\ell} \quad .$$

This involves the use of the matrix elements $K_{mj\ell}$.

6. Use the numbers $V(\ell,\sigma,p)$ to form

(6.25) $$V(\ell,\sigma,p) \leftarrow \{B(V(\cdot,\cdot,p)) + C(V(\cdot,\cdot,p))\}_{\ell,\sigma}$$

($\ell = 1,2,\ldots,2N+5$; $\sigma = 1,2,\ldots,4N+2$; $p = 3,4,\ldots,2N+3$; and where the evaluation of $B(\cdot)$ and $C(\cdot)$ is explained in Sec. 6.2), and then set

(6.26) $$W(\ell,\sigma,p) \leftarrow W(\ell,\sigma,p) - \frac{i}{k} V(\ell,\sigma,p) \quad .$$

7. Solve the equations

(6.27) $$W(\ell,\sigma,p) = \int_{P_{\sigma\ell}} F_p(\rho,\theta)dr$$

for F_p, $p = 3,4,\ldots,2N+3$.

8. The "surfaces" F_p can now be "grey-scaled" and viewed in 3 dimension [8]

6.2. The Evaluation of B and C.

Let us now describe in greater detail the formation of $B(\cdot)$ and $C(\cdot)$ in step 6 of algorithm II and III above. Since the computations are planar for the case of $B(\cdot)$, we suppress the subscript p, and we consider forming

$$B_{\ell\sigma} \equiv B(W(\ell,\sigma)) \tag{6.28}$$

where ℓ and σ are integers in the range $1 \leq \ell \leq 2N+1$, $1 \leq \sigma \leq 4N+2$. By equation (3.10), this involves the following 5 coordinates $(W(\ell-2,\sigma+2);\eta_1)$, $(W(\ell-1,\sigma+1),\eta_2)$, $(W(\ell,\sigma),\eta_3)$, $(W(\ell+1,\sigma-1),\eta_4)$, $(W(\ell+2,\sigma-2),\eta_5)$, this being the W data received along 5 consecutive parallel rays. Before assigning values to the η_i, let us consider what happens when the subscripts fall outside of the above indicated ranges. To this end, we have

$$\begin{aligned} W(\ell,\sigma) &= W(\ell,4N+2+\sigma) \quad \text{if} \quad \sigma \leq 0 \\ W(\ell,\sigma) &= W(\ell,\sigma-(4N+2)) \quad \text{if} \quad \sigma > 4N+2 \\ W(\ell,\sigma) &= 0 \quad \text{if} \quad \ell \leq 0 \\ W(\ell,\sigma) &= W(4N+2-\ell,\sigma+\ell) \quad \text{if} \quad \ell > 2N+1 \quad . \end{aligned} \tag{6.29}$$

The first two cases of (6.29) follow readily from the cyclic occurrance of the data on the cylinder. The third case is a case of a source and a detector without a body in between. In the fourth case, assume for the moment that the source enters the cylinder at $ae^{i(\sigma+\ell)\Delta}$ and is detected at $ae^{i\sigma\Delta}$, where ℓ and σ are within their proper ranges. We have assumed in our algorithm that the detected data $W(\ell,\sigma)$ takes the same value $W(\ell,\sigma)$ when the source is at $ae^{i\sigma\Delta}$ and the detector at $ae^{i(\sigma+\ell)\Delta}$. Our algorithm does not detect data when the source is at $ae^{i\sigma\Delta}$ and the detector is at $ae^{i(\sigma+\ell)\Delta}$. The fourth equation in (6.29) enables us to compensate for this "omission" of data collection by taking advantage of the symmetry relation of source and detector.

Let us next discuss the computation of the η_i in Eq. (3.11). Clearly $\eta_3=\rho_\ell$, where the ρ_j are defined in Eq. 5.1. Hence, if the subscripts of ρ are in the range $1 \leq j \leq 2N+1$, $(\eta_1,\eta_2,\eta_3,\eta_4,\eta_5) = (\rho_{\ell-4},\rho_{\ell-2},\rho_\ell,\rho_{\ell+2},\rho_{\ell+4})$. However, the subscripts fall out of their defined ranges if $\ell \leq 2$ or if $\ell \geq 2N$ It is therefore convenient to define numbers d_j and e_j by

$$
\begin{aligned}
& d_1 = 2a-\rho_1 && e_1 = 2a-\rho_1 \\
& d_2 = a && e_2 = a \\
& d_j = \rho_{2j-5}\ ,\ j = 3,4,\ldots,N+3 && e_j = \rho_{2j-4}\ ,\ j = 3,4,\ldots,N+2 \\
& d_j = -\ d_{2N+6-j}\ ,\ j = N+4,N+5 && e_j = -\ e_{2N+5-j}\ ,\ j = N+3,N+4
\end{aligned}
\tag{6.30}
$$

We then define η_j by

$$
\begin{aligned}
& \eta_j = d_{j+s} && \text{if } \ell = 2s+1 && (\ell \text{ odd}) \\
& \eta_j = e_{j+s-1} && \text{if } \ell = 2s && (\ell \text{ even})\ .
\end{aligned}
\tag{6.31}
$$

The computations for B can now be carried out the divided difference via the standard method and then substituting into (3.10).

The evaluation of the coefficient C in step 6 of algorithms II and III is much simpler. Here have the explicit expression

$$C(V(\sigma,\ell,p)) = \frac{1}{2h^2}\,(\delta_p^2 - \frac{1}{12}\,\delta_p^4)V(\sigma,\ell,p)$$

$$(6.32) \qquad = (-V(\ell,\sigma,p+2) + 16V(\ell,\sigma,p+1) - 30V(\ell,\sigma,p)$$

$$+ 16V(\ell,\sigma,p-1) - V(\ell,\sigma,p-2))/24h^2 \quad .$$

7. REMARKS ON THE COMPUTER PROGRAM OF THE ALGORITHMS IN SEC. 6.

The coefficients $J^0_{mj\ell}$, $J^1_{mj\ell}$ and $J^2_{mj\ell}$ may be written in the form

$$(7.1)\quad \begin{cases} J^0_{mj\ell} = e^{im\ell\Delta/2}\, H^0_{mj\ell} \\ J^1_{mj\ell} = e^{im\ell\Delta/2}[u_\ell H^0_{mj\ell} + iH^1_{mj\ell}] \\ J^2_{mj\ell} = e^{im\ell\Delta/2}[u_\ell^2 H^0_{mj\ell} + 2iu_\ell H^1_{mj\ell} + H^2_{mj\ell}] \end{cases}$$

where the $H^q_{mj\ell}$ are real. This enables us to save on computer storage, since we need only store the H's . Although the algorithm works, we are presently experiencing ill conditions in the solution triangular system (1.9), which can lead to gross errors in the regions near the center of the cylinder.

Another error occurs as a consequence of the asymptotics. While the asymptotic relative error is of order k^{-2} , the actual theoretical relative error of the algorithm in Sec. 6 is roughly $(\frac{a}{k})^2 \cong .003$ if $a = 25$ cm. , $k = 454$/cm. ; this combined with the instability of the solution of (1.9) leads to further errors. The author is presently studying ways of correcting these errors.

REFERENCES

[1] Ball, J., S. A. Johnson and F. Stenger, *Explicit Inversion of the Helmholtz Equation for Ultrasound Insonification and Spherical Detection,* to appear in Proc. of Houston Conference on Acoustical Imaging 9 (1980).

[2] Bleistein, N. and R. A. Handelsman, *Asymptotic Expansion of Integrals,* Holt, Rinehart and Winston (1975).

[3] Greenleaf, J. F., S. A. Johnson, W. F. Samoyoa and F. A. Duck, *Reconstruction of Spacial Distributions of Retractive Indices in Tissue from Time of Flight Profiles, Image Processing for* 2D *and* 3D *Reconstruction from Projections: Theory and Practice in Medicine and the Physical Sciences. A Digest of Technical Papers.* August 4-7 (1975), Stanford, Cal., pp. MA2-1-MA2-4.

[4] McNamee, J., F. Stenger and E. L. Whitney, *Whittaker's Cardinal Function in Retrospect,* Math. Comp. 25 (1971), pp. 141-154.

[5] Morse, P. M. and K. U. Ingard, *Theoretical Acoustics,* McGraw-Hill, N.Y. (1965).

[6] Mueller, R. K., M. Kaveh and G. Wade, *Reconstructive Tomography and Applications to Ultrasonics,* Proc. IEEE 67 (1979) pp. 567-587.

[7] Olver, F. W. J., *Asymptotics and Special Functions,* Academic Press, N.Y. (1974).

[8] Rawson, E. G., *Vibrating Verifocal Mirrors for* 3-D *Imaging.* IEEE, Spectrum 6 (1969) pp. 37-43.

[9] Stenger, F. and S. Johnson, *Ultrasonic Transmission Tomography Based on the Inversion of the Helmholtz Wave Equation for Plane and Spherical Wave Insonification,* Appl. Math. Notes 4 (1979) pp. 102-127.

ON THE NUMERICAL APPROXIMATION OF SECONDARY BIFURCATION PROBLEMS

BY

H. WEBER

Abteilung Mathematik
Universität Dortmund
D-4600 Dortmund 50

ON THE NUMERICAL APPROXIMATION OF SECONDARY BIFURCATION PROBLEMS

Helmut Weber

We discuss stable numerical methods for the approximation of the solutions of a nonlinear parameter - dependent equation near a non-trivial bifurcation point. The problem of finding the bifurcation point is reformulated as a well-posed equation of higher dimension. The nearby branches can be calculated in a stable manner after applying a certain transformation having its origin in the Lyapunov - Schmidt theory. We also treat the perturbed bifurcation problem and present numerical results.

1. INTRODUCTION

Let X and Y be real Banach spaces and (.,.) a continuous inner product defined on X. We consider the nonlinear equation

(1) $$F(u,\lambda) = 0$$

where $F \in C^3(X\times\mathbb{R},Y)$. We also write equation (1) in the form

$$F(z) = 0$$

with $z = (u,\lambda) \in W := X\times\mathbb{R}$, $F: W \to Y$. For the following let us assume that the point $z_0 = (u_0,\lambda_0) \in W$ satisfies

(A1) $$F(z_0) = 0$$

(A2) $$\dim N(F'(z_0)) = 2,\quad N(F'(z_0)) = \operatorname{span}\{p,q\},\quad p,q \in W$$

(A3) $$\operatorname{codim} R(F'(z_0)) = 1,\quad R(F'(z_0)) = \{y\in Y \mid \langle\psi,y\rangle = 0\},\quad 0 \neq \psi \in Y'$$

(A4) $$\tau = \alpha\gamma - \beta^2 < 0$$

where

(2) $$\alpha = \langle\psi,F''(z_0)pp\rangle,\quad \beta = \langle\psi,F''(z_0)pq\rangle,\quad \gamma = \langle\psi,F''(z_0)qq\rangle .$$

Here N and R are the nullspace respectively the range of a linear operator. Y' denotes the dual space of Y and $\langle \cdot,\cdot \rangle$ the duality pairing between Y' and Y.

In the applications, e.g. to differential eqution or integral equations, we will often have the following setting: $X \subset Y \subset H$, where H is a Hilbert space.

In §2 we will show that under the above hypotheses the point z_0 is a "simple" bifurcation point. That means: there are exactly two solution curves of (1) which (locally) only intersect at z_0. Our result gives us a constructive way for computing these solution curves in the vicinity of z_0. In section 3 we discuss a "direct" method for computing simple bifurcation points. It has relations to those presented by Seydel [14,15], Weber [17] and Moore [12]. For a survey of numerical methods for bifurcation problems we refer to Mittelmann, Weber [11]. It should be pointed out that we derive well-posed formulations of both problems, for finding the bifurcation point as well as for moving on the nearby branches. Remarks on the application of discretization algorithms to the equations formulated in §2 and §3 can be found in §4. In the following section we analyze an algorithm for calculating the solutions of a perturbed bifurcation problem. We use recent results of Beyn [1,2]. Finally two numerical examples give an impression of the accuracy obtainable by our methods.

2. A BIFURCATION THEOREM

The following result is an extension of a well known fundamental theorem of Crandall and Rabinowitz [4].

Theorem 1. Under the assumptions (A1-4) the point z_0 is a simple bifurcation point: there is a neighbourhood U of z_0 in W such that the solution set of $F|_U$ exactly consists of two smooth curves z_i: $I(\varepsilon_i) \to W$, $I(\varepsilon_i) = [-\varepsilon_i, +\varepsilon_i]$, which are crossing only at z_0, $z_i(0) = z_0$ $(i=1,2)$.

Proof: Without loss of generality we assume that $\alpha \neq 0$. If $\alpha = \gamma = 0$ we drop the following transformation. We transform p and q into p' and q' by

$$p' = \frac{-\beta + \sqrt{-\tau}}{\alpha}\, p + q\,, \qquad q' = \frac{-\beta - \sqrt{-\tau}}{\alpha}\, p + q.$$

Then for α', β' and γ' ($\alpha' = \langle\Psi, F''(z_0)p'p'\rangle$, etc.) we find

$$(3) \qquad \alpha' = 0, \quad \beta' = 2\mathcal{T}/\alpha, \quad \gamma' = 0, \quad \tau' = \alpha'\gamma' - \beta'^2 < 0.$$

In the following we will write again α, β, γ, τ, p, q for the transformed numbers and vectors.

Define a continuous inner product $(.|.)$ on W. Then perturbation theory implies that the solutions of $F(z) = 0$ near z_0 have the form

$$(4) \qquad z(\varepsilon) - z_0 = \varepsilon\{a(\varepsilon)p + b(\varepsilon)q\} + \varepsilon^2 v(\varepsilon)$$

with

$$(5) \qquad a(\varepsilon)^2 + b(\varepsilon)^2 = 1, \quad (v(\varepsilon)|p) = (v(\varepsilon)|q) = 0,$$

ε being a small perturbation parameter. A Taylor expansion of F yields

$$F(z) = F_0'(z - z_0) + \frac{1}{2}F_0''(z - z_0)^2 + R(z),$$

where $F_0^{(i)} := F^{(i)}(z_0)$ and R is of third order in $z - z_0$. Inserting (4) and (5) in this expression and dividing by ε^2 leads us to the following nonlinear system of equations:

$$(6) \qquad \begin{aligned} &F_0'v(\varepsilon) + \tfrac{1}{2}F_0''[a(\varepsilon)p + b(\varepsilon)q + \varepsilon v(\varepsilon)]^2 + \varepsilon^{-2}R(z(\varepsilon)) = 0 \\ &a(\varepsilon)^2 + b(\varepsilon)^2 - 1 = 0 \\ &(v(\varepsilon)|p) = (v(\varepsilon)|q) = 0 \end{aligned}$$

for $v(\varepsilon)$, $a(\varepsilon)$ and $b(\varepsilon)$. We rewrite (6) in the form

$$(7) \qquad \mathfrak{F}(x,\varepsilon) = 0,$$

where $x = x(\varepsilon) = (v(\varepsilon), a(\varepsilon), b(\varepsilon))$, $\mathfrak{F}: V\times\mathbb{R} \to Z$, $V = W\times\mathbb{R}\times\mathbb{R}$, $Z = Y\times\mathbb{R}\times\mathbb{R}\times\mathbb{R}$, $\|x\|_V = \|(x_1, x_2, x_3)\| = \sup\{\|x_i\|\}$, $\|y\|_Z = \|(y_1, y_2, y_3, y_4)\| = \sup\{\|y_i\|\}$. V and Z are Banach spaces.

Since $R(z) = O(\|z - z_0\|^3)$ equation (7) is simplified for $\varepsilon \to 0$ into

$$(8) \qquad \begin{aligned} &F_0'v(0) + \tfrac{1}{2}F_0''[a(0)p + b(0)q]^2 = 0 \\ &a(0)^2 + b(0)^2 - 1 = 0 \\ &(v(0)|p) = (v(0)|q) = 0\,. \end{aligned}$$

From (A3) and (A4) it follows that

(9) $$2\beta a(0)b(0) = 0, \quad a(0)^2 + b(0)^2 - 1 = 0$$

is necessary for the solvability of (8). The solutions of (9) are

(10) $$\xi_1 = (a_1,b_1)^T = (1,0)^T, \ \xi_2 = (a_2,b_2)^T = (0,1)^T, \ \xi_3 = -\xi_1, \ \xi_4 = -\xi_2.$$

Henceforth we are only concerned with the linearly independent pair of solutions (ξ_1,ξ_2). Given $a_i(0) = a_i$, $b_i(0) = b_i$, the solution $v_i(0)$ of (8) is uniquely determined, $i = 1,2$.

We want to apply the implicit function theorem in Banach space (cf. Dieudonné [5]) to $\mathfrak{F}(x,\varepsilon) = 0$ at the point

$$x^{(i)} = (v_i(0),a_i(0),b_i(0)), \quad \varepsilon = 0 \quad (i=1,2).$$

Thus we have to prove that the Fréchet derivative $D_x\mathfrak{F}(x^{(i)},0)\in\mathcal{L}(V,Z)$ is a linear homeomorphism between V and Z. We write down the equation $D_x\mathfrak{F}(x^{(i)},0)h = 0$, $h = (\tilde{v},\tilde{a},\tilde{b})$ componentwise:

(11) $$\begin{aligned} &F_0'\tilde{v} + \tilde{a}a_iF_0''pp + \tilde{a}b_iF_0''pq + \tilde{b}a_iF_0''pq + \tilde{b}b_iF_0''qq = 0 \\ &2(\tilde{a}a_i + \tilde{b}b_i) = 0 \\ &(\tilde{v}|p) = (\tilde{v}|q) = 0 \ . \end{aligned}$$

a) <u>Injectivity</u>: Necessary for the solvablity of (11) is

$$\tilde{a}b_i\beta + \tilde{b}a_i\beta = 0, \quad \tilde{a}a_i + \tilde{b}b_i = 0.$$

From (10) we conclude $\tilde{a} = \tilde{b} = 0$ for $i=1,2$. It remains $F_0'\tilde{v} = 0$, $(\tilde{v}|p) = (\tilde{v}|q) = 0$. (A2) implies $\tilde{v} = 0$.

b) <u>Surjectivity</u>: $(i=1)$ Consider the equation

(12) $$\begin{aligned} F_0'\tilde{v} + \tilde{a}F_0''pp + \tilde{b}F_0''pq &= g \in Y \\ 2\tilde{a} &= g_1 \in \mathbb{R} \\ (\tilde{v}|p) &= g_2 \in \mathbb{R} \\ (\tilde{v}|q) &= g_3 \in \mathbb{R} \ . \end{aligned}$$

Inserting $\tilde{a} = g_1/2$ yields

$$F_0'\tilde{v} = g - \tfrac{1}{2}\, g_1F_0''pp - \tilde{b}F_0''pq.$$

A necessary condition for the solvability of this equation is ($\alpha = 0$) $\langle\Psi, g\rangle - \tilde{b}\beta = 0$. With $\tilde{b} = \langle\Psi, g\rangle/\beta$ we get $F_0'\tilde{v} = \bar{g} \in R(F_0')$ with an appropriate $\bar{g}$. If $\hat{v}$ is a solution of this equation, then also $\tilde{v} = cp + dq + \hat{v}$, $c, d \in \mathbb{R}$. The third and fourth equation of (12) lead to

$$\begin{aligned}(p|p)c + (p|q)d &= g_2 - (\hat{v}|p)\\ (p|q)c + (q|q)d &= g_3 - (\hat{v}|q)\ .\end{aligned}$$

This 2x2 - system is uniquely solvable since p and q are linearly independent.

The case i=2 is proved in a similar manner.

From the bijectivity of $D_x\mathfrak{F}(x^{(i)}, 0) \in \mathcal{L}(V, Z)$ we conclude with the aid of the closed graph theorem that this linear operator is a homeomorphism from V onto Z. The implicit function theorem then yields the existence of continuous families $(v_i(\varepsilon),\ a_i(\varepsilon),\ b_i(\varepsilon))$ of solutions of (6). Together with (4) this proves the existence of intersecting solution branches of $F(z) = 0$. The uniqueness is easily shown by an application of the Crandall - Rabinowitz theorem ([4]).

q.e.d.

Remarks.

1. The nonlinear system (6) gives us a well-posed formulation of the problem of computing the bifurcating branches near the branch point. Of course this approach is only applicable, if good approximations to z_0, p and q are available. A numerical method for finding those is treated in the next paragraph.

2. Theorem 1 holds also under the relaxed smoothness condition $F \in C^2(W, Y)$, as easily can be seen.

3. COMPUTING THE BIFURCATION POINT

Let us assume that in addition to (A1-4) at least one of the branches crossing at z_0 is smoothly parametrizable by λ, the original bifurcation parameter. This means

(A5) $$F_\lambda(z_0) \in R(F_u(z_0)).$$

Then we can choose the elements p and q spanning $N(F_0')$ in the special form

$$p = (r_0, 0), \quad q = (s_0, 1),$$

where r_0, $s_0 \in X$, $(r_0, s_0) = 0$, $(r_0, r_0) = 1$. This choice uniquely determines p and q (up to sign). We further assume that we know an element

(13) $$0 \neq \varphi \in Y, \ \langle \psi, \varphi \rangle \neq 0.$$

Consider the following nonlinear system of equations:

(14)
$$\begin{aligned} &F(u,\lambda) + \mu\varphi = 0 \\ &F_u(u,\lambda)r = 0 \\ &F_u(u,\lambda)s + F_\lambda(u,\lambda) = 0 \\ &(r,r) - 1 = 0 \\ &(r,s) = 0. \end{aligned}$$

μ is a real imperfection parameter. We rewrite (14) as equation

(15) $$G(w) = 0, \quad w = (u,r,s,\lambda,\mu) \in H$$

in the Banach space $H = X \times X \times X \times \mathbb{R} \times \mathbb{R}$, endowed by the maximum product norm.

<u>Theorem 2.</u> Assume that the solution $w_0 = (u_0, r_0, s_0, \lambda_0, 0)$ of $G(w) = 0$ satisfies (A1-5) and (13). Then w_0 is an <u>isolated</u> solution of (15), i.e. the Fréchet derivative $G'(w_0) = D_w G(w_0) \in \mathcal{L}(H,H)$ is a homeomorphism.

Proof: We consider the equation $G'(w_0)\tilde{w} = c$ with $\tilde{w} = (\tilde{u}, \tilde{r}, \tilde{s}, \tilde{\lambda}, \tilde{\mu}) \in H$ and $c = (c_1, c_2, c_3, \mu_1, \mu_2) \in H$. It has the form $(F_u^0 := F_u(z_0), \ldots)$

(15)
$$\begin{aligned} &F_u^0 \tilde{u} + F_\lambda^0 \tilde{\lambda} + \tilde{\mu}\varphi = c_1 \in X \\ &F_{uu}^0 r_0 \tilde{u} + F_{u\lambda}^0 r_0 \tilde{\lambda} + F_u^0 \tilde{r} = c_2 \in X \\ &F_{uu}^0 s_0 \tilde{u} + F_{u\lambda}^0 s_0 \tilde{\lambda} + F_u^0 \tilde{s} + F_{u\lambda}^0 \tilde{u} + F_{\lambda\lambda}^0 \tilde{\lambda} = c_3 \in X \\ &2(r_0, \tilde{r}) = \mu_1 \in \mathbb{R} \\ &(r_0, \tilde{s}) + (\tilde{r}, s_0) = \mu_2 \in \mathbb{R}. \end{aligned}$$

The application of ψ onto the first row of (15) implies

$$\tilde{\mu} = \langle \psi, c_1 \rangle / \langle \psi, \varphi \rangle \quad , \text{ since (A5) holds.}$$

Thus with $\bar{c}_1 = c_1 - \frac{\langle\psi, c_1\rangle}{\langle\psi,\varphi\rangle}\varphi \in R(F_u^0)$ it follows the equation

$$F_u^0\tilde{u} + F_\lambda^0\tilde{\lambda} = \bar{c}_1.$$

It has the general solution $(\tilde{u},\tilde{\lambda}) = (u_1 + \delta r_o + \kappa s_o, \kappa)$, $\delta,\kappa \in \mathbb{R}$, $u_1 \in N(F_u^0)^{\perp} = \{x \in X \mid (x, r_o) = 0\}$. Inserting $\tilde{u}$ into the second respectively third row of (15) then leads us to the 2×2 - system

$$0 = \langle\psi, c_2 - F_{uu}^0 r_o u_1\rangle - \delta\langle\psi, F_{uu}^0 r_o r_o\rangle - \kappa\langle\psi, F_{uu}^0 r_o s_o + F_{u\lambda}^0 r_o\rangle$$

$$0 = \langle\psi, c_3 - F_{uu}^0 s_o u_1 - F_{u\lambda}^0 u_1\rangle - \delta\langle\psi, F_{uu}^0 s_o r_o + F_{u\lambda}^0 r_o\rangle$$
$$- \kappa\langle\psi, F_{uu}^0 s_o s_o + 2F_{u\lambda}^0 s_o + F_{\lambda\lambda}^0\rangle \quad .$$

From (2) it follows

$$\text{(16)} \qquad \begin{aligned} \alpha\delta + \beta\kappa &= \langle\psi, c_2 - F_{uu}^0 r_o u_1\rangle \\ \beta\delta + \gamma\kappa &= \langle\psi, c_3 - F_{uu}^0 s_o u_1 - F_{u\lambda}^0 u_1\rangle. \end{aligned}$$

Since $\alpha\gamma - \beta^2 < 0$ (A4), (16) is uniquely solvable, say by (δ_o, κ_o). Now $(\tilde{u},\tilde{\lambda}) = (u_1 + \delta_o r_o + \kappa_o s_o, \kappa_o)$ is uniquely determined and the second and third row are put in the form

$$F_u^0\tilde{r} = \bar{c}_2 \in R(F_u^0), \qquad F_u^0\tilde{s} = \bar{c}_3 \in R(F_u^0).$$

This implies $r = r_1 + \chi r_o$, $r_1 \in N(F_u^0)$, $\chi \in \mathbb{R}$, and
$s = s_1 + \xi r_o$, $s_1 \in N(F_u^0)$, $\xi \in \mathbb{R}$.

From the last two rows of (15) we conclude $\chi = \mu_1/2$ and $\xi = \mu_2 - (r_1, s_o)$. Hence for given $c \in H$ the equation $G'(w_o)\tilde{w} = c$ has a unique solution. An application of the closed graph theorem completes the proof.

q.e.d.

<u>Remark.</u> A modification of the nonlinear system of equations (14), which does not need the element φ is to replace the first row of (14) by

$$\text{(17)} \qquad F(u,\lambda) + \mu r = 0.$$

If the hypotheses of Theorem 2 are satisfied and if in addition zero is a simple eigenvalue of F_u^0, i.e. $\langle\psi, r\rangle \neq 0$, then w_o is an isolated solution of the modified system.

4. APPLICATION OF DISCRETIZATION METHODS

The computational solution of bifurcation problems is not a trivial problem at all. For instance, it has been shown by Beyn [1,2], Brezzi - Rappaz - Raviart [3] and Yamaguti - Fuji [21] that discretization is in general bifurcation - destroying. That means: a discretized version of equation (1), say

$$F_h(u_h,\lambda) = 0, \quad F_h: \mathbb{R}^{n(h)} \times \mathbb{R} \longrightarrow \mathbb{R}^{n(h)},$$

generally has no true bifurcation point "near" the bifurcation point (u_0,λ_0) of the original problem, which is assumed to exist. The bifurcation diagram splits up into two different non - intersecting branches, see fig. 1.

Fig. 1

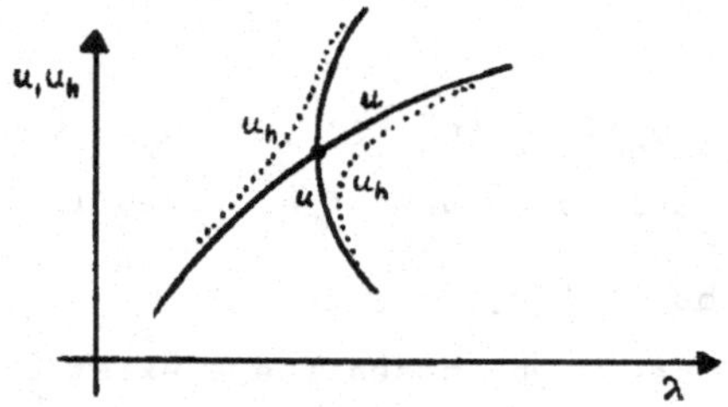

A criterion for the occurence of this effect is the non - vanishing of

$$\langle \psi, e \rangle ,$$

where e is the main component of the local discretization error of F_h at the bifurcation point, cf. [1]. Note that e.g. in the situation of branching from the trivial solution at a simple eigenvalue this behaviour does not occur. The reason is, that the trivial solution is also the exact solution of the equation $F_h(u_h,\lambda) = 0$. For further information see e.g. [9,19]. So the numerical approximation of the solution branches near a secondary bifurcation point is a difficult task.

It turns out that the order of application of the discretization and the Lyapunov - Schmidt reduction (or of a similar process) is crucial. In the author's opinion it is often advantageous to compute first a good approximation of the branch point, for instance by discretizing equation (14). The next step is then the derivation of a well-posed formulation of the problem of calculating the branches near z_o. This was done in section 2, especially in equation (6). The last step consists of a discretization of this reformulated equation, i.e. of (6), where the results of the first step are incorporated.

We discuss this approach in more detail. Equation (14,15): $G(w)=0$ is discretized using a <u>stable</u> discretization method of order p ($p \geq 1$) of <u>consistency</u> (see Stetter [16], Keller [7]). This yields a sequence of smooth problems

$$(18) \qquad G_h(w_h) = 0, \quad G_h: E_h \to E_h^o, \quad E_h \cong E_h^o \cong \mathbb{R}^{N(h)}$$

with the properties

$$(19) \qquad \|G_h(\Delta_h w_o) - \Delta_h^o G(w_o)\|_h^o = O(h^p), \quad h \to 0 \text{ (consistency)}$$

and

$$(20) \qquad \|w_h' - w_h''\|_h \leq C \, \|G_w(w_h') - G_h(w_h'')\|_h^o \quad, h \leq h_o,$$

C independent of h (stability)

for all w_h', w_h'' in a neighbourhood of $\Delta_h w_o$.

It is well known that under these hypotheses there exist unique solutions w_{oh} of (18) for sufficiently small h with

$$(21) \qquad \|\Delta_h w_o - w_{oh}\|_h = O(h^p), \quad h \to 0.$$

Here $\Delta_h, \Delta_h^o, \|.\|_h, \|.\|_h^o$ are the usual discretization mappings from $H \to E_h$, $H \to E_h^o$ and norms on E_h and E_h^o. (21) gives us approximations

$$\bar{u}_h, \bar{r}_h, \bar{s}_h, \bar{\lambda}_h \quad \text{to the} \quad u_o, r_o, s_o, \lambda_o$$

of order h^p.

Inserting $\bar{u}_h$, $\bar{r}_h$, $\bar{s}_h$ (eventually prolongations of these quantities) and $\bar{\lambda}_h$ into (6) yields a slightly perturbed family of smooth problems

$$(22) \qquad \mathfrak{F}^h(x,\varepsilon) = 0$$

with

$$\| \bar{\mathfrak{F}}^h(x,\varepsilon) - \mathfrak{F}(x,\varepsilon) \|_Z = O(h^p).$$

Since (7) has isolated solutions for small $|\varepsilon|$ the perturbation (22) only amounts in a perturbation of order h^p in the solutions $x(\varepsilon)$. (22) is discretized by a stable and consistent scheme of order q:

$$(23) \qquad \bar{\mathfrak{F}}^h_k(x_k,\varepsilon) = 0, \quad \bar{\mathfrak{F}}^h_k: Q_k \times \mathbb{R} \longrightarrow Q^o_k, \quad Q_k \cong Q^o_k \cong \mathbb{R}^{M(k)}.$$

The parameter-dependent systems of equations (23) may be solved numerically via Newton's method and ε - imbedding.

The standard convergence theorem now states: for $k \leq k_0$ we obtain unique smooth families $x^h_k(\varepsilon,i)$ of solutions of (13) which satisfy

$$(24) \qquad \| x^h_k(\varepsilon,i) - \Delta_k x^{(i)}(\varepsilon) \|_k = O(h^p+k^q), \; h,k \to 0, \; |\varepsilon| < \varepsilon_i, i=1,2.$$

Hence we have constructed approximate branching solutions

$$u^h_k(\varepsilon,i) \;, \; \lambda^h_k(\varepsilon,i), \; i=1,2 \quad \text{of order } h^p + k^q$$

of $F(u,\lambda) = 0$ near (u_0,λ_0), which intersect in $(\bar{u}_h,\bar{\lambda}_h)$, see fig.2.

Fig. 2

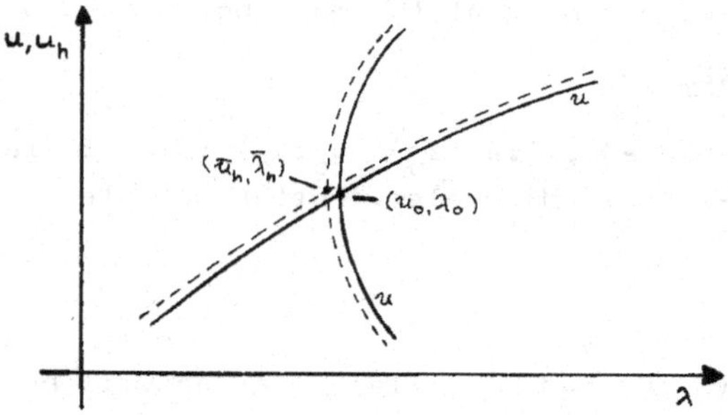

In the above approach the bifurcation is not destroyed by discretization. We want to point out that the proof of the stability of a concrete discretization scheme applicable to an equation of the form (15) or (22)is by no means trivial since we have to consider non-standard discretizations. However the main supposition necessary for such proofs, namely that the solution of the continuous problem is isolated, is satisfied in both cases. Thus one has not to encounter general difficulties. For some classes of problems there are special tricks which lead to standard discretizations, cf. Weber [17] for ordinary differential equations.

Finally let us make some remarks on the numerical treatment of the system (14) in the finite - dimensional case $X = Y = \mathbb{R}^n$. Then (14) represents a system of nonlinear equations of dimension 3n+2. The application of Newton's method to it makes it necessary to solve a linear system of dimension 3n+2 in each step of the iteration. To reduce the computational effort it is possible to apply a certain block elimination technique to the linear systems under consideration. so one can solve one 3n+2 - system with aid of solving six n+1 - systems, some of which have the same matrix of coefficients. This may be seen by a thorough inspection of the structure of the Jacobian of (14). For similar results in this direction we refer to Moore [12], Weber - Werner [18].

5. TREATMENT OF PERTURBED PROBLEMS

Here we consider the equation

(25) $$F(u,\lambda,\tau) = F(z,\tau) = 0,$$

where $F: W \times \mathbb{R} \to Y$ is a C^3- mapping, $W = X \times \mathbb{R}$. τ denotes a perturbation parameter. We assume that the mapping

$$\hat{F}(z) := F(z,0)$$

satisfies the hypotheses (A1-4), i.e. $z_0 = (u_0,\lambda_0)$ is a simple bifurcation point of the non-perturbed problem. Furthermore let

(A6) $$\langle \psi, F_\tau^0 \rangle \neq 0$$

be valid, where $F_\tau^0 := F_\tau(z_0,0)$ and a similar notation will be used for other derivatives at the point $(z_0,0)$.

It will be convenient to normalize p, q and the functional $\psi \in Y'$ in such a way that

(A7) $$\alpha = \gamma = 0, \quad \beta = 1, \quad \langle \psi, F_\tau^0 \rangle = -1.$$

In the following only the case $\tau > 0$ is considered since the results for $\tau < 0$ can be obtained from the equation $F(z,-\tau) = 0$ after redefining p, q and ψ. It is well known, at least for the special case of branching from the trivial solution, that under the conditions

(A1-4) and (A6) the perturbation ($\tau \neq 0$) has the effect of destroying the bifurcation (see e.g. Keener, Keller [6]).

The aim of this paragraph is to advocate a constructive approach for numerically computing the non-intersecting branches of the perturbed problem. We use a modification of Beyn's [1] "hyperbolic" ansatz

$$(26)\quad \begin{aligned} & z(t,\varepsilon) = z_0 + t\{\varepsilon p + \varepsilon^{-1} q + g n_\varepsilon (p+q)\} + t^2 v, \quad n_\varepsilon = (\varepsilon + \varepsilon^{-1})/2 \\ & \tau = t^2, \quad (v|p) = (v|q) = 0, \quad g = g(t,\varepsilon), \quad v = v(t,\varepsilon), \quad \varepsilon > 0 \end{aligned}$$

for the branches in question. t,ε and g are real parameters.

F is developed according to

$$(27)\quad \begin{aligned} F(z,\tau) = {} & F_z^0(z-z_0) + F_\tau^0 \tau + \frac{1}{2} F_{zz}^0 (z-z_0)^2 + \frac{1}{2} F_{\tau\tau}^0 \tau^2 \\ & + F_{z\tau}^0 (z-z_0)\tau + R(z,\tau), \end{aligned}$$

where R is of third order in $(z - z_0)$ and τ. Putting (26) into (27) then easily yields the system of equations

$$(28)\quad \begin{aligned} & F_z^0 v + F_\tau^0 + \frac{1}{2} F_{zz}^0 [\varepsilon p + \varepsilon^{-1} q + g n_\varepsilon (p+q) + tv]^2 + \frac{1}{2} F_{\tau\tau}^0 t^2 \\ & \quad + F_{z\tau}^0 t[\varepsilon p + \varepsilon^{-1} q + g n_\varepsilon (p+q) + tv] + t^{-2} R(z,t^2) = 0, \\ & (v|p) = (v|q) = 0, \quad \varepsilon > 0. \end{aligned}$$

For shortness we write (28) as

$$(29)\quad \mathcal{T}(m;\varepsilon,t) = 0, \quad \varepsilon > 0,$$

$m = (v,g)$, $\mathcal{T}: E \times \mathbb{R}^2 \to Z$, $E = W \times \mathbb{R}$, $Z = Y \times \mathbb{R} \times \mathbb{R}$. E and Z are Banach spaces when supplied with the maximum product norms.

We inspect the smooth mapping $\mathcal{T}$ for $t \to 0$. It has the form

$$(30)\quad \begin{aligned} & F_z^0 v + F_\tau^0 + \frac{1}{2} F_{zz}^0 [\varepsilon p + \varepsilon^{-1} q + g n_\varepsilon (p+q)]^2 = 0, \\ & (v|p) = (v|q) = 0, \quad \varepsilon > 0. \end{aligned}$$

From the hypothesis (A7) we conclude that a necessary condition for the solvabilty of (30) is

$$g = g(0, \varepsilon) = 0$$

(or $g = -2$, which brings no essential new solutions). Now one can easily prove that the solution $v(0,\varepsilon) = v_\varepsilon^0$ of (30) uniquely exists for $g(0,\varepsilon) = 0$.

The Fréchet - derivative $D_m\mathcal{T}((v_\varepsilon^0,0);\varepsilon,0)$ is a linear homeomorphism from E onto Z, since (A1-4) hold. This is easily seen using the techniques of §2 and §3. We then apply a quantitative version of the implicit function theorem (cf. [2]). This implies the following result:

<u>Theorem 3.</u> Under the hypotheses of this paragraph there is a smooth family of isolated solutions $(v(t,\varepsilon), g(t,\varepsilon))$ of equation (28) from which we can construct solutions of (25) near $(z_0,0)$ with the aid of the ansatz (16) for $\tau > 0$.

<u>Remark.</u>

1. p and q are the tangent vectors of the intersecting branches of $F(z,0) = 0$ at the bifurcation point, see fig. 3.

<u>Fig. 3</u>

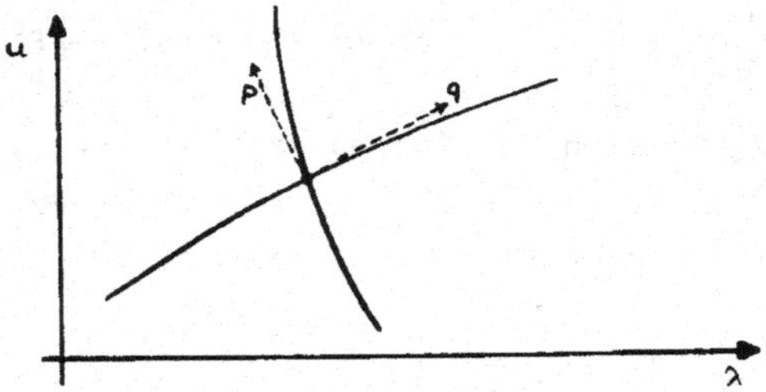

2. Equation (28,29) gives us a constructive way for numerically computing the non - intersecting branches for $\tau \neq 0$. Of course a discretization of this <u>regular</u> problem will be necessary before in most cases. The same arguments as in §4 apply here, too.
3. Note that this approach yields approximations to <u>all</u> branches near z_0, whereas the methods of Keener, Keller [6] are applicable only for branches exhibiting turning points.

6. NUMERICAL EXAMPLES

Example 1.

$X = Y = \mathbb{R}^2$, $$F(x,y,\lambda) := \begin{pmatrix} x + \lambda(x^3 - x + xy^2) \\ 10y - \lambda(y + 2x^2y + y^2) \end{pmatrix}.$$

The equation $F(x,y,\lambda) = 0$ (cf. Crandall, Rabinowitz [4]) has the trivial solution $x = y = 0$ for all real λ and a nontrivial branch

$$\begin{pmatrix} x_{\pm}(\lambda) \\ y(\lambda) \end{pmatrix} = \begin{pmatrix} \pm\sqrt{(\lambda-1)/\lambda} \\ 0 \end{pmatrix}, \quad \lambda \geq 1.$$

The point $(x_0, y_0, \lambda_0) = (\sqrt{3}/2, 0, 4)$ is a secondary bifurcation point which satisfies (A1-5). The direct method (14) with modification (17) from §3 was used for computing the vector w_0 incorporating x_0, y_0 and λ_0. The following table gives representative numerical results. They were obtained by applying Newton's method in 5 iteration steps.

	Starting value	w_{num}	w_0 (exact)
u_0	1	0.86602543	0.86602540
	0.2	0	0
λ_0	3.5	4	4
r_0	-0.1	0	0
	1.2	1	1
s_0	0.1	$0.36084380_{10}-1$	$0.36084312_{10}-1$
	0.1	0	0
μ_0	0	0	0

Example 2.

The integral equation (cf. [13])

$$(31) \qquad \lambda u(s) = \frac{2}{\pi}\int_0^{\pi} (3\sin t \sin s + 2\sin 2t \sin 2s)(u(t)+u(t)^3)\,dt, \quad 0 \leq s \leq \pi,$$

has the trivial solution $u^0 = 0$ and the nontrivial solutions

$$u^1(\lambda;t) = \pm\frac{2}{\sqrt{3}}\sqrt{\frac{\lambda}{3}-1}\,\sin t,$$

$$u^2(\lambda;t) = \pm\frac{2}{\sqrt{3}}\sqrt{\frac{\lambda}{2}-1}\,\sin 2t,$$

$$u_+^3(\lambda;t) = \frac{2}{3}\sqrt{\frac{2}{3}\lambda - 1}\ \sin t \ \pm \frac{2}{3}\sqrt{\frac{\lambda}{6} - 1}\ \sin 2t\ ,$$

$$u_-^3(\lambda;t) = -\frac{2}{3}\sqrt{\frac{2}{3}\lambda - 1}\ \sin t \ \pm \frac{2}{3}\sqrt{\frac{\lambda}{6} - 1}\ \sin 2t\ .$$

This problem has two primary bifurcation points: $\lambda = 2$ and $\lambda = 3$. We are interested in the secondary bifurcation at $(u_0, \lambda_0) = (\frac{2}{\sqrt{3}}\sin t, 6)$.

(31) can be transformed easily into the nonlinear boundary - value problem

$$\text{(32)} \quad \begin{aligned} u_1' &= \frac{6}{\pi\lambda}\cos t(u_2 + u_3) + \frac{8}{\pi\lambda}\cos 2t(u_4 + u_5) \\ u_2' &= (u_1 + u_1^3)\sin t \\ u_3' &= -(u_1 + u_1^3)\sin t \\ u_4' &= (u_1 + u_1^3)\sin 2t \\ u_5' &= -(u_1 + u_1^3)\sin 2t \\ u_1(0) &= u_2(0) = u_3(\pi) = u_4(0) = u_5(\pi) = 0. \end{aligned}$$

The bifurcation point (u_0, λ_0) satisfies the conditions (A1-5), so that we can formulate the equations (14,17). The last two normalizing equations are transformed into differential equations and appropriate boundary conditions (cf. [17]). For λ and μ we add two trivial differential equations. This leads to a nonlinear boundary value problem of dimension 19, the solution of which is isolated. Its numerical solution was determined by means of the shooting method - for the convergence of this algorithm see Weiss [20]. The initial value problems were integrated by the classical 4-th order Runge-Kutta scheme with uniform stepsize h. The following table presents some typical results.

	$h = \pi/50$	$h = \pi/100$	$h = \pi/200$
$u_3(0)$	3.6275734	3.6275971	3.6275986
λ	5.9999802	5.9999985	5.9999999
μ	$1.9517438_{10}{-5}$	$1.2394881_{10}{-6}$	$7.6438107_{10}{-8}$

For computing the branches u^1 and u_+^3 near the bifurcation point the nonlinear system (6) is formulated as a boundary value problem of dimension 10 (cf. [17]). The shooting method was applied, using Runge-Kutta integration with stepsize $h = \pi/100$ and the results of the above calculations with $h = \pi/200$.

Results for branch u^1:

ε	-0.1	-0.01	0	0.01	0.1
$\lambda(\varepsilon)$	5.8997736	5.9899974	5.9999999	6.0099974	6.0997757
$u_1(\varepsilon,\frac{\pi}{2})$ (num.)	1.1352482	1.1527740	1.1547004	1.1566231	1.1737454
$u(\lambda(\varepsilon),\frac{\pi}{2})$ (ex.)	1.1352481	1.1527739	1.1547005	1.1566230	1.1737453

Results for branch u_+^3:

ε	-0.1	-0.01	0	0.01	0.1
$\lambda(\varepsilon)$	6.0022132	6.0000217	5.9999999	6.0000217	6.0022003
$u_1(\varepsilon,\frac{\pi}{2})$ (num.)	1.1549844	1.1547032	1.1547004	1.1547032	1.1549827
$u(\lambda(\varepsilon),\frac{\pi}{2})$ (ex.)	1.1549845	1.1547032	1.1547005	1.1547033	1.1549828

The branches were computed for $\varepsilon \in [-1,1]$ without difficulties, using Newton's method.

All computations were performed on an IBM 370/158 computer in Fortran, single precision.

REFERENCES

1. W.-J. Beyn, On discretizations of bifurcation problems, pp. 46-73 in [10], 1980.

2. W.-J. Beyn, Zur Approximation von Lösungsverzweigungen nichtlinearer Randwertaufgaben mit dem Differenzenverfahren, Manuskript, Konstanz 1980.

2. F. Brezzi, J. Rappaz, P.-A. Raviart, Finite-dimensional approximation of nonlinear problems, Part III: Bifurcation points, manuscript, 1980.

4. M.G. Crandall, P.H. Rabinowitz, Bifurcation from simple eigenvalues, J. Functional Anal. 8(1971), 321-340.

5. J. Dieudonné, Foundations of modern analysis, Academic Press, New York 1960.

6. J.P. Keener, H.B. Keller, Perturbed bifurcation theory, Arch. Rational Mech. Anal. 50 (1974), 159-175.

7. H.B. Keller, Approximation methods for nonlinear problems with applications to two-point boundary value problems, Math. Comp. 29(1975), 464-474.

9. F. Kikuchi, Finite element approximation of bifurcation problems, Theoretical and Applied Mechanics 26(1976), 37-51, University of Tokyo Press.

10. H.D. Mittelmann, H. Weber (eds.), Bifurcation problems and their numerical solution, Workshop Dortmund Jan. 15-17, 1980, Intern. Ser. Numer. Math. Vol. 54, Birkhäuser, Basel 1980.

11. H.D. Mittelmann, H. Weber, Numerical methods for bifurcation problems - a survey and classification, pp. 1-45 in [10], 1980.

12. G. Moore, The numerical treatment of non-trivial bifurcation points, Techn. report Na/6, University of Bath, 1980, submitted for publication.

13. G.H. Pimbley, Eigenfunction branches of nonlinear operators, and their bifurcations, Lecture Notes in Maths. Vol. 104, Springer-Verlag, Berlin 1969.

14. R. Seydel, Numerical computation of branch points in ordinary differential equations, Numer. Math. 32(1979), 51-68.

15. R. Seydel, Numerical computation of branch points in nonlinear equations, Numer. Math. 33(1979), 339-352.

16. H.J. Stetter, Analysis of discretization methods for ordinary differential equations, Springer-Verlag, Berlin-Heidelberg-New York, 1970.

17. H. Weber, Numerische Behandlung von Verzweigungsproblemen bei gewöhnlichen Randwertaufgaben, pp. 176-190 in: J. Albrecht, L. Collatz, K. Kirchgässner (eds.), Constructive methods for nonlinear boundary value problems and nonlinear oscillations, Intern. Ser. Numer. Math. Vol. 48, Birkhäuser, Basel 1979.

18. H. Weber, W. Werner, On the numerical determination of nonisolated solutions of nonlinear equations, Preprint Nr. 32(1979/80), Univ. Dortmund, submitted for publication.

19. R. Weiss, Bifurcation in difference approximations to two-point boundary value problems, Math. Comp. 29(1975), 746-760.

20. R. Weiss, The convergence of shooting methods, BIT 13(1973), 470-475.

21. M. Yamaguti, H. Fujii, On numerical deformation of singularities in nonlinear elasticity, pp. 267-278 in: R. Glowinski, J.L. Lions (eds.), Computing methods in applied sciences and engineering, Lecture Notes in Maths. 704, Springer - Verlag, Berlin 1979.

Helmut Weber
Abteilung Mathematik
Universität Dortmund
Postfach 50 05 00
D-4600 Dortmund 50
Bundesrepublik Deutschland

SOME IMPROVEMENTS OF CLASSICAL ITERATIVE METHODS FOR THE SOLUTION OF NONLINEAR EQUATIONS

BY

W. WERNER

Fachbereich Mathematik
Universität Mainz
D-6500 Mainz

SOME IMPROVEMENTS OF CLASSICAL ITERATIVE METHODS FOR THE SOLUTION OF NONLINEAR EQUATIONS

Wilhelm Werner

Fachbereich Mathematik der
Johannes Gutenberg-Universität

65 MAINZ, GERMANY

INTRODUCTION

Let $F: D \subseteq X \to Y$ sufficiently smooth, D open and convex, X,Y real Banach spaces; one of the basic problems in numerical analysis is to solve the nonlinear equation

$$F(x) = 0 .$$

For smooth functions F several iterative methods based on truncation of the Taylor series are well known; the most famous algorithm of this type certainly is

(1) Newton's method:

given $x_o \in D$, compute x_{i+1} from

$$0 = F(x_i) + F'(x_i)(x_{i+1}-x_i)$$

$i=0,1,2,3,\ldots$.

Methods using higher than first derivatives are by far not as important in applications as (1); for special types of problems - such as nonlinear integral equations - however they may be advantageous if it is not particularly expensive or onerous to evaluate the derivatives which are involved in these methods. A well known representant of this type of algorithm is

(2) given $x_o \in D$, compute y_i, x_{i+1} from

$$0 = F(x_i) + F'(x_i)(y_i-x_i)$$

$$0 = F(x_i) + \left[F'(x_i)+ \frac{\alpha}{2} F''(x_i)(y_i-x_i)\right](x_{i+1}-x_i) + \frac{1-\alpha}{2} F''(x_i)(y_i-x_i)^2$$

with some fixed $\alpha \in \mathbb{R}$, $i=0,1,2,3,\ldots$.

For $\alpha = 0$ (2) is called <u>Chebyshev's method</u> or <u>method of tangent parabola</u>, $\alpha = 1$ corresponds to <u>Halley's method</u> or <u>method of tangent hyperbola</u> (for references concerning these methods see [8]);

the case $\alpha = 2$ corresponds to the application of a Newton step to the nonlinear equation

$$0 = F(x_i) + F'(x_i)(x-x_i) + \tfrac{1}{2}F''(x_i)(x-x_i)^2$$

with initial value y_i. From the numerical point of view the preferable choice of α is $\alpha = 0$: then one has to solve the same linear system twice with different inhomogenities.

For scalar equations $F: D \subseteq \mathbb{R} \to \mathbb{R}$ there are numerous algorithms which have no multivariate analogue, e.g. the

(3) third-order Newton method:

given $x_0 \in D$, compute x_{i+1} from

$$0 = F(x_i) + F'(x_i)(x_{i+1}-x_i) + \tfrac{1}{2}F''(x_i)(x_{i+1}-x_i)^2$$

(<u>if</u> this quadratic equation has a real root which is locally at least true for simple roots of F; x_{i+1} then is chosen as that root lying nearest to $x_i - F(x_i)/F'(x_i)$).

In the sequel we are concerned with several iterative methods using the same function evaluations per step as (1), (2) or (3) but yet have higher order of convergence for simple roots.

1. Higher order iterative methods; general case

Let us assume in the following that

(i) $F \in C^4(D,Y)$

(ii) there exist $L_i \in \mathbb{R}_+$, $i=1,2,3$, such that for all $x,y \in D$, $h \in X$:

$$\|F^{(i)}(x)h^i - F^{(i)}(y)h^i\| \leq L_i \|x-y\| \|h\|^i .$$

Starting point for the development of the announced methods will be Taylor's series with remainder

$$F(x) = F(x_i) + F'(x_i)(x-x_i) + \int_0^1 (1-t)F''(x_i+t(x-x_i))dt\ (x-x_i)^2$$

$$= F(x_i) + \left[F'(x_i) + \alpha \int_0^1 (1-t)F''(x_i+t(x-x_i))dt\ (x-x_i)\right](x-x_i)$$

$$+ (1-\alpha) \int_0^1 (1-t)F''(x_i+t(x-x_i))dt\ (x-x_i)^2$$

For $\alpha = 0$ a natural approach to solve $F(x)=0$ is to replace

$$\int_0^1 (1-t)F''(x_i+t(x-x_i))dt\ (x-x_i)^2$$

by $\int_0^1 (1-t)F''(x_i+t(y_i-x_i))dt\ (y_i-x_i)^2 = F(y_i)$

where $y_i := x_i - F'(x_i)^{-1}F(x_i)$. This leads to a well known third order modification of Newton's method due to TRAUB :

(4) given $x_0 \in D$, compute y_i, x_{i+1} from

$$0 = F(x_i) + F'(x_i)(y_i-x_i)$$

$$0 = F(y_i) + F'(x_i)(x_{i+1}-y_i)$$

$i=0,1,2,3,...$.

Approximation of the terms $\int_0^1 (1-t)F''(x_i+t(x-x_i))dt\ (x-x_i)^j$ $(j=1,2)$ by $\frac{1}{2}F''(x_i)(y_i-x_i)^j$ leads to (2). However a locally at least better approximation of $\int_0^1 (1-t)F''(x_i+t(x-x_i))dt$ is $\frac{1}{2}F''(\frac{2}{3}x_i + \frac{1}{3}y_i)$ as is shown in the following

LEMMA 1

Let $x \in D$ be a simple root of F; then there are constants c_1, c_2 such that for any x_i sufficiently close to x

(i) $\left\| \int_0^1 (1-t)F''(x_i+t(x-x_i))dt - \frac{1}{2}F''(x_i) \right\| \leq c_1 \|x_i-x\|$

(ii) $\left\| \int_0^1 (1-t)F''(x_i+t(x-x_i))dt - \frac{1}{2}F''(\frac{2}{3}x_i+\frac{1}{3}y_i) \right\| \leq c_2 \|x_i-x\|^2$

where $y_i := x_i-F'(x_i)^{-1}F(x_i)$.

The proof easily follows by Taylor expansion.

Therefore one is led to the following improved version of (2):

(5) given $x_o \in D$, compute y_i, z_i, x_{i+1} from

$$0=F(x_i)+F'(x_i)(y_i-x_i)$$

$$0=F(x_i)+\left[F'(x_i)+\frac{\alpha}{2}F''(\tfrac{2}{3}x_i+\tfrac{1}{3}y_i)(y_i-x_i)\right](z_i-x_i)+\frac{1-\alpha}{2}F''(\tfrac{2}{3}x_i+\tfrac{1}{3}y_i)(y_i-x_i)^2$$

$$0=F(x_i)+\left[F'(x_i)+\frac{\alpha}{2}F''(\tfrac{2}{3}x_i+\tfrac{1}{3}y_i)(z_i-x_i)\right](x_{i+1}-x_i)+\frac{1-\alpha}{2}F''(\tfrac{2}{3}x_i+\tfrac{1}{3}y_i)(z_i-x_i)^2$$

$\alpha \in \mathbb{R}$ fixed, $i=0,1,2,3,\ldots$.

Note that one step of (5) requires the same function evaluations as the third order method (2). Concerning the order of convergence of (5) however we have

PROPOSITION 2

If the assumptions of lemma 1 are fulfilled then there exists a neighborhood U of x such that for any starting point $x_o \in U$ the sequences $\{x_i\}$, $\{y_i\}$, $\{z_i\}$ which are generated by (5) converge to x with Q-order 4 at least.

REMARK

The concept of Q- and R-order is discussed in detail in [6].

Proof of proposition 2:

The proof of convergence of (2) (cf. [8], p.163 f.) may be used to show that for some constants c_1, $c_2 > 0$:

$$\|y_i-x\| \le c_1 \|x_i-x\|^2 , \quad \|z_i-x\| \le c_2 \|x_i-x\|^3 \quad .$$

From

$$0=F(x_i)+\left[F'(x_i)+\frac{\alpha}{2}F''(\tfrac{2}{3}x_i+\tfrac{1}{3}y_i)(z_i-x_i)\right](x-x_i)$$
$$-\frac{\alpha}{2}F''(\tfrac{2}{3}x_i+\tfrac{1}{3}y_i)(z_i-x_i)(x-x_i)+\int_0^1(1-t)F''(x_i+t(x-x_i))dt\ (x-x_i)^2$$

and

$$0=F(x_i)+\left[F'(x_i)+\frac{\alpha}{2}F''(\tfrac{2}{3}x_i+\tfrac{1}{3}y_i)(z_i-x_i)\right](x_{i+1}-x_i)$$
$$-\frac{\alpha}{2}F''(\tfrac{2}{3}x_i+\tfrac{1}{3}y_i)(z_i-x_i)^2 \qquad + \frac{1}{2}F''(\tfrac{2}{3}x_i+\tfrac{1}{3}y_i)(z_i-x_i)^2$$

one concludes that

$$\left[F'(x_i)+\frac{\alpha}{2}F''(\tfrac{2}{3}x_i+\tfrac{1}{3}y_i)(z_i-x_i)\right](x_{i+1}-x\ ) \quad =$$

$$\int_0^1(1-t)\left[F''(x_i+t(x-x_i))-F''(\tfrac{2}{3}x_i+\tfrac{1}{3}y_i)\right]dt\ (x-x_i)^2 \quad +$$

$$+ \frac{1}{2} F''(\tfrac{2}{3}x_i + \tfrac{1}{3} y_i)(x-z_i)(x-2x_i+z_i) - \frac{\alpha}{2} F''(\tfrac{2}{3}x_i + \tfrac{1}{3} y_i)(z_i-x_i)(x-z_i) \quad .$$

Since $F'(x)$ is continuously invertible

$$\left[F'(x_i)+\frac{\alpha}{2} F''(\tfrac{2}{3}x_i+\tfrac{1}{3} y_i)(z_i-x_i)\right]^{-1}$$

is uniformly bounded for any x_i, y_i, z_i in a sufficiently small neighborhood of x; the conclusion then follows from the last equation and lemma 1 .

REMARK

The idea that led to the improvement of (2) may also be applied to a general class of iterative methods due to EHRMANN [3]; these methods use the values $F(x_i)$, $F'(x_i),\dots,F^{(n)}(x_i)$ in each step. Their order is n+1 in the case of simple roots; our improved version of these methods has order n+2 for $n \geq 2$. Let us mention the case $n=3$, $\alpha=0$:

given $x_o \in D$, compute u_i, v_i, w_i, x_{i+1} from

$$0=F(x_i)+F'(x_i)(u_i-x_i)$$

$$0=F(x_i)+F'(x_i)(v_i-x_i)+\frac{1}{2} F''(x_i)(u_i-x_i)^2$$

$$0=F(x_i)+F'(x_i)(w_i-x_i)+\frac{1}{2} F''(x_i)(v_i-x_i)^2+\frac{1}{6}F'''(\tfrac{3}{4}x_i+\tfrac{1}{4}v_i)(v_i-x_i)^3$$

$$0=F(x_i)+F'(x_i)(x_{i+1}-x_i)+\frac{1}{2} F''(x_i)(w_i-x_i)^2+\frac{1}{6}F'''(\tfrac{3}{4}x_i+\tfrac{1}{4}v_i)(w_i-x_i)^3$$

$i=0,1,2,3,\dots$.

Methods of this type are however of little practical importance. We suggest that the order n+2 for a method using the data F, F', ..., $F^{(n)}$ $(n \geq 2)$ once per step cannot be increased.

EXAMPLES

(a) The matrix eigenvalue problem

find $x \in \mathbb{R} \setminus \{0\}$, $\lambda \in \mathbb{R}$ such that $Ax = \lambda x$

can be solved by looking for zeros of $F: \mathbb{R}^{n+1} \to \mathbb{R}^{n+1}$,

$$F(x,\lambda) := \begin{pmatrix} Ax - \lambda x \\ x^T x - 1 \end{pmatrix}$$

(cf. [2], p. 257 f.); in this case

$$\frac{1}{2} F''(x,\lambda)\begin{pmatrix} h \\ \mu \end{pmatrix}\begin{pmatrix} h \\ \mu \end{pmatrix} = \begin{pmatrix} -\mu h \\ h^T h \end{pmatrix} \quad .$$

(b) An approximate solution of the nonlinear Hammerstein equation

$$x(t) + \int_0^1 K(t,s)f(s,x(s))\,ds = g(t), \quad 0 \le t \le 1,$$

on the grid $0 \le t_1 < t_2 < \dots < t_{n-1} < t_n \le 1$ can be computed from the system of nonlinear equations

$$(*) \qquad x_j + \sum_{i=1}^{n} \omega_i K(t_j,t_i) f(t_i,x_i) = g(t_j),\ j=1,2,\dots,n,$$

if $\sum_{i=1}^{n} \omega_i h(t_i) \doteq \int_0^1 h(t)\,dt$ is a suitable quadrature formula .

Introducing the matrix $\mathbb{K} := (\omega_i K(t_j,t_i))_{i,j=1(1)n}$ (*) reads:

$$F(x_1,\dots,x_n) = \begin{pmatrix} x_1 \\ \vdots \\ x_n \end{pmatrix} + \mathbb{K} \begin{pmatrix} f(t_1,x_1) \\ \vdots \\ f(t_n,x_n) \end{pmatrix} - \begin{pmatrix} g(t_1) \\ \vdots \\ g(t_n) \end{pmatrix} = 0 \ .$$

Hence $F'(x_1,\dots,x_n) = I + \mathbb{K}\mathrm{diag}(f_x(t_1,x_1),\dots,f_x(t_n,x_n))$,

$$\tfrac{1}{2}F''(x_1,\dots,x_n)hh = \tfrac{1}{2}\mathbb{K} \begin{pmatrix} f_{xx}(t_1,x_1)h_1^2 \\ \vdots \\ f_{xx}(t_n,x_n)h_n^2 \end{pmatrix} \quad \text{if } h = \begin{pmatrix} h_1 \\ \vdots \\ h_n \end{pmatrix} \in \mathbb{R}^n \ .$$

In the following table we list the computational costs (per step of the iteration) for three fourth order methods and one method of order $2+\sqrt{5}$, which will be introduced later:

(A) method (5) with $\alpha = 0$

(B) two steps of Newton's method (considered as one step of a fourth order method)

(C) the modified fourth order Newton method (one Newton step followed by two modified Newton steps)

(D) a method of order $2+\sqrt{5} \doteq 4.24$, which will be introduced later (method (11) with m=4)

Three types of computational costs arise in this example:

1. function evaluation of f, f_x, f_{xx}
2. algebraic operations necessary to evaluate F, F', F''
3. algebraic operations necessary to solve linear systems of equations

In 2., 3. we only count multiplications and divisions .

<u>Table 1</u> Comparison of the computational costs per step to solve a nonlinear Hammerstein equation for 4 iterative methods

method	computations necessary to evaluate						algebraic operations necessary to solve the occuring linear systems	total number of algebraic operations per step
	F		F'		F"			
	evaluations of f	algebraic operations	evaluations of f_x	algebraic operations	evaluations of f_{xx}	algebraic operations		
(A)	n	n^2	n	n^2	n	$2n^2$	$\frac{1}{3}n^3 + 3n^2$	$\frac{1}{3}n^3 + 7n^2$
(B)	2n	$2n^2$	2n	$2n^2$	-	-	$\frac{2}{3}n^3 + 2n^2$	$\frac{2}{3}n^3 + 6n^2$
(C)	3n	$3n^2$	n	n^2	-	-	$\frac{1}{3}n^3 + 3n^2$	$\frac{1}{3}n^3 + 7n^2$
(D)	3n	$3n^2$	n	n^2	-	-	$\frac{1}{3}n^3 + 4n^2$	$\frac{1}{3}n^3 + 8n^2$

A conclusion concerning the efficiency of method (A) relative to that of (B), (C) depends on the relative costs for evaluating f and f_{xx}; if one evaluation of f_{xx} is "cheap" relative to 2 evaluations of f then (A) should be preferred. (D) should be always preferred to (C) since it has higher than fourth order without relevant additional computational costs.

REMARK

A more detailed discussion of the higher order iterative methods introduced in this section can be found in [12].

2. Higher order iterative methods; scalar case

Our motivation for the higher order iterative methods mentioned so far was to solve (approximately at least) the quadratic equation

$$(6)\quad 0=F(x_i)+F'(x_i)(x-x_i)+\int_0^1(1-t)F''(x_i+t(y_i-x_i))dt\,(x-x_i)^2\ .$$

In the scalar case, i.e. $X=\mathbb{R}$, this can be done explicitly since

$$(\alpha)\qquad \int_0^1(1-t)F''(x_i+t(y_i-x_i))dt = F(y_i)/(y_i-x_i)^2\ ;$$

one could, however, also use the approximation

$$(\beta)\qquad \int_0^1(1-t)F''(x_i+t(y_i-x_i))dt \doteq \frac{1}{2}\,F''(\tfrac{2}{3}x_i+\tfrac{1}{3}y_i)$$

instead as was done in the general case.

In any case that solution of (6) should be chosen which lies next to y_i; if (6) has no real solution then it is reasonable to determine x_{i+1} as the extremum of the parabola defined by the right side of (6). Note, however, that in the case of a simple root x, (6) locally always has two real roots.

Let us define the function $\phi:\mathbb{R}\to\mathbb{R}_+$ by

$$\phi(t) := \begin{cases} \dfrac{2}{1+\sqrt{1-2t}} & \text{if } t\le\frac{1}{2} \\ \dfrac{1}{t} & \text{if } t\ge\frac{1}{2} \end{cases},$$

then the algorithm described above can be written in the following comprehensive notation:

(7) given $x_0\in\mathbb{R}$, compute y_i, x_{i+1} as follows:

$$y_i = x_i-F(x_i)/F'(x_i)$$

$$x_{i+1} = x_i+\phi(t_i)(y_i-x_i)$$

where

$$t_i := \begin{cases} 2\, F(y_i)/F(x_i) & \text{in case } (\alpha) \\ F''(\tfrac{2}{3}x_i+\tfrac{1}{3}y_i)F(x_i)/F'(x_i)^2 & \text{in case } (\beta) \end{cases}$$

$i=0,1,2,3,\ldots$.

REMARK

(a) Note that $(7)_\beta$ mainly requires the same computational work per step as the third order Newton method (3) which is mentioned in many textbooks on numerical analysis.

(b) Two steps of Newton's method which asymptotically yield the same accuracy as one step of $(7)_\alpha$ require two evaluations of F' per step instead of one for $(7)_\alpha$.

(c) If (α) is used, and (6) is approximately solved by one Newton step with initial value y_i, then one gets a method suggested by OSTROWSKI [8], Appendix G.

PROPOSITION 3

(i) If x is a simple root of $F \in C^4(\mathbb{R})$, then $(7)_\alpha$, $(7)_\beta$ locally converge to x with Q-order 4 at least.

(ii) If x is a double root of $F \in C^4(\mathbb{R})$, then $(7)_\alpha$, $(7)_\beta$ locally converge to x with Q-order 3/2 at least.

(iii) If x is a root of multiplicity $p > 2$ of $F \in C^4(\mathbb{R})$, then $(7)_\alpha$, $(7)_\beta$ locally converge Q-linear to x; it is

$$\lim_{i\to\infty} \frac{x_{i+1}-x}{x_i-x} = \begin{cases} 1-\frac{1}{2}(1-\frac{1}{p})^{-p}/p & \text{in case } (\alpha) \\ 1-(1-\frac{1}{p})^{-1}(1-\frac{1}{3p})^{2-p}/p & \text{in case } (\beta) \end{cases}$$

Proof. The proof of (i) is similar to the proof of proposition 2 and therefore is omitted. We only give a proof for (ii), case (α):

Let x be a double root of F: $F(x)=F'(x)=0$, $F''(x)\neq 0$; then

$$\begin{aligned} x_{i+1}-x_i &= (x_i-2F'(x_i)^{-1}F(x_i)-x) + (2-\phi(t_i))F'(x_i)^{-1}F(x_i) \\ &= \tfrac{1}{2}(2-\phi(t_i))(x_i-x) + O(\|x_i-x\|^2)\ . \end{aligned}$$

The order of convergence therefore is determined by the asymptotic behaviour of the null sequence $\{2-\phi(t_i)\}$.

(a) If $t_i \le \frac{1}{2}$ then $\phi(t_i) = 2/(1+\sqrt{1-4F(y_i)/F(x_i)})$ whence by Taylor expansion: $2-\phi(t_i) = O(\|x_i-x\|^{1/2})$.

(b) If $t_i \ge \frac{1}{2}$ then $\phi(t_i)=\frac{1}{2}F(x_i)/F(y_i)$ so that $2-\phi(t_i) = O(\|x_i-x\|)$.

Some generalizations of (7) and related algorithms are discussed in [13].

3. Iterative methods using first derivatives only

Let $F \in C^3(D,Y)$, $F(x)=0$, $x \in D$, $F'(x)$ continuously invertible; then

$$F(x) = F(x_i) + \int_0^1 F'(x_i+t(x-x_i))dt\,(x-x_i) = 0 .$$

The approximation of $\int_0^1 F'(x_i+t(x-x_i))dt$ by $\frac{1}{2}F'(\frac{x_i+y_i}{2})$ where

$$y_i := x_i - F'(x_i)^{-1}F(x_i)$$

then leads to a third order method:

(8) given $x_0 \in D \subseteq X$, compute y_i, x_{i+1} from

$$0 = F(x_i) + F'(x_i)(y_i-x_i)$$

$$0 = F(x_i) + F'(\tfrac{1}{2}(x_i+y_i))(x_{i+1}-x_i)$$

$i=0,1,2,3,\ldots$

(cf. [9], p. 164, [5]).

The main disadvantage of (8) in the case of systems of nonlinear equations is the necessity of computing the derivative twice per step; this difficulty however can be overcome (for the price of a reduced asymptotic order of convergence) if the method is modified in the following way:

(9) given x_0, $y_0 \in D \subseteq X$, compute x_{i+1}, y_{i+1} from

$$0 = F(x_i) + F'(\tfrac{1}{2}(x_i+y_i))(x_{i+1}-x_i)$$

$$0 = F(x_{i+1}) + F'(\tfrac{1}{2}(x_i+y_i))(y_{i+1}-x_{i+1})$$

$i=0,1,2,3,\ldots$

which is a method of R-order $1+\sqrt{2}$ under the above assumptions. Method (9) was proposed by the author in [10]; later it was pointed out to the author that R.F.KING [4] had previously suggested the method in different notation ("method (I,II)" in [4]). It seems however that the utility of this method had not been fully recognized then.

As was mentioned in [10] one may replace the sample point $\frac{1}{2}(x_i+y_i)$ by $x_i+\lambda_i(y_i-x_i)$ without reduction of the asymptotic order of convergence if the sequence $\{\lambda_i\}$ converges to 1/2 "fast enough". Here we are concerned with a reasonable choice of λ_i for the case of finite systems of nonlinear equations; we propose to perform the method in the following way:

(10) given $x_0, y_0 \in D \subseteq \mathbb{R}^n$, compute y_{i+1}, x_{i+1} as follows:

$$x_{i+1} = x_i + t_i \quad \text{where} \quad 0 = F(x_i)+F'(\tfrac{1}{2}(x_i+y_i))t_i$$

$$y_{i+1} = x_{i+1} + \mu_i s_i$$

$$\text{where} \quad 0 = F(x_{i+1})+F'(\tfrac{1}{2}(x_i+y_i))s_i$$

$$\text{and} \quad \mu_i = t_i^T t_i/(t_i-s_i)^T t_i$$

$i=0,1,2,3,\ldots$.

Note that (10) could also be written similarly to (9) with variable λ_i ; if μ_i as defined above differs "too much" from 1 it should be set 0.

The motivation for this construction is the following: the auxiliary point y_{i+1} should be as good an approximation for the solution x as is computable from the data $F(x_i)$, $F(x_{i+1})$, $F'(\frac{1}{2}(x_i+y_i))$, x_i, x_{i+1}. Therefore a rank one correction of $F'(\frac{1}{2}(x_i+y_i))$ is applied to improve the approximation:

$$0 = F(x_{i+1}) + \left[F'(\tfrac{1}{2}(x_i+y_i))+uv^T\right](y_{i+1}-x_{i+1})$$

where $u,v \in \mathbb{R}^n$ are determined from the quasi-Newton-equation

$$\left[F'(\tfrac{1}{2}(x_i+y_i))+uv^T\right](x_{i+1}-x_i) = F(x_{i+1})-F(x_i):$$

$$u= F(x_{i+1}), \quad v=t_i/t_i^T t_i \ .$$

By the SHERMAN-MORRISON-inversion formula (cf. [6], p. 50) therefore

$$y_{i+1} = x_{i+1} + \left[t_i^T t_i/(t_i-s_i)^T t_i\right] s_i \ .$$

In practical computations it is often useful to keep the Jacobian constant for some steps of the iteration (see e.g. method (4)); this leads to the following generalization of (9) :

(11) given $x_o^{(m-1)}$, $x_o^{(m)}$ ε $D \subseteq X$, compute $x_{i+1}^{(j)}$ $(j=1,\ldots,m)$ from

$$x_{i+1}^{(o)} := x_i^{(m-1)}$$

$$0 = F(x_{i+1}^{(j-1)}) + F'(\tfrac{1}{2}(x_i^{(m)}+x_i^{(m-1)}))(x_{i+1}^{(j)} - x_{i+1}^{(j-1)})$$

$i=0,1,2,3,\ldots$.

It is evident that (11) also can be combined with quasi-Newton methods as was done in (10); we do not go into these details here.

PROPOSITION 4

Let (i) $F \varepsilon C^3(D,Y)$, $D \subseteq X$ open, convex;

(ii) $F(x) = 0$ for some $x \varepsilon D$;

(iii) $\sup_{z \varepsilon D} \|F'(z)^{-1}\| \leq \Gamma < \infty$

(iv) there exist $L_i \varepsilon \mathbb{R}_+$, $i=1,2$, such that for any $x,y \varepsilon D$, $h \varepsilon X$:

$$\|F^{(i)}(x)h^i - F^{(i)}(y)h^i\| \leq L_i \|x-y\| \|h\|^i \;;$$

(v) $x_o^{(m)}$, $x_o^{(m-1)}$ ε $S(x,r) := \{z \varepsilon X \mid \|x-z\| < r\} \subset D$,

where r satisfies: $Ar^2+Br < 1$, $3Br < 1$ with $A := \Gamma L_2/24$, $B := \Gamma L_1/2$.

Then the sequences $\{x_i^{(j)}\}_{i \varepsilon \mathbb{N}}$ $(j=0,1,\ldots,m-1)$ generated by (11) converge to x with R-order

$$\frac{m}{2} + \sqrt{\frac{m^2}{4} + 1} \quad \text{at least .}$$

For the proof of this proposition we need the following

LEMMA 5

Under the above assumptions (i)-(iv) one has

(a) $\|F(x)-F(y)-F'(\tfrac{1}{2}(x+y))(x-y)\| \leq L_2/24 \|x-y\|^3$

(b) $\|F(x)-F(y)-F'(z)(x-y)\| \leq L_1/2 \|x-y\| \{\|x-z\| + \|y-z\|\}$

for any $x,y,z \varepsilon D$.

Now let us set $z_{i+1} := \frac{1}{2}(x_i^{(m)}+x_i^{(m-1)})$; then

$$
\begin{aligned}
x_{i+1}^{(1)} - x &= x_{i+1}^{(o)} - x - F'(z_{i+1})^{-1}F(x_{i+1}^{(o)}) \\
&= F'(z_{i+1})^{-1}\left[F'(z_{i+1})(x_{i+1}^{(o)} - x) - F(x_{i+1}^{(o)})\right] \\
&= F'(z_{i+1})^{-1}\left[F'(\tfrac{1}{2}(x_{i+1}^{(o)}+x))(x_{i+1}^{(o)}-x)-F(x_{i+1}^{(o)})\right] \\
&\quad + F'(z_{i+1})^{-1}\left[F'(z_{i+1})-F'(\tfrac{1}{2}(x_{i+1}^{(o)}+x))\right](x_{i+1}^{(o)}-x)
\end{aligned}
$$

implies

$$
(\alpha) \quad \|x_{i+1}^{(1)}-x\| \leq A\,\|x_{i+1}^{(o)}-x\|^3 + B\,\|x_i^{(m)}-x\|\cdot\|x_{i+1}^{(o)}-x\| .
$$

Furthermore

$$
\begin{aligned}
x_{i+1}^{(j+1)}-x &= x_{i+1}^{(j)} - x - F'(z_{i+1})^{-1}F(x_{i+1}^{(j)}) \\
&= F'(z_{i+1})^{-1}\left[F'(z_{i+1})(x_{i+1}^{(j)} - x) - F(x_{i+1}^{(j)})\right]
\end{aligned}
$$

whence

$$
(\beta) \quad \|x_{i+1}^{(j+1)}-x\| \leq B\,\|x_{i+1}^{(j)}-x\|\{\|x_{i+1}^{(o)}-x\| + \|x_i^{(m)}-x\| + \|x_{i+1}^{(j)}-x\|\}.
$$

(v) then implies that all iterates $x_i^{(j)}$ remain in $S(x,r) \subset D$ and converge to x.

It remains to prove the statement concerning the asymptotic order of convergence:

(β) implies that

$$
\|x_{i+1}^{(j+1)}-x\| \leq 3B\,\|x_{i+1}^{(j)}-x\|\,\|x_{i+1}^{(o)}-x\| \leq (3B)^j\,\|x_{i+1}^{(1)}-x\|\,\|x_{i+1}^{(o)}-x\|^j;
$$

therefore from

$$
\|x_{i+1}^{(1)}-x\| \leq A\,\|x_{i+1}^{(o)}-x\|^3 + 3B^2\|x_i^{(m-1)}-x\|\cdot\|x_i^{(o)}-x\|\cdot\|x_{i+1}^{(o)}-x\|
$$

we may conclude

$$
\begin{aligned}
\|x_{i+2}^{(o)}-x\| &\leq C_o\,\|x_{i+1}^{(o)}-x\|^{m-2}\,\|x_{i+1}^{(1)}-x\| \\
&\leq C_1\,\|x_{i+1}^{(o)}-x\|^{m+1} + C_2\,\|x_{i+1}^{(o)}-x\|^m\,\|x_i^{(o)}-x\| \\
&\leq C_3\,\|x_{i+1}^{(o)}-x\|^m\,\|x_i^{(o)}-x\|
\end{aligned}
$$

with some suitable constants $C_1, C_2, C_3 > 0$; thus $\{x_i^{(o)}\}$ converges to x with R-order of convergence $(m+\sqrt{m^2+4})/2$ at least.

REMARK

(a) The corresponding modifications of Newton's method which use the same function evaluations per step (one evaluation of F' and m-1 evaluations of F) have order m only; for m=3 see (4) .
(b) The above result is mentioned without proof in [11] where several additional results on (9) can be found.

REFERENCES

[1] COLLATZ, L., Näherungsverfahren höherer Ordnung für Gleichungen in Banach-Räumen, Arch.Rat.Mech.Anal.2, 66-75 (1958)

[2] COLLATZ, L., Funktionalanalysis und Numerische Mathematik. Springer, Berlin-Heidelberg-New York (1964)

[3] EHRMANN, H., Konstruktion und Durchführung von Iterationsverfahren höherer Ordnung, Arch.Rat.Mech.Anal. 4, 65-88 (1959)

[4] KING, R.F., Tangent methods for nonlinear equations, Num.Math. 18, 298-304 (1972)

[5] KLEINMICHEL, H., Stetige Analoga und Iterationsverfahren für nichtlineare Gleichungen in Banachräumen, Math.Nachr. 37, 313-344 (1968)

[6] ORTEGA, J.M., RHEINBOLDT, W.C., Iterative solution of nonlinear equations in several variables, Academic Press, New York-London (1970)

[7] OSTROWSKI, A.M., Solution of equations in Euclidean and Banach spaces, Academic Press, New York-London (1973)

[8] SCHWETLICK, H., Numerische Lösung nichtlinearer Gleichungen, R.Oldenbourg Verlag, München-Wien (1979)

[9] TRAUB, J.F., Iterative methods for the solution of equations, Prentice Hall, Englewood Cliffs (1964)

[10] WERNER, W., Über ein Verfahren der Ordnung $1+\sqrt{2}$ zur Nullstellenbestimmung, Num.Math. 32, 333-342 (1979)

[11] WERNER, W., Some supplementary results on the $1+\sqrt{2}$ order method for the solution of nonlinear equations, submitted for publication

[12] WERNER, W., On higher order iterative methods for the sol solution of nonlinear equations, submitted for publication

[13] WERNER, W., Some efficient algorithms for the solution of a single nonlinear equation, to appear in Int.J.Comp.Math., Ser. B, (1981)